Studies in Systems, Decision and Control

Volume 495

Series Editor

Janusz Kacprzyk, Systems Research Institute, Polish Academy of Sciences, Warsaw, Poland

The series "Studies in Systems, Decision and Control" (SSDC) covers both new developments and advances, as well as the state of the art, in the various areas of broadly perceived systems, decision making and control–quickly, up to date and with a high quality. The intent is to cover the theory, applications, and perspectives on the state of the art and future developments relevant to systems, decision making, control, complex processes and related areas, as embedded in the fields of engineering, computer science, physics, economics, social and life sciences, as well as the paradigms and methodologies behind them. The series contains monographs, textbooks, lecture notes and edited volumes in systems, decision making and control spanning the areas of Cyber-Physical Systems, Autonomous Systems, Sensor Networks, Control Systems, Energy Systems, Automotive Systems, Biological Systems, Vehicular Networking and Connected Vehicles, Aerospace Systems, Automation, Manufacturing, Smart Grids, Nonlinear Systems, Power Systems, Robotics, Social Systems, Economic Systems and other. Of particular value to both the contributors and the readership are the short publication timeframe and the world-wide distribution and exposure which enable both a wide and rapid dissemination of research output.

Indexed by SCOPUS, DBLP, WTI Frankfurt eG, zbMATH, SCImago.

All books published in the series are submitted for consideration in Web of Science.

Preface

This book contains eleven original and survey scientific research articles that arose from invited talks given at International Workshop on Image Processing and Inverse Problems, held in Beijing Computational Science Research Center, Beijing, China, April 21–24, 2018.

The purpose of the conference was to bring together international researchers to exchange ideas, recent achievements on various aspects of image processing and inverse problems. Conference topics cover image reconstruction, image restoration, image registration and inverse problems and so on. Deep learning, PDE, Statistics based methods and techniques were discussed. The newest developments on mathematical analysis, numerical algorithm and applications were presented. This book aims to collect presentation papers which introduce new research trends and show improved results. It should be a good reference for people working on related problems, as well as for people working on computer vision and visualization, inverse problems, image processing and medical imaging.

To ensure the scientific quality of the book, each contributed paper was carefully reviewed. Special thanks go to all contributors and referees. Without their efforts, this book would not be possible.

Finally, we wish to thank the conference organizers and supports which were partially given by the National Nature Science Foundation of China. Many thanks also go to Springer-Verlag colleagues, Daniel Wang, Banu Dhayalan and Zongren Peng. Their help and collaboration are kind and effective.

Kowloon Tong, Hong Kong — Xue-Cheng Tai
Beijing, China — Suhua Wei
Beijing, China — Haiguang Liu

Contents

Chapter 1
Basic Characteristics of Dynamical Systems

1.1 Cornerstone

Everywhere in the nature occur dynamical systems producing rhythms and vibrations in the forms of a strong, regular, more or less repeating in time, changes of motion, lightning, or sound. Natural environments generally relate to the outdoor environments (rivers, forests, grasslands, mountains, lakes, oceans) in which human tribes and other living species evolved, but they also contain human-made indoor or outdoor environments (buildings, cities, rural settlements, planes, cars, motorways). The environment in which an organism lives and the tasks it performs makes it possible for it to survive and reproduce shape of the design of its perceptual, cognitive and motor systems through evolution and experience. In contrast to the natural environment appears the manufactured environment made by humans and/or animals. Manufactured environments are places where humans have fundamentally transformed landscapes to the forms of urban settings and agricultural land conversion with animal habitats. In this way the natural environment is greatly and systematically changed into a friendly human environment. However, even acts which seem less extreme, such as building a mud hut or a photovoltaic system in the desert, form the modified environment becoming an artificial one. Fortunately, using physics, chemistry, biology, chronology and mathematics, we are able to induce a qualitative and quantitative understanding of the principal cognitive areas concerning appropriate fields.

In this book, we will repeatedly exploit the term "system", defined as "a configuration of components functioning together as a whole and their relationships". Putting aside the impartial discussion of this definition, we can check its effectiveness immediately on the basis of real examples from the relevant field. The key point here is that most systems are hierarchical; they are composed of smaller sets of systems made of smaller interacting parts. For instance, an ecosystem is a type of systems, in which the components are living things like plants, animals, and microbes plus a habitat formed of natural, urban, and agricultural environments. Taking into

Z. Trzaska, *Mathematical Modelling and Computing in Physics, Chemistry and Biology*, Studies in Systems, Decision and Control 495,
https://doi.org/10.1007/978-3-031-39985-5_1

account all the relationships among these component parts, with an emphasis on the reciprocities between the living parts of the system then their interactions manifest themselves in, for example, food webs in which plants feed herbivores and herbivores nourish carnivores. Systems that contain a large number of components interacting in multiple ways (like an ecosystem, mentioned above, or the power systems distributed through a large territory) are often said to be complex. The term "complex" may have an obvious and general meaning adopted from everyday use such as "of course it is complex! there are lots of components and relationships!" but in the matter considered here the medics, physicists, geoscientists, ecologists, engineers and social scientists mean something specific, namely: they are referring to ways that different complex systems, starting from human cardiovascular system, ocean food webs, to the global climate system, and to the ecosystem of a dairy farm, display common types of behavior related to their complexity [1]. The natural system instantaneous state exhibits the idea that formal and inter-component relations within a structure are most important for its proper operation. This is due to the fact that components within a structure have multiple interactions, and equilibrium-building or blanking of conflict processes drive respective structural action (Fig. 1.1).

Activity of systems can be identified as a process which is observed with respect to a particular set of changes experienced by a system. The same system may be portrayed in terms of neither a change which can be neither increasing (such as pressing) nor decreasing (such as color). Another system may be described in terms of a change perceived as increasing (such as a runner heart rhythm), whilst it may also be described in terms of another property which is decreasing (such as car wheel tire pressure).

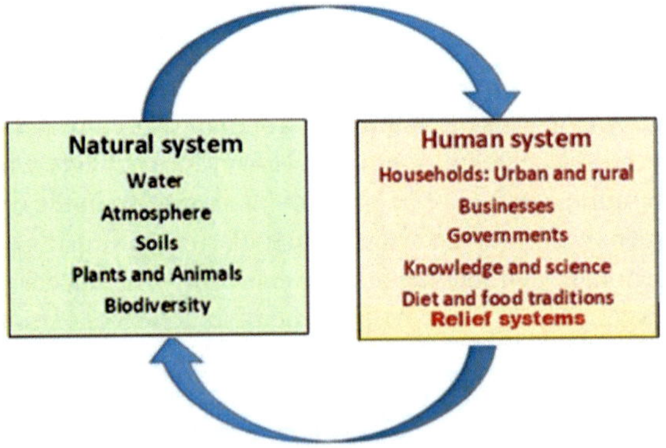

Fig. 1.1 A simple illustration of coupled natural and human systems with reference to each day situation. The list of natural system components (water, atmosphere, soils, plants and animals, biodiversity) and human (households: urban and rural, businesses, governments, knowledge and science, diet and food traditions, belief systems) on each side of the diagram is not exhaustive, and the diagram may be completed throughout the subject matter. The reciprocal arrows represent the mutual effects of each subsystem on the other, and are highly schematic although they can denote specific impacts or feedbacks

Complex processes forever change into new forms without repetition. Hence a given state of a system passing along a complex process will be different to all past and future states of that system. The changes that the system passes along may occur in a predictable (but infinite) sequence or they may be random and unpredictable (chaotic or noisy) [2].

In general, the dynamical system can be studied in terms of three system attributes: *openness*, *integration* and *complexity*.

Openness relates to whether a system is isolated from other systems in the sense that whether it can or cannot exchange matter and/or energy with other systems. An open system is not isolated in that it exchanges matter and/or energy with other systems. The closed system is isolated and cannot exchange mass and/or energy with other systems at all. The law of entropy establishes that energy cannot be recycled; therefore, "high quality" energy (such as visible light falling on the surface of the object) is degraded to "lower quality" energy (such as waste heat). Therefore, any system that does not have a restoring energy supplied from outside source will eventually over time discontinue to being active.

Integration refers to the strength of the interactions among the parts of the system. For instance, the human body is a highly integrated system whose cells are interdependent and stay in close communications. The loss of certain cells such as those composing the heart or brain can result in the death of all other cells in the whole organism because the cells are so interdependent. At the other extreme side are systems with very weak integration, such as the cells in a colony of single-celled organisms called the gold algae.

Complexity is often found as how many kinds of partners a system exhibits. This definition conforms to our intuition: A tiny insect seems more complex to us than a large rock because it has many more "parts". The insect has more complex molecules, as well as more different types of molecules, and these are used to construct cells and organs. This example also illustrates that complexity is often hierarchical, with smaller components being used to construct larger ones. Resilience is a sort of self-regulation of complex systems which they often tend in a self-organized way to be opposite to changes, like the way by which human body attempts to always maintain a temperature of order 37 °C.

Studying complex systems often occurs to be difficult because they have many parts that often interact mutually in different ways. Systems theory and chaos theory, for example, try to produce general laws of complexity. Such laws not only make complex systems more understandable, but also allow predicting more accurately how these systems behave. How important such prediction could be in a complex system it is easy check with such cases as the plane intercontinental trip or the space ship flight to the moon. Systems theory was one of the first widely used attempts to find "laws" of complex systems. It treats a complex system as a "black box" with inputs and outputs. Such a system is kept at equilibrium by negative feedback processes, defined as processes that counteract perturbations. For instance, in a thermostat that turns on a furnace to produce heat when a house is cold and turns off air conditioning when it is hot. In contrast, positive feedback processes amplify perturbations. This means that the positive feedback occurs when part of a system responds to change in a way that magnifies the initial change. In nontechnical terms, positive feedback is often referred to as a "snowball effect" or a "vicious circle". An important aspect of such an effect is that the more advanced the process is, the more difficult it is to stop it. Treating a system as a "black box", the theory omits many of the details of how the system operates. Therefore, present studies of complex systems focuses on more mathematical and rigorous description of them. Much attention has getting one theory that is chaos theory. In accord to chaos theory even simple systems, such as several atoms, often have chaotic properties [3, 4].

Chaotic systems are extremely sensitive to even the slightest change that can become greatly amplified through positive feedback. The classic example is the weather, as first described by E. Lorenz who created a set of equations that precisely described atmospheric conditions and showed how even tiny changes in one of parameters could cause a massive alterations of the weather in a few days. This is often called the "butterfly effect" (Fig. 1.2). Consequently, predicting a precise behavior of chaotic systems very far into the future is nearly impossible. In a similar manner it always is impossible to accurately predict precise future behavior in complex systems, because they have even more potential for chaos than simple systems. Nevertheless, by applying chaos theory, patterns of regularity can be discerned and investigated [5].

Presently, complex and chaotic nonlinear dynamics systems constitute a growing and increasingly important area that comprises advanced research activities and strongly interdisciplinary approaches. This area is of a fundamental interest in many sciences, including physics, chemistry and biology.

Fig. 1.2 **a** The Lorenz butterfly which is actually the appearance of a three-dimensional diagram of the Lorenz attractor [6], **b** Lorenz attractor pattern from nature

1.2 Concise Characterization of Complex Dynamical Systems

In general, most real-world dynamic systems are non-linear, and time-varying. Although they appear mainly in the world of physics, it is worth noticing that their dynamics indicates on impressive analogies with other types of dynamical systems, such as chemical, medical, biological, and even appearing in some social processes or market exposure capital. Presently, complex and chaotic nonlinear dynamics systems constitute a growing and increasingly important area that comprises advanced research activities and strongly interdisciplinary approaches. This area is of a fundamental interest in many sciences, and above all related to main environmental problems of our time [7, 8].

The functioning of the whole system results not only from the functioning of each of its parts separately, but also from the mutual relations and interactions of these parts with each other. Very often, the complex dynamical systems are finely composed grand and this is why they are prone to often unexpected influences of one component on other portions of the system. Thus it is valuable at the outset distinguish basic parts or elements from other components, viz.- parts without which a work in any particular system cannot even come into existence. Thanks to this, it is possible to build systems with a much richer behavior and functioning than it would only result from their components. Thus it is extremely beneficial to use a systems approach in analyzing global complex dynamical systems issues.

In this book a formal means of classifying physical processes for virtual and physical interpreting responses of complex dynamical systems is presented. Additionally, the classification of physical processes facilitates the analysis of existing mathematical works based on presented descriptions of dynamic properties. This is to be preferred over a classification based on a mathematical description's static structural features, because these features do not necessarily reflect the system's inherent dynamism. Moreover, the characterization of physical processes is useful in the synthesis of virtual dynamical systems, since it establishes a basis from which studied processes may be described algorithmically. It should be emphasized that all physical processes may be classified in terms of a handful of different classes. Moreover, more broadly based research assembling descriptions of physical processes with attempts to explain how they run also exist, especially when linking physics, chemistry, biology, mathematics and intellectual processes. These form a basis from which a process may be represented algorithmically.

Still in antiquity Aristotle [9] and Plato [10] have each made a considerable contribution in contemporary understanding of the meaning of process. Aristotle has much established about the processes which govern living systems. On the other hand, Plato's writings, among many other problems, highlight the importance of the technology of the day in determining the processes (and terms) by which humankind attempts to understand (and describe) the events which occur in its environment. It is worth to mention and there is no doubt that many writers and philosophers from Eastern countries have made valuable contributions to the philosophy of the

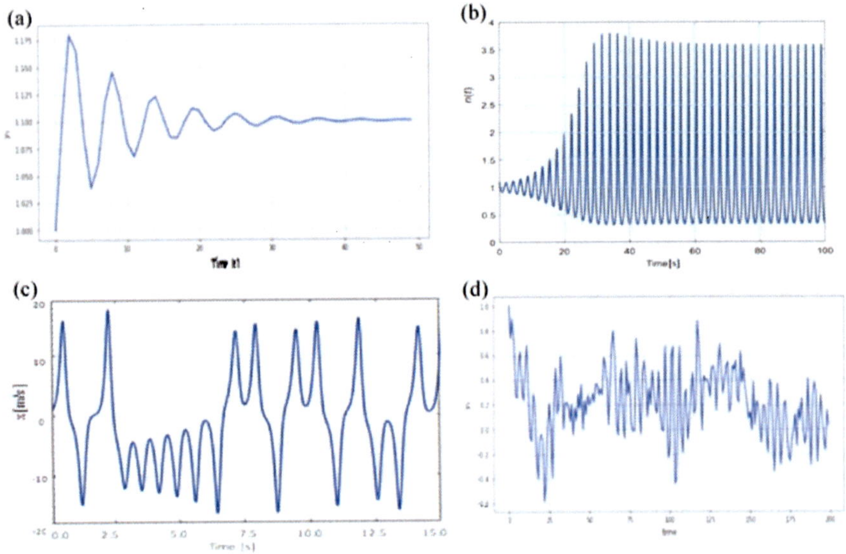

Fig. 1.3 Representative quantities varying over time in specific categories of dynamic systems with: **a** homogeneous state, **b** limit cycle, **c** strange attractor, **d** localized complex pattern

dynamics and process. For instance, Wolfram's [11] categorization of the behavior of continuous complex dynamic systems is divided into four categories, namely:

1. movement into a homogeneous state, i.e., exists a limit-point (Fig. 1.3a);
2. movement into simple, separated, periodic structures, i.e., exists a limit-cycle (Fig. 1.3b);
3. producing chaotic non-periodic patterns, i.e., exist strange attractors (Fig. 1.3c);
4. producing complex patterns of localized structures (Fig. 1.3d).

Complex processes forever change into new forms without repetition. Hence a given state of a system passing along a complex process will be different to all past and future states of that system. The changes that the system undergoes may occur in a predictable (but infinite) sequence or they may be random and unpredictable (chaotic or noisy).Categorizing physical processes from an algorithmic standpoint is needed for the same reasons that anything is categorized: for drawing out similarities between things; for noting their differences; and so it gives the possibility to learn more about the way they operate. Since each of these categories of processes is characterized with its specific language, then the universality of their dynamic behavior is clearly and entirely exhibited at the level of their mathematical description, which in turn leads to the interpretation of the solutions of the respective nonlinear differential equations. Therefore, the current achievements in the field of dynamics of complex nonlinear systems constitute a significant help in the analysis of processes occurring in real systems [12–14].

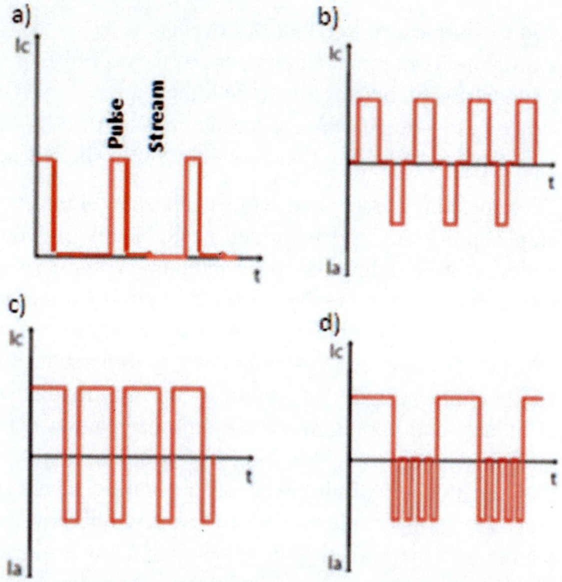

Fig. 1.4 Pulse waveforms: **a** periodic positive, **b** periodic reverse with time lag, **c** unsymmetrical reverse periodic, **d** periodic with reverse segment set

It is worth to emphasize that a process is something which is perceived as changes experienced by natural and/or artificial observers from within observer-dependent frames of reference. For instance, 'movement' and 'expansion' are perceived with respect to a 'stationary' or 'fixed' standard. Both concepts 'perception' as well 'processes' are entwined with the experience of time. It is clear that the temporal dimension is necessary for perception. Time may be perceived as moving rapidly, or it may seem to run at an agonizing rate. If it is observed at all it is through the observation of a change, a process. The first category of process is a Pulse (Fig. 1.4). A Pulse is a repeating set of experiences, a rhythmic course of events. Pulses such as the regular pumping of the heart or the musical pulse [15] define the temporal domain.

The spacing between events comprising a Pulse may be less in amount to create what is perceived as a continuous, uniform Stream. Individual locations within this continuum are not distinguishable from one another. For instance, two snapshots of the system taken at different times in the Stream look identical. The entropy of a system undergoing a Stream process does not change. Streams, although they are forever mobile, are timeless. They are unique because they may not be directly perceived. A marking of the Stream process by the super-position of a Pulse makes it possible to be perceived. For a sequence of events to be perceived as a Pulse, their spacing must be large enough that a continuum is not perceived, and short enough that the space/time between events may be evaluated by an observer. Note that the notion of a Stream makes clear that although a system may appear static, an underlying process may be perceived if the Stream is marked. Thus, the Stream preserves the apparently static form from moment to moment. In view of the above statements, the five categories of physical process can be distinguished, namely:

1. Pulse—repeating sequence of events;
2. Stream—continuum (may be marked by a Pulse or Complex process);
3. Increase—increasing (in some dimension);
4. Decrease—decreasing (in some dimension);
5. Complex—changing (in some dimension which is not increasing or decreasing).

The meanings associated with Increase and Decrease, with Streams, Pulses and Complex processes, may be sometimes subjective, but their properties may be recommended as a guideline for those intending to employ them in their activity. Each physical process to be realized may be seen as a mixture of these five process types. Taking them into account makes it much easier to complete the task.

Scientists are trying systematically to understand the inherent laws underlying these phenomena over the centuries through mathematical modeling. Mathematical models and paradigms describing the relation between system state variables and their derivatives with taking into account exciting terms are stated by differential equations. A driving mechanism in most physical and biological phenomena is conditioned by nonlinearity. It is worth underline that differential equations are essential for a mathematical description of nature—they lie at the core of many physical theories. Between most famous them there are: Newton and Lagrange equations for classical mechanics, Maxwell equations for classical electro-magnetism, Nernst equation as a chemical thermodynamically relationship in electro-chemistry, wave equation describing the propagation of energy and/or informations, Boltzmann equation relating the entropy of a macrostate to the number of microstates corresponding to that macrostate, Schrödinger equation for quantum mechanics, and Einstein equation for the general theory of gravitation, to mention a few. These equations usually reflect a physical process or some aspects of the natural environment from the viewpoint of the global analytical approach which is developed in this work. It is worth to point out that the superposition principle cannot be applied for nonlinear equations. This is one of the reasons why nonlinear equations are harder to solve analytically, except for a few special equations. Throughout the book, emphasis has been given to exposition of the subject mathematically and then explaining the dynamics of systems physically. In either case the mathematics must conform to measurements [16–33].

Note that system dynamics and its representations with differential equations are inextricably related to mathematics. In the modern approach to nonlinear complex systems, not having frequently explicit solution in terms of Newton's analytical methods, the emphasis is put on qualitative dynamics rather than strictly quantitative solutions. When the differential equations of a dynamical system are nonlinear, then their analytical solution is often unknown and they must be solved numerically with effective computer assistance. Hence, both the development of computers and the practical dissemination of numerical computations contribute significantly to the progress in research on the dynamics of all nonlinear systems.

1.3 Relevance and Emergence of Natural Systems

The popularization of large-scale systems and inexpensive robust installations leads to the need to treat them as extensive systems. For this reason, it is advisable to focus on the operation of individual devices in the system or even entire subsystems. This leads to the complex object being treated as a modular system that can be easily configured and maintained in proper operation compared to a monolithic system. If a combination of a large number of simple elements as one object is presented, then its operation is more complex than the operation of only the elements. The functioning of the whole system results not only from the functioning of each of its parts separately, but also from the mutual relations and interactions of these parts with each other. Thanks to this, it is possible to build systems with a much richer behavior and functioning than it would only result from their components. This coming to light of qualitatively new forms and behaviors from the interaction between simpler elements is called emergence. Generally, this term is explained by saying that the effect of the action of the whole is not only the sum of the effects of the action of its parts—as a whole, there are usually new properties and phenomena that do not appear in its components. A system is defined as emergent if in some sense it cannot be represented by describing only its component parts. This means, in the strict scientific sense, that the given object is irreducible. In this case, the simplified description of the lower level object known to us is not sufficient to describe the higher level object. In others words an emergent system is one where: (a) the final system has properties and effects which exceed the properties and effects of its component parts, and (b) whose final properties and effects cannot be predicted from merely examining the component parts. There are different types of emergence and related effects. In the case of emergence related to size scale differences, the reason is usually the ordering and strengthening of some specific interactions between elements, causing effects that cover the entire system. Such a phenomenon is usually impossible to predict on the basis of observation of only the elements themselves. Air can serve as an example, as none of the physical features of its molecules suggest that their large number will form an object (air) capable of transmitting sound waves. Another example is the so-called the "wetness" of water (H_2O). It results from the fact that water has a certain set of properties, which we commonly refer to when we say that water is "wet", although its components—hydrogen and oxygen—are not. Well, this "wetness" of water comes from the fact that the hydrogen and oxygen atoms in the water molecule interact so that an elongated structure is formed, one end of which has a positive electric charge and the other negative one. Hence the water molecule is an electric dipole. Thanks to these electrically charged extremities, water molecules can interact in a special way with each other and with molecules of other substances, in a way that we attribute to the "wet" quality. However, at least 16 molecules H_2O must combine into a single system for this "wet" property to occur. In the case of separate molecules of hydrogen (H_2) and oxygen (O_2), their charges are evenly distributed, that is, they are not dipoles and therefore are "dry". One of the reasons why the emergent behavior of systems is difficult to predict is the number of interactions between elements that should be considered when simulating them. Often it is sufficient to consider only very simple interactions: for example, simply identifying the reaction of each locust insect to the behavior of its neighbors is sufficient to simulate the movement of an entire locust colony. The same is not possible with air, however, due to the astronomical number of molecules to be simulated. However, a multiplicity of interactions alone is not enough to produce the emergent behavior of a system. If they are not properly related, they can mutually cancel out, generating only noise on a larger scale. Systems with emergent properties can appear to be non-compliant with the second law of thermodynamics because they self-order without a central source of control. In reality, however, this principle is not broken by them, because their operation always takes place at the expense of releasing entropy to the environment. In a frictional system individual atoms do not create friction, because all the forces they are subject to are conservative. Only the proper arrangement of atoms into a crystal structure enables the conversion of mechanical energy into heat, which we define as friction. Similar descriptions apply to other emergent systems that exhibit properties such as viscosity, elasticity, strength, etc. In classical mechanics, the laws of systems can be described as emergent, and resulting from averaging the principles of quantum mechanics over all elements of a macroscopic object. This is of particular importance as quantum mechanics is usually considered more complex, while in most cases we assume that lower-level rules are less complex than the resulting emergent properties. A similar dependence occurs in the case of the laws of thermodynamics. They are relatively simple, although they result from complex interactions between elementary molecules of the system. It should be emphasized that the emergent behavior of the system is the result of nonlinear interactions of elements causing deterministic processes. Such behavior should be distinguished from accidental, as their nature is different. Various forms of nonlinearity are the cause of emergent states of practical systems. Three of the possible ones are the most important, namely: technological limitations in the production of high-capacity components, feedback and signal transmission delays. The emergent system can operate in a wide range of dynamic states, including the undesirable steady states, oscillations and chaos. Therefore, the first step in analyzing such cases of systems is to identify whether any of these particular states are present in the system under study. Let the process of supplying parts on the assembly line in a certain car factory serve as an illustration of this issue. We assume that the assembly takes place on the production line and the appropriate sensors are installed at points that are critical for assembly. The result of their operation is the measurement of the assembly delay time caused by the delivery of parts. More than 1600 units coming off the belt over a 14-day period were analyzed. The measured delay times are shown in Fig. 1.5. The axis of ordinates shows the delay times on a logarithmic scale, depending on the production progress. It can be clearly seen changes in lag times on a scale by three orders of magnitude. Delay times greater than 2 h occur almost regularly, but shorter delays are not regular. Such dependencies make it possible to identify potential cases of emergent production systems. If emergent operation of a given system is not desired, then an appropriate strategy of its control should be adopted. One of them is to eliminate nonlinear interactions from the system, which is often not feasible. Another way is to cause

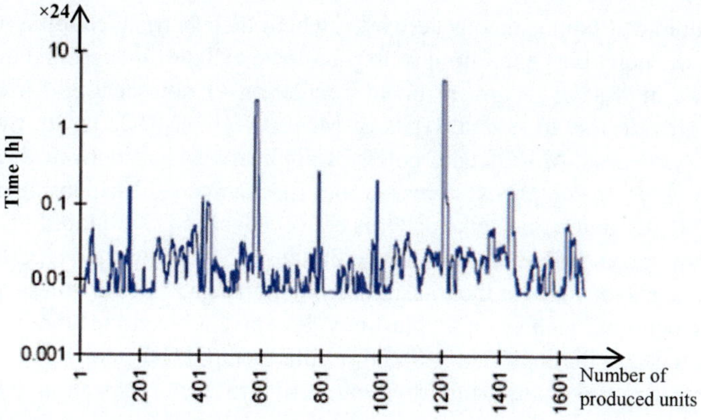

Fig. 1.5 Dependence of the delay time on the number of units produced

any oscillation in the system to be dampened by limiting its operation well below the optimality level. In the case of automation, emergent states of systems appear to be a consequence of modern architecture of control systems.

Designers and producers of appropriate technologies must take certain steps to make it easy to identify, evaluate and modify the resulting systems. This provides opportunities to reduce potential damage due to unpredictable system behavior. Consequently, a system is obtained which is more efficient and resistant to undesirable states. Diverse neuronal oscillators arising in a mixed population of neurons with different excitability properties produce mixed mode oscillations (MMOs). Various MMOs, including MMBOs (mixed mode bursting oscillations) and synchronized tonic spiking appear in a randomly connected network of neurons, where a fraction of them is in a quiescent (silent) state and the rest in self-oscillatory (firing) states [34]. MMOs and other patterns of neural activity depend on the number of oscillatory neighbors of quiescent nodes and on electrical coupling strengths. For weak neuron couplings, MMOs appear due to the desynchronization of a large number of quiescent neurons in the networks. The quiescent neurons together with the firing neurons generate oscillations of high frequency and bursting mobility. The overriding goal is to detect favorable network architecture of neurons and corresponding parameter spaces where Izhikevich model [35] of neurons produces diverse responses laying from MMOs to tonic spiking. In the absence of neuron coupling, the activity of the considered neuronal population reveals two types of dynamical states (or excitabilities), ranging from spike-bursting by subthreshold to quiescent states. The coupling parameter space and the ratio of mixed populations where MMOs and fast tonic spiking exist create basic factor to form emergence. At weak couplings of neurons and a diluted random network setting, the desynchronized subthreshold neurons exhibit MMOs. With the increase of the coupling, all subthreshold neurons fire in a mixed-mode state. In both cases, MMOs are not prominent in oscillatory neurons and eventually disappear as the coupling strength increases. Consequently, neural subpopulations emerge as synchronous clusters exhibiting tonic spiking behavior. For diluted random and homogeneous networks, where the electrical coupling strength is constant, the neighbors exhibiting self-sustained oscillations determine the structural patterns of MMOs. A given mixed population of quiescent and oscillatory nodes can provide rise to several types of MMOs and MMBOs in the two types of neuronal networks. MMOs take potential applications in biophysical, medical and other systems. In complex systems, various mechanisms subsist during different oscillatory phases generating spike patterns between fast and slow amplitude motion together with spikes and subthreshold oscillations. There exists a suggestion that MMOs can be responsible for the transition from high firing rates to quiescent states by reducing neuronal gain synaptic plasticity. It is also noticed [34] the impacts of small amplitude oscillations/subthreshold oscillations (STOs) on diverse neuronal responses such as spike clustering, rhythmic activities, synchronization and others as emergent processes.

1.4 Unusual Behavior of Complex Nonlinear Dynamical Systems

In the modern approach to nonlinear complex systems, not having frequently explicit solution in terms of Newton's analytical methods, the emphasis is put on qualitative dynamics rather than strictly quantitative solutions. It is well known that the dynamics of a system may be expressed either as a continuous-time or as a discrete-time-evolutionary process. The simplest mathematical models of continuous systems are those consisting of first-order linear differential equations. The dynamics of first-order linear autonomous system (explicit in time), belongs to a very restrictive class of system since it courses along the real line. On the other hand, even in simple nonautonomous cases of linear systems, their dynamics can be very rich.

Nonlinear systems are greatly difficult (if not sometime impossible) to solve than linear systems, because the latter follow the superposition principle and their solution can be divided into parts. Each part can be solved individually and then appropriately combining them together gives the vital result.

The modeling of a complex nonlinear dynamical system requires much attention considering the large number of components or active particles, the multiple nonlinear interactions, and the emerging collective behaviors [35–37]. In such a case the goal is to develop models which capture the essence of various interactions allowing the outcome to be more fully exposed. Thus, it is important to find which details are really relevant and must be incorporated into a model.

The continuous-time dynamical systems may be described mathematically as follows:

Let

$$x = x(t) \in \mathrm{R}^{\mathrm{n}};\, t \in \mathrm{I} \subseteq \mathrm{R} \tag{1.1}$$

be the vector representing the dynamics of a continuous system (continuous-time system). The mathematical representation of the system may be written as

$$\frac{d\boldsymbol{x}}{dt} = \dot{\boldsymbol{x}} = \boldsymbol{f}(\boldsymbol{x}, t) \tag{1.2}$$

where $\boldsymbol{f}(\boldsymbol{x},t)$ is a sufficiently smooth function defined on some subset $U \subset \mathbb{R}^n \times \mathbb{R}$: Schematically, this can be shown as

$$\mathbb{R}^n_{state\,space} \times \mathbb{R}_{time} = \mathbb{R}^{n+1}_{space\,of\,motion} \tag{1.3}$$

The variable t is usually interpreted as time and the function $\boldsymbol{f}(\boldsymbol{x},t)$ is generally nonlinear. The time interval may be finite, semi-finite or infinite. On the other hand, the discrete system is related to a discrete map (given only at equally spaced points of time) such that from a point $\boldsymbol{x}_0$, one can obtain a point $\boldsymbol{x}_1$ which in turn maps into $\boldsymbol{x}_2$, and so on. In other words, $\boldsymbol{x}_{n+1} = g(\boldsymbol{x}_n) = g(g(\boldsymbol{x}_{n-1}))$; etc. This is also written in the form $\boldsymbol{x}_{n+1} = g(\boldsymbol{x}_n) = g^2(\boldsymbol{x}_{n-1}) = \ldots$

Equation (1.2), in a developed form, with n differential equations takes the configuration

$$\begin{aligned} \dot{x}_1 &= f_1(x_1, \cdots, x_n, t) \\ &\vdots \\ \dot{x}_n &= f_n(x_1, \cdots, x_n, t) \end{aligned} \tag{1.4}$$

It is accepted that in a dynamical system "time" is the independent variable. The dependent variables x_k, are spanned in $\mathbb{R}^n$ which is called the phase space or the state space. Whereas space $\mathbb{R}^n \times \mathbb{R}$ is called the system space.

The term solution of the system (1.2) means a map $\boldsymbol{x}$, from an interval $I \subset \mathbb{R}$ to $\mathbb{R}^n$. It is called also a flow. Geometrical interpretation of the map $\boldsymbol{x}$ leads to a curve in $\mathbb{R}^n$, and $\dot{\boldsymbol{x}} = \boldsymbol{f}(\boldsymbol{x},t)$ is the tangent vector at each point of the curve. Thus, $\dot{\boldsymbol{x}} = \boldsymbol{f}(\boldsymbol{x},t)$ defines a vector field because the solution $\boldsymbol{x}$ for each point is a curve in $\mathbb{R}^n$, for which the speed is given by $\boldsymbol{f}(\boldsymbol{x})$. In addition, the space $\mathbb{R}^n$ of independent variables determines the phase space of $\dot{\boldsymbol{x}} = \boldsymbol{f}(\boldsymbol{x},t)$. The geometry of "solution-curves" in the phase-space forms the basis to determine properties of the studied system. The set $\{t, \boldsymbol{x}(t): t \in I\}$ is the trajectory of $\boldsymbol{x}$ contained in the system space.

Such problems as, among others, the bursting dynamics in Taylor–Couette flow, the compartmental Wilson–Callaway model of the dopaminergic neurotransmitters, stellate cell dynamics, the famous Hodgkin–Huxley model of neuron dynamics, the coupled calcium oscillator model, the complex plasma instability model, and the surface oxidation reaction and autocatalytic chemical reaction (for such examples see [6, 15, 38–48]) are described by singularly perturbed systems of nonlinear ordinary differential equations (ODEs) with three (or more) state variables changing at different time scales (usually two or three).

The Duffing system $\ddot{x} + b\dot{x} - x + x^3 = c \cdot \cos(t)$ can be written as

$$
\begin{aligned}
\dot{x}_1 &= x_2 \\
\dot{x}_2 &= -bx_2 - x_1 + x_1^3 - c \cdot \cos(x_3) \\
\dot{x}_3 &= 1
\end{aligned}
\tag{1.5}
$$

The Duffing system models a metallic plate moving between magnet poles (Fig. 1.6a). The plate is under an external periodic driving force. The Eq. (1.5) is a nonlinear one due to the third power term. There are not a universal recipe for its solutions. So that, there is possible only to investigate its behavior using tools from the qualitative theory of differential equations. The system response (state variables) and the phase trajectory are obtained by modeling and simulation the Eq. (1.5) in MATLAB/Simulink. Using the MATLAB function ODE23 to solve the equation gives plots of state variables and phase space of the trajectory shown in Fig. 1.6b, c, respectively. This looks complicated, but in fact, most of the plot shows the final range of time during which the course of state variables is diverging from its initial course which is much simpler. The early course is called an "initial transient". There is no sign of the data settling down to a periodic orbit. The system course is chaotic. A more detailed study shows that the period doubling transition to chaos occurs just as in the logistic map with the same value of the Feigenbaum constant γ [1].

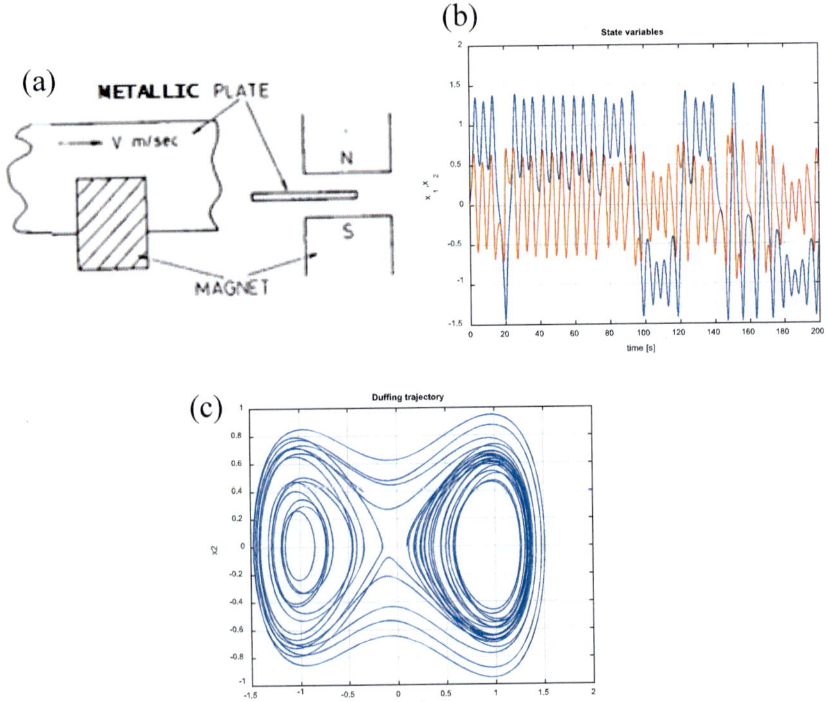

Fig. 1.6 Duffing system: **a** scheme of a metallic plate moving between magnet poles, **b** time-varying state variables, **c** system trajectory

Hence, both the development of computers and the practical dissemination of numerical computations contribute significantly to the progress in research on the dynamics of all nonlinear systems. Performing computer calculations, one can get not only numerical solutions to various problems, but also discover completely unexpected phenomena. In support of this statement, an excellent example can be cited, which was the observation made in 1963 by Lorenz [49], who identified the unpredictable long-term evolution of solutions to only three nonlinear differential equations. These studies dealt with the dynamics of atmospheric phenomena. It is worth emphasizing that this discovery initiated the intensive development of research into deterministic chaos, characterized by complex, nonperiodic dynamics generated by a completely deterministic non-linear dynamical system, without the occurrence of stochastic processes.

1.5 Entropy of Complex Processes in Dynamical Systems

Amongst the large number of different methods for nonlinear dynamical systems studies, entropy analysis has gained broad attention, and, thereby, has witnessed its wide suitability for studies of state variables with time-courses of limited length, short length, or even extremely short length. Entropy commonly thought of as a measure of disorder, counts the number of ways an object's internal parts can be rearranged without any change in its overall state. This systems theoretical characteristic is appropriate for dealing with macroscopic complex systems as a whole from nonreductionistic viewpoints, without the equivocal and nondefinitive results inherent in many prevalent studies, such as e.g. cryptology, electrochemistry and nonlinear physiology. It is worth underlying that the entropy concept lies at the core of the Second Law of thermodynamics. Process state variables, entropy flow, and entropy production can be quantified by the use of energetic data and physical methods. There are three ways to form a concept or idea of entropy, namely: entropy as an extensive thermodynamic quantity of physical systems, entropy as a measure for information contained in ergodic processes, and entropy as a means for statistical inference on multinomial Bernoulli processes [41]. From the physical point of view the entropy measure follows from fundamental laws of thermodynamics.

The First Law of thermodynamics is concerned about energy. Note that the energy concept is well known and has been extensively employed in the physics and chemistry as well as in social sciences, and even in our daily lives. However, entropy itself cannot be measured and calculated for biological systems, even for very small organisms. In the biological sciences, such approaches as bioenergetics, energy budget, biocalorimetry, and ecological energetics, among others, are examples of studies using the energy concept.

First law of thermodynamics sets out the principle of conservation of energy, which states that there must exist a physical quantity having the property of a state function, called internal energy (U). It has a total differential dU. Furthermore, let the internal energy of a system be increased by dU and if a certain amount of heat (dQ) is introduced into the system, and/or if certain work (dW) is done on the system then

$$\mathrm{dU} = \mathrm{dQ} + \mathrm{dW} \tag{1.6}$$

In physics and chemistry, the law of conservation of energy states that the total energy of an isolated system remains conserved over times. For instance, chemical energy is converted to kinetic energy when a stick of dynamite explodes.

The Eq. (1.6) contains the essence of the First Law of thermodynamics. Both the differentials dQ, as well as dW, are reflections of a change of energy. However, according to the Second Law of thermodynamics, heat (Q), as a physical quantity, differs from all other forms of energy because it possesses a specific property: *any form of energy can be completely transformed into heat, but heat itself can only partly be transformed into work.*

Entropy (S) makes its appearance in phenomenological thermodynamics as a kind of measure of the quality of heat and for a quasi reversible process it is defined as follows:

$$\mathrm{dS} = \mathrm{dQ}/\mathrm{T} \tag{1.7}$$

where T denotes temperature of the system.

The differential of work (dW) can be expressed as a sum of products of sort of work coordinates, extensive quantities indicating a kind of measure of the performed work, and work coefficients, intensive quantities, reflecting the effort required for this. For example, the work ($\mathrm{dW_p}$) which is done when a gas is compressed by a pressure p (the work coefficient) resulting in a volume alteration dV (the work coordinate) is determined by

$$\mathrm{dW_p} = -\mathrm{pdV} \tag{1.8}$$

The sign of this equation depends on the definition of the work differential. A positive dW means that there is an increase in the work done in favor of the system. In this case work is achieved through the compression, i.e. a negative differential of the volume.

In accord to the Second Law of thermodynamics an isolated system changes spontaneously its state towards a maximum in its entropy. When this state is achieved, then it is in thermodynamic equilibrium. In the same way, the decrease of the free energy down to a minimum can be considered as the way towards the equilibrium in the sense of the Second Law. Any change as a result of energy transformation leads to an increase in the entropy of the system or its environment. The term entropy measure ($\sigma = \mathrm{dS/dt}$) has been introduced to characterize this process. The entropy measure is always positive, but can approach zero asymptotically. The condition: $\sigma = 0$ would mean an idealized reversible process. Thermodynamically, a process is defined as being reversible if it can be repeated an arbitrary number of times without requiring the supply of additional energy.

The concept of entropy as a measure for information named Shanon entropy and contained in an ergodic process shares some intuition with Bernoulli's, and some of the mathematics developed in information theory turns out to have relevance in statistical mechanics.

In the discrete theory of information the main concern refers to real functions H defined on the family of sequences (p_1,..., p_n) such that $p_i \geq 0$ and $\sum p_i = 1$. There, a very significant role is played by the Shannon entropy, given by

$$H(p_1, \ldots, p_n) = -(p_1\log(p_1) + \ldots + p_n\log(p_n)) \tag{1.9}$$

where 'log' stands for the logarithm to the basis of two. Any change toward equalization of p_1,..., p_n leads to an increase of H, which reaches its maximum, log[n], for $p_1 = \cdots = p_n = 1/n$. Furthermore, $H(p_1, ..., p_n) = 0$ iff all p_i but one equal zero.

It is the only symmetric (i.e. independent of the order of p_i) continuous function of such sequences that is normalized by $H(1/2, 1/2) = 1$ and satisfies the following grouping axiom

$$\begin{aligned} &H(a_1p_1, \cdots, a_kp_1, b_1p_2, \cdots, b_lp_2, \cdots, c_1p_n, \cdots, c_mp_n) \\ &= H(p_1, p_2, \cdots, p_n) + p_1H(a_1, \cdots, a_k) \\ &\quad + p_2H(b_1, \cdots, b_r) + \cdots + p_nH(c_1, \ldots, c_m) \end{aligned} \tag{1.10}$$

The shape of the grouping axiom, leads to think about entropy as a value assigned to transformations, divisions or partitions, say from a number p_1, to its partition a_1p_1,..., a_kp_1, where a_i sum up to 1. In fact, extending H to nonnegative sequences, states $H(a_1p_1,..., a_kp_1) = p_1H(a_1,..., a_k)$ and satisfies the relation

$$\begin{aligned} &H(a_1p_1, \ldots, a_kp_1, b_1p_2, \ldots, blp_2, \ldots, c_1p_n, \ldots, c_mp_n) \\ &= H(p_1, p_2, \ldots, pn) \\ &\quad + H(a_1p_1, \ldots, a_kp1) \\ &\quad + H(b_1p_2, \ldots, b_lp_2) + \cdots + H(c_1p_n, \ldots, c_mp_n) \end{aligned} \tag{1.11}$$

whenever a_i, b_i, and c_i sum up to 1.

Moreover, the Shannon entropy is the unique (continuous) function that satisfies the following key properties:

(i) Continuity, meaning that $H(p_1, ..., p_n)$ is continuous in all its arguments p_1, ..., p_n;
(ii) Additivity, statting that $H(p_1q_1, p_1q_2, ..., p_nq_k) = H(p_1, ..., p_n) + H(q_1, ..., q_k)$;
(iii) Monotonicity, meaning that for any arbitrary k, n ∈ N: with k > n, we have $H(1/k, ...,1/k) > H(1/n, ...,1/n)$;

(iv) Branching, measuring of information independently on division of the process into parts, i.e. $H(p_1, \ldots, p_n) = H(p_A, p_B) + p_A H(p_1/p_A, \ldots, p_s/p_A) + p_B H(p_{s+1}/p_B, \ldots, p_n/p_B)$ where $p_A = (p_1, \ldots, p_s)$ and $p_B = (p_{s+1}, \ldots, p_n)$;

(v) Bit normalisation, meaning that the average information gained for two equally likely variables is one-bit ('binary digit'): $H(1/2,1/2) = 1$.

For a modern survey of characteristics of Shannon entropy, its relatives and the various conditions concatenated with their definition, see, among others [50].

The Shannon entropy can be generalized to the continuous case systems. Let p(x) be a space distribution probability density. The continuous Shannon entropy is

$$H(p) = -\int_a^b p(x)\log[p(x)]dx \tag{1.12}$$

with $a, b \in \boldsymbol{R}$. The generalization of (1.12) to densities of n variables $x_1, \ldots, x_n$ is straightforward. Usually we are interested in entropy differences rather than in absolute values, and then taking into account that the entropy differences are coordinate independent, therefore continuous Shannon entropy can be used to identify differences in information. Note that there is a fundamental connection between dynamical systems theory and information theory. In dynamical systems theory when chaotic state with deterministic yet irregular and unpredictable or even random behavior occupies centre stage, the Kolmogorov-Sinai entropy (in brief KS-entropy), also known as the metric entropy, appears as a convenient analytical tool. The KS-entropy describes the average information loss (respectively, increase of information) with respect to the time development of a trajectory. The KS-entropy can be seen as a value measuring the creation of information at each of iterations under the action of a chaotic map. Generally speaking, the KS-entropy is positive for chaotic systems and zero for regular systems. In general, in higher dimensions the KS-entropy rather than the Lyapunov exponent measures the loss of information.

In what follows we propose a method to determine the KS-entropy directly from a time-varying signal, i.e., a state variable. It can be determined in a following manner. Assuming an invariant probability measure μ of some flow $f(x)$ on the defined phase space X, the KS-entropy denoted as $H_{KS}(\mu)$ can be determined as follows. Suppose that there is an attractor in phase space and that the trajectory $x = x(t)$ is located in the basin of attraction. Consider a bounded region $M \subseteq X$ such that $\mu(M) = 1$ and be invariant under transformation of the flow $f(x)$. Let M consist of k disjoint partitions such that

$$M = M_1 \cup M_2 \cup \cdots \cup M_k \tag{1.13}$$

Thus, the entropy of the partition $\{M_i\}$ can be written as

$$H(\mu, \{M_i\}) = -\sum_{i=1}^{k} \sum \mu(M_i)\log(\mu(M_i)) \tag{1.14}$$

Since the flow $f(x)$ evolves in time, it produces a series of intersections $\{M(n)\}$ of the form $M_j \cap g^{-1}(M_i)$, $j, i = 1, 2, \ldots, k$ such that for n iterations the refined partitions $\{M_j(n)\}$ are given as

$$M_{j1} \cap f^{-1}(M_{j2}) \cap f^{-2}(M_{j3}) \cap \cdots \cap f^{-(n-1)}(M_{jn}) \tag{1.15}$$

with $j1, j2, \ldots, jn = 1, 2, \ldots, k$. Thus the entropy for the partitions $\{M_i\}$ can be written as

$$\overline{H}(\mu, \{M_i\}) = \lim_{n\to\infty} \frac{1}{n} H(\mu, \{Mj(n)\}) \tag{1.16}$$

From Eq. (1.15) can be seen, that the entropy depends on the original partition $\{M_i\}$.

Now, the KS-entropy can be formally defined as the supremum of Eq. (1.16) over all initial partitions $\{M_i\}$.

$$H_{KS}(\mu) = \sup_{\{Mi\}} \overline{H}(\mu, \{M_i\}) \tag{1.17}$$

In other words, the KS-entropy is the remaining uncertainty of the next outcome x_{n+1}, if all the past outcomes $x_n, x_{n-1}, \ldots, x_0$ with a certain uncertainty are known.

It can be easily proved that the KS-entropy does not exceed the sum of all positive Lyapunov exponents λ_p. Note, that entropy allows portraying the probability distributions of the possible state of a system, and therefore the information contained in the temporal dynamics. Its high value indicates on states of high complexity, while its very low level is synonymous with regular, "robust" behavior of the system. Moreover, its unquestionable advantage as a tool for assessing patients' health is the fact that it can be relatively easily calculated on the basis of state variables models, and as an individual number it is easily interpretable.

In order to establish the correct diagnosis with regard to the instantaneous state of a complex dynamical system and apply high-quality control, it is suitable to consider that all of the different possible internal states of it can be described by entropy following purely in terms of universal thermodynamic quantities like energy and temperature. Entropy measures complexity, or irregularity and unpredictability, of state variables which is extraordinarily relevant to the investigation of deterministic chaotic dynamics, quantification of the complexity of fractal dynamics and so on. It triggered an avalanche of new work, including cardiovascular studies as demonstrated by wide studies up to date performed. Note that the complexity can be estimated, though hardly, by classical methods, including time-, frequency-domain analysis, and time–frequency analysis that are the common built-in possibilities in commercialized standard and statistical software, but still obtained results based upon these traditional approaches are not quite satisfactory.

1.6 Instantaneous States of Dynamical Systems

Dynamical systems are extremely widespread in the natural world and in science. They are frequently used to model physical and technological phenomena. The goal is to develop models which take into account every single process and capture the essence of crucial interactions within the system components allowing their outcome to be more fully identified and predicted in their behavior with the time. Very often mathematical models are autonomous, i.e., they do not entail an explicit dependence on time. Such a definition can be extended to include as well systems with periodic excitation, since time can be added as an extra-variable with constant time-derivative.

Creating a dynamical system is to specify the rule for the time evolution of the system's state variables. The evolution rule of dynamical systems is an implicit relation that gives the state of the system for only a short time into the future. This rule must be defined to make the state variables be a complete description the state of the system in the following sense: the value of the state variables at a particular time must completely determine the system's evolution to all future states. In a wide variety of research fields such as mathematics, physics, biology, chemistry, engineering, economics, history, and medicine, a dynamical system is described as a "particle or ensemble of particles whose state varies over time and thus obeys differential equations involving time derivatives". The definition of particle has been expanded to also include the entities derived from living matter, e.g., human cells, virus, pedestrians, and swarms. A particle is not a mere entity but is now assumed to be able to perform a strategy, function, or interaction and thereby acquire the denomination of being 'active' or, even less accurate, 'intelligent'. In order to make a prediction about the system's future behavior, an analytical solution of such equations or their integration over time through computer simulation is realized. An instantaneous state is taken to be characterized by the instantaneous values of the variables considered crucial for a complete description of the state. In a continuous dynamical system the state of the system evolves through continuous time. In this context, the term evolution equation can be considered as a general framework whose solution is a function describing the time evolution of a microscopic, mesoscopic, or macroscopic quantity related to the system. One can think of the state of the system as flowing smoothly through the state space. As time evolves, the state $x(t)$ at time t can be thought of as a point that moves through the state space (Fig. 1.7). The evolution rule will specify how this point $x(t)$ moves by giving its velocity, such as through a function $v(t) = F(x(t))$, where $v(t)$ is the velocity of the point at time t. In this case, starting with an initial state $x(0)$ at time $t = 0$, the trajectory $x(t)$ of all future times will be a curve through the state space. Often the function $F(x(t))$ is deterministic, that is, for a given time interval only one future state follows from the current state. However, some systems are stochastic, in that random events also affect the evolution of the state variables.

To determine the state for all future times requires iterating the system's relation many times—each advancing time a small step. The iteration procedure is referred to as *solving the system* or *integrating the system*. If the system can be solved and

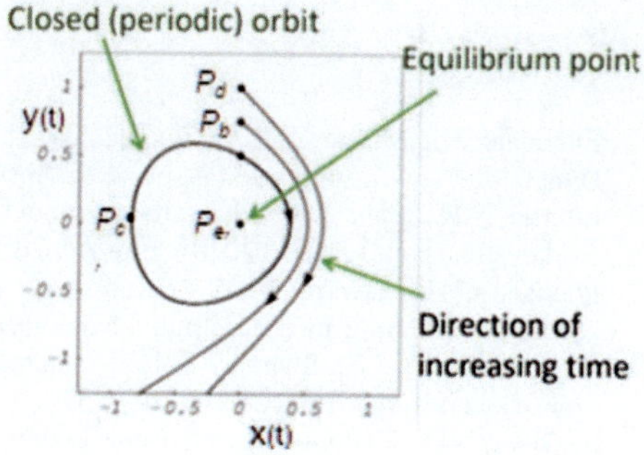

Fig. 1.7 Illustrative trajectories of a nonlinear system with two state variables $x(t)$ and $y(t)$ for different initial conditions P_e, P_c, P_b and P_d

an initial point is fixed, then it is possible to determine all its future answers, in the form of a trajectory or orbit.

A steady state is characterized by recurrent action, i.e., a particular point in the phase space takes a steady state if the system, after sufficient time, returns enough close to that point [37]. This definition includes fixed points, limit cycles, quasiperiodic, and chaotic steady states.

A dynamical system is in a transient state if its mapping point in the phase space does not meet the steady state conditions. Both steady and transient states are typical of dissipative dynamical systems with time invariant parameters and constant or periodic excitations in time, i.e., autonomous or periodically-forced systems, which therefore can be called *steady systems.* Transient systems are characterized by state variables which change in time. Note, that the theory of dynamical systems is well elaborated to deal with steady states problems. Fixed points, limit cycles, basins of attraction, etc., are all features of steady problems, even if they can be applied to study transient trajectories. However, meticulous attention must be paid to systems which are transient because they are subjected to nonperiodic excitations.

This problem especially concerns dynamical systems with parameters varying in time. They are subject to a transient excitation that is neither constant nor periodic. As mentioned above these systems belong to the transient systems, and are distinguished from steady systems in which parameters are constant. The latter can represent earthquake vibrations, traffic of cars in the city, vibrations of power lines under the influence of wind gusts, crossing the storm zone during the flight of the plane, etc. Therefore transient systems can be seen as an intermediary evolution of an ordinary steady system into another ordinary steady system, for both of which the classical theory of dynamical systems holds. The evolution from one steady system to the other is excited by a transient force, which states a control between the two steady systems.

References

1. Edelman, M., Macau, E.E.N., Sanjuan, M.A.F. (eds.): Chaotic, Fractional, and Complex Dynamics: New Insights and Perspectives. Springer, Cham, Switzerland (2018)
2. Lumley, S.F., et al.: Antibody status and incidence of SARS-CoV-2 infection in health care workers. N. Engl. J. Med. (2020). https://doi.org/10.1056/nejmoa2034545
3. Podhaisky, H., Marszalek, W.: Bifurcations and synchronization of singularly perturbed oscillators: an application case study. Nonlinear Dyn. **69**, 949–959 (2012)
4. Wolf, A., Swift, J.B., Swinney, H.L., Vastano, J.A.: Determining Lyapunov exponent from a time series. Physica **16D**, 285–317 (1985)
5. de Oliveira, O.: The implicit and inverse function theorems: easy proofs. Real Anal. Exchange. **39**(1): 214–216. https://doi.org/10.14321/realanalexch.39.1.0207. S2CID 118792515 2013
6. Solinski, M., Gieraltowski, J.: Different Calculations of Mathematical Changes of the Heart Rhythm. Center of Mathematics Applications, Warsaw University of Technology, Warsaw (2015)
7. Ion, S., Marinoschi, G., Popa, C.: Mathematical Modeling of Environmental and Life Sciences Problems. Editura Academiei Romane, Bucuresti (2006)
8. Caly, L., et al.: Isolation and rapid sharing of the 2019 novel coronavirus (SARS-CoV-2) from the first patient diagnosed with COVID-19 in Australia. Med. J. Aust. (2020). https://doi.org/10.5694/mja2.50569
9. Ross, W.D. (ed.): De Anima. In: The Works of Aristotle Translated into English, 12 vols. Oxford University Press, Oxford, 1908–1952
10. .Cornford, F.M.D.: Plato's cosmology: the Timaeus of Plato. Translated with a Running Commentary. Hackett Publishing Company, Inc., Indianapolis (1997)
11. Wolfram, S.: Universality and complexity in cellular automata. Physica **10** (1984)
12. Petras, I.: Fractional-Order Nonlinear Systems Modeling. Analysis and Simulation. Higher Education Press, Beijing (2011)
13. Binti, H.F.A., Lau, C., Nazri, H., Ligot, D.V., Lee, G., Tan, C.L., et al.: CoronaTracker: World-wide COVID-19 outbreak data analysis and prediction [preprint]. Bull World Health Organ. E-pub: 19 Mar 2020. https://doi.org/10.2471/BLT.20.255695
14. Babu, G.R., Ray, D., Bhaduri, R., Halder, A., Kundu, R., Menon, G.I., Mukherjee, B.: COVID-19 pandemic in India: through the lens of modeling. Global Health Sci. Pract. **9**(2) (2021)
15. Pilipchuk, V.N.: Between Linear and Impact Limits. Springer, Berlin (2010)
16. Wu, K., Darcet, D., Wang, Q., Sornette, D.: Generalized logistic growth modeling of the COVID-19 outbreak: comparing the dynamics in the 29 provinces in China and in the rest of the world. Nonlin. Dyn. https://doi.org/10.1007/s11071-020-05862-6
17. Cao, L., Liu, Q.: COVID-19 modeling: a review.https://doi.org/10.1145/nnnnnnn.nnnnnnn
18. Coronavirus disease (COVID-19) outbreak situation (2020). https://www.who.int/emergencies/diseases/novel-coronavirus (2019)
19. Zhu, N., et al.: A novel coronavirus from patients with pneumonia in China, 2019. N Engl J Med **382**, 727–733 (2019)
20. Fuk-Woo, C.J., et al.: Novel coronavirus indicating person-to-person transmission: a study of a family cluster. Lancet **2019**, 395 (2020)
21. Cohen, J., Normile, D.: New SARS-like virus in China triggers alarm. Science **367**, 234–235 (2020)
22. Tang, B., Wang, X., Li, Q., Bragazzi, N.L., Tang, S., Xiao, Y., et al.: Estimation of the transmission risk of the 2019-ncov and its implication for public health interventions. J. Clin. Med. **9**(2), 462 (2020)
23. Chen, Y., Liu, Q., Guo, D.: Emerging coronaviruses: genome structure, replication, and pathogenesis. J. Med. Virol. **92**, 418–423 (2020)
24. Zhou, P., et al.: A pneumonia outbreak associated with a new coronavirus of probable bat origin. Nature **579**, 270–273 (2020)
25. Centers for disease control and prevention. Coronavirus disease 2019 (COVID-19) (2020). https://www.cdc.gov/coronavirus/2019-ncov/index.html

26. Jingyuan, W., Ke, T., Kai, F., Weifeng, L.: High temperature and high humidity reduce the transmission of COVID-19 (2020). https://doi.org/10.2139/ssrn.3551767
27. Khajanchi, S., Sarkar, K., Mondal, J.: Dynamics of the COVID-19 pandemic in India (2020). arXiv preprint arXiv: 2005.06286
28. Event horizon—COVID-Coronavirus COVID-19 global risk assessment (2020). http://rocs.hu-berlin.de/corona/relative-import-risk
29. National Centre for Disease Control: COVID-19 outbreak in China (2020). https://ncdc.gov.in/WriteReadData/l892s/34827556791580715701.pdf
30. Bhargava, B., Sudan, P.: Prepared for the coronavirus. The Hindu (2020). https://www.thehindu.com/opinion/op-ed/prepared-for-the-coronavirus/article30785312.ece
31. Ferguson, N.M., et al.: Impact of non-pharmaceutical interventions (NPIs) to reduce COVID-19 mortality and healthcare demand. In: Imperial college COVID-19 response team, pp. 1–20 (2020). https://www.imperial.ac.uk/media/imperialcollege/medicine/sph/ide/gida-fellowships/Imperial-College-COVID19-NPImodelling-16-03-2020.pdf
32. Khajanchi, S., Bera, S., Roy, T.K.: Mathematical analysis of the global dynamics of a HTLV-I infection model, considering the role of cytotoxic T-lymphocytes. Math. Comput. Simulation **180**, 354–378 (2021)
33. Anderson, R.M., Heesterbeek, H., Klinkenberg, D., Hollingsworth, T.D.: How will country-based mitigation measures influence the course of the COVID-19 epidemic? Comment **395**, 931–934 (2020)
34. Trzaska, Z.: Study of mixed mode oscillations in a nonlinear cardiovascular system. Nonlinear Dyn. **100**(3), 2635–2656 (2020)
35. Izhikevich, E.M.: Dynamical Systems in Neuroscience: The Geometry of Excitability and Bursting. MIT Press, Cambridge (2007)
36. Heldt, T., Chang, J.L., Chen, J.J.S., Verghese, G.C., Mark, R.G.: Cycle-averaged dynamics of a periodically-driven, closed loop circulation model. Control Eng. Pract. **13**(9), 1163–1171 (2005)
37. Parlikar, T.A., Heldt, T., Verghese, G.C.: Cycle-averaged models of cardiovascular dynamics. IEEE Trans. Circuits Syst. I Fundam. Theory Appl. **53**(11), 2459–2468 (2006)
38. Kozlovskaya, I.B., Grigoriev, A.I.: Russian system of countermeasures on board of the International Space Station (ISS): the first results. Acta Astronaut. **55**, 233–237 (2004). j.actaastro.2004.05.049
39. Adamec, I., Junakovic, A., Krbot, M., Habek, M.: Association of autonomic nervous system abnormalities on head-up tilt table test with joint hypermobility. Eur. Neurol. **79**(5–6), 319–324 (2018)
40. Rajzer,M., Kawecka-Jaszcz, K.: Arterial compliance in arterial hypertension. From pathophysiology to clinical relevance. Arter. Hypertens. **6**(1), 61–73 (2002)
41. Dimitrakopoulos, E.G.: Nonsmooth analysis of the impact between successive skew bridge-segments. Nonlinear Dyn. **74**, 911–928 (2013)
42. Heldt, T., Verghese, G.C., Mark, R.G.: Mathematical modeling of physiological systems. In: Batzel, J.J., Bachar, M., Kappel, F. (eds.) Included in Mathematical Modeling and Validation in Physiology: Applications to the Cardiovascular and Respiratory Systems. Springer, Berlin (2013)
43. Subramaniam, B., Khabbaz, K.R., Heldt, T., Lerner, A.B., Mittleman, M.A., Davis, R.B., Goldberger, A.L., Costa, M.D.: Blood pressure variability: can nonlinear dynamics enhance risk assessment during cardiovascular surgery. J. Cardiothorac. Vasc. Anesth. **28**(2), 392–397 (2014)
44. Trzaska, Z.: Nonsmooth analysis of the pulse pressured infusion fluid flow. Nonlinear Dyn. **78**, 525–540 (2014)
45. Trzaska, Z.: Dynamical processes in sequential-bipolar pulse sources supplying nonlinear loads. Electr. Rev. **90**(3), 147–152
46. Trzaska, Z.: Properties and applications of memristors—memristor circuits with innovation in electronics. In: Czyz, Z., Maciag, K. (eds.) Contemporary Problems of Electrical Engineering and Development and Evaluation of Technological Processes, pp. 76–93. Publisher TYGIEL, Lublin (2017)

47. Giusti, A., Mainardi, F.: A dynamic viscoelastic analogy for fluid-filled elastic tubes. Meccanica **51**(10), 2321–2330 (2016)
48. Colombaro, I., Giusti, A., Mainardi, F.: On the propagation of transient waves in a viscoelastic Bessel medium. Z. Angew. Math. Phys. **68**(3), 62 (2017)
49. Trzaska, Z.: Mixed Mode Oscillations (MMOs). Springer, Cham, Switzerland (2021)
50. Fedorowski, A., Stavenow, L., Hedblad, B., Berglund, G., Nilsson, P.M., Melander, O.: Orthostatic hypotension predicts all-cause mortality and coronary events in middle-aged individuals. Eur. Heart J. **31**, 85–91 (2010)

Chapter 2
Basic Ingredients of Nonlinear Systems Dynamics

2.1 Introduction

By and large, most real-world phenomena are dynamic, non-linear, and time-variable. Despite the fact that they appear mainly in the world of physics, it should be emphasized that their dynamics indicates on impressive analogies with other types of dynamical systems, such as chemical, medical, biological, and even appearing in some social processes or market exposure capital. System science is the study of the dynamic behavior of systems, based on interactions of its components and their interactions with other systems. The basis of the method is the recognition that the structure of any system is often just as important in determining its behavior as the individual components themselves. Examples are chaos theory and social dynamics. It is also asserted that there are often properties-of-all that cannot be found amongst the properties-of-elements. In some cases, the behavior of the whole cannot be explained in terms of the behavior of the parts. The interactions that have place in different systems, such as a driver trying to control a car safely or a production supervisor trying to optimize the manufacturing process, exhibit interesting analogies. For instance, the systems theories being developed for better control and optimization of chemical and mechanical production processes applied judiciously may create benefits in other fields, such as politics, law, sociology, psychology, medicine, and marketing. Practically all systems of human investigation interest, whether mechanical, ecological, or social, are systems which evolve over time; that is, they are *time dynamic* systems.

A natural process is described by a set of natural variables that depend on a single independent variable. For example, natural variables can be the position, speed or acceleration of a car as it drives along a racetrack. In most cases, natural processes change over time according to laws or natural principles involving natural variables, so that *time* is almost always the *independent variable.* Natural variables whose given values at an instant, together with the natural laws of the process, uniquely

Z. Trzaska, *Mathematical Modelling and Computing in Physics, Chemistry and Biology*, Studies in Systems, Decision and Control 495,
https://doi.org/10.1007/978-3-031-39985-5_2

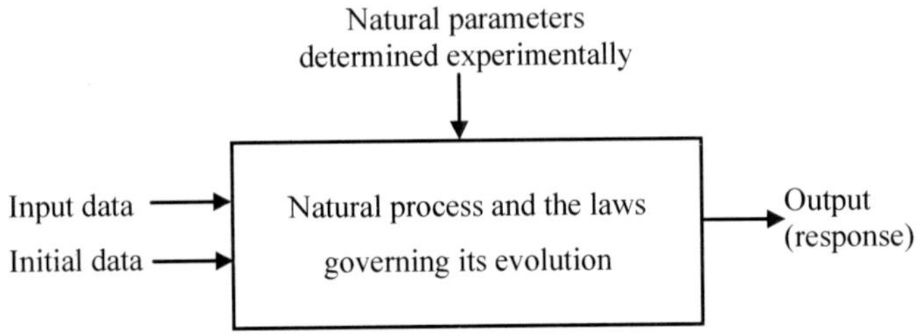

Fig. 2.1 Diagram of the operation of a dynamical system

determining the state of the dynamical system at that instant are called *state variables*. A natural process or its mathematical representation known as *mathematical model*, described by state variables is referred as *dynamical system*. Factors in the external environment that affect a dynamical system are *input data* to the system. They form *driving terms* or *source terms*. Systems without the driving terms are called *undriven* or *autonomous systems* and those affecting by such terms are called *driven* or *nonautonomous*. Values of the state variables at fixed initial time are *initial data*.

A collection of the state variables and input data that follow the natural laws of the process and respond to initial data is the response called also output of the dynamical system. Figure 2.1 shows a diagram of the operation of a dynamic system with focusing primary attention on the basic ingredients. After selecting state variables for a dynamical system, and applying natural laws governing the activity of the system, a mathematical model consisting of a system of ODEs and initial conditions is created.

2.2 Nonlinear Phenomena and Nonlinear Equations

2.2.1 Preliminaries

Everywhere in nature occur phenomena manifesting in the form of rhythms and vibrations leading to more or less regular patterns of movement, light or sound, which can sometimes be strong. In general, most of them are dynamic, non-linear, and time-varying. Though they appear mainly in the world of physics, it is worth noticing that their dynamics indicates on considerable analogies with other kinds of systems such as chemical, medical, biological, and even appearing in some social processes or production market capital.

The prevalence in nature of large-scale systems and inexpensive robust installations in practice leads to the need to investigate them as extensive systems. For this reason, it is advisable to focus the study attention on the operation of individual devices in the system or even complete subsystems. This leads to the complex object being treated as a modular system that can be easily configured and maintained in proper operation compared to a monolithic system.

If a combination of a large number of simple elements as one object is presented, then its operation is more complex than the operation of only the elements. The functioning of the whole system results not only from the functioning of each of its parts separately, but also from the mutual relations and interactions of these parts with each other. Thanks to this, it is possible to build systems with a much richer behavior and functioning than it would only result from their components. This coming to light of qualitatively new forms and behaviors from the interaction between simpler elements is called emergence. In general, the reason for this statement is that the effect of the action of the whole is not just the sum of the effects of the action of its parts—overall, there are usually new properties and phenomena that do not show up in its components. If a system cannot be in any sense described only by its component parts, then it is defined as emergent. This means, in the strict scientific sense, that the given object is irreducible. In this case, the simplified description of the lower level object known to us is not sufficient to describe the higher level object. In others words an emergent system is one where:

- the final system has properties and effects which exceed the properties and effects of its component parts, and
- whose final properties and effects cannot be predicted from merely examining the component parts only.

Recall that local interactions between components of an initial disordered system create a process of self-organization leading to some form of overall order. The resulting organization is wholly decentralized, distributed over all the components of the system.

It is worth emphasizing that the process can be spontaneous when sufficient energy is available, not needing control by any external agent. It is often triggered by seemingly random fluctuations, amplified by positive feedback [1].

Let us now specify four fundamental conditions of self-organization:

- strong dynamical non-linearity, often necessarily involving positive and negative feedback,
- balance of exploitation and exploration,
- abundant interplay,
- accessibility of energy to overpass natural tendency promoting increase of entropy, or disorder.

There are different types of emergence and related effects. It should be emphasized that the emergent behavior of the system is the result of non-linear interactions of elements causing deterministic processes. Such behavior should be distinguished from accidental, as their nature is different. Various forms of nonlinearity are the cause of emergent states of practical systems. Three of the possible ones are the most important, namely: technological limitations in the production of high-capacity components, feedback and signal transmission delays. The emergent system can operate in a wide range of dynamic states, including the undesirable steady states, periodic oscillations and chaos. If emergent operation of a given system is not desired, then an appropriate strategy of its control should be adopted. One of them is to eliminate nonlinear interactions from the system, which is often not feasible. Another way is to cause any oscillation in the system to be dampened by limiting its operation well below the optimality level. In the case of automation, emergent states of systems appear to be a consequence of modern architecture of control systems [2, 3]. Designers and producers of appropriate technologies must take adequate measures so that the resulting systems can be easily identified, evaluated and modified. When the microstructural dimension approaches nano size and/or the interfacial boundary achieves space-charging behavior, the traditional linear driving force and mobility terms become insufficient to describe physical observations. This situation involves non-linear and higher order phenomena, which if properly analyzed can provide new insights and allow better predictions of small dimensional phenomena.

Before proceeding to the particular investigation of physical phenomena, it is helpful to distinguish among and define three classes of linear and/or nonlinear models.

- *Fully Linear Models*

In this case, both the static equilibrium problem (usually trivially so) and the dynamic variability problem may be formulated by linear models. Moreover, changes in time of the dynamic system's variables are sufficiently small that there are no significant nonlinear effects due to system's phenomena evolutions.

- *Dynamically Linear Models*

For these models, the static problem must be investigated by a nonlinear model, but the system dynamic evolution about this static equilibrium may be investigated by a linear dynamic model. The solution, in sequence, of first the nonlinear static model and then the dynamically linear model usually has major conceptual and computational advantages over a fully nonlinear model. The above means that if the behavior of a natural system is to be modeled as dynamically linear, the dynamic changes of system's variables must remain sufficiently small or to say it another manner, the study concerns those phenomena that involve small dynamic changes. Some differential equations have solutions that can be written in an exact and closed form.

- *Fully Nonlinear Models*

In this category of models, the dynamic changes themselves are so large that a nonlinear dynamic model must be used not considering advantages to solving the static and dynamic problems separately and consecutively.

A system behavior pattern is said to be linear if the ratio of change in an output due to a change in input is constant. In other words, the output varies in a straight line for the variation of its input. Note that the vast majority of systems concepts to which researchers are engaged are linear since there exist for them elegant analytical and algorithmic solutions to allow linear problems to be easily solved [1, 4–11]. The concept of linear system can mean at least three different, but highly related ideas,

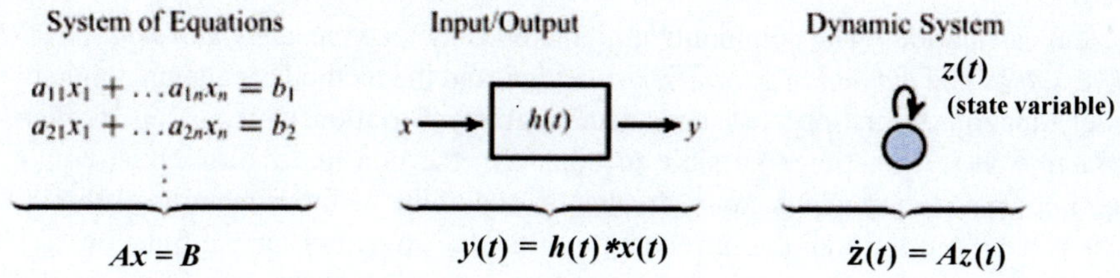

Fig. 2.2 Ilustration of different concepts of linear systems

illustrated in Fig. 2.2. The distinction between these three approaches depends on context, but it is not recommended for mixing them. For the time being, the attention is focused on linear dynamic systems. In all cases, the fundamental principle of superposition applies: For a system subject to the superposition principle, its response to a sum of multiple inputs is exactly the sum of responses to the individual inputs.

Nonlinear systems are more common in the nature, such as the growth of a car speed with increase of the fuel supply or the attraction force of a piece of iron by a magnet with the distance between them. It is worth emphasizing that significant research has been undertaken worldwide in the fields of large-scale, nonlinear problems, so a great deal is in fact known, however the analysis of such systems is really very challenging [10, 12–21].

For instance, to explain phenomena in structural mechanics, nanotechnology, chemistry and physics, very often nonlinear dynamic models with integer or non-integer nonlinearity are evident in engineering applications. Nowadays, the relevance of nanotechnology is well recognized, so that new developments and applications based on non-linear dynamics are achieved in an interdisciplinary setting. Experimental investigation conducted in materials engineering showed that the stress–strain properties of the material are strong nonlinear. Therefore, the stress–strain diagrams designated are mathematically approximate with a one polynomial term whose coefficient and sequence correspond to the experimental data obtained.

In the following, dynamically linear and fully non-linear dynamic models will be expended. Fully linear models will not be considered.

2.2.2 *Paradigmatic Nonlinear Models*

For many years, linear models, both theoretical and experimental, have served the researchers and practitioners extraordinarily well. However, most non-linear differential equations cannot be solved by hand, and for these equations it is necessary to rely either on advanced theory or numerical methods. On the other hand, there are special types of ordinary differential equations which can be solved analytically. Several suitable techniques can be applied in practice for nonlinear dynamic system calculations. The commonly used methods include the *method of separation of variables*, the *method of parameter variation*, and the method of reducing higher-order equations to first-order systems with parameter variation, which is called *order reduction*. It is sometimes possible to transform the non-linear differential equation into an equation which can be treated by one of the methods mentioned above. This is achieved through the introduction of another function for the function $x(t)$ in an appropriate manner. This technique lies at the core of the *simplifications by substitution method*.

In this book we consider a different research frontier for dynamic processes, that is, the study of nonlinear dynamical systems. Of course, it has been known for many years that nonlinear effects in natural and manmade systems may be important. Indeed "nonlinear" effects are often appealed as a possible explanation for any difference between theory and experiment. Many times these nonlinear effects are small, hence the major successes of linear models. However, sometimes nonlinear effects are more important, and, very often they are crucial. Of course, there are many natural phenomena that have defied rational modeling, at least until recently, and often this is thought to be due to important nonlinear effects. Among those that are of significant interest are: earthquakes, climate changes, spaceflights, military security, electromagnetic terrorism, achievements of artificial intelligence, etc.

There are two basic limitations of linearization. First, since linearization is an approximation in the neighborhood of an operating point, it can only predict the "local" behavior of the nonlinear system in the vicinity of that point. It cannot predict the "nonlocal" behavior far from the operating point and certainly not the "global" behavior throughout the state space. Second, the dynamics of a nonlinear system are much richer than the dynamics of a linear system. There are "essentially nonlinear phenomena" that can take place only in the presence of nonlinearity; hence, they cannot be described or predicted by linear models. The following are examples of essentially nonlinear phenomena:

- *Finite escape time.* The state of an unstable linear system goes to infinity as time approaches infinity; a nonlinear system's state, however, can go to infinity in finite time.
- *Multiple isolated equilibria.* A linear system can have only one isolated equilibrium point; thus, it can have only one steady-state operating point that attracts the state of the system irrespective of the initial state. A nonlinear system can have more than one isolated equilibrium point. The state may converge to one of several steady-state operating points, depending on the initial state of the system.
- *Limit cycles.* For a linear time-invariant system to oscillate, it must have a pair of eigenvalues on the imaginary axis, which is a nonrobust condition that is almost impossible to maintain in the presence of perturbations. Even if we do, the amplitude of oscillation will be dependent on the initial state. In real life, stable oscillation must be produced by nonlinear systems. There are nonlinear systems that can go into an oscillation of fixed amplitude and frequency, irrespective of the initial state. This type of oscillation is known as a limit cycle.

- *Subharmonic, harmonic, or almost-periodic oscillations.* A stable linear system under a periodic input produces an output of the same frequency. A nonlinear system under periodic excitation can oscillate with frequencies that are submultiples or multiples of the input frequency. It may even generate an almost-periodic oscillation; an example is the sum of periodic oscillations with frequencies that are not multiples of each other.
- *Chaos.* A nonlinear system can have a more complicated steady-state behavior that is not equilibrium, periodic oscillation, or almost-periodic oscillation. Such behavior is usually referred to as chaos. Some of these chaotic motions exhibit randomness, despite the deterministic nature of the system [19, 22–26].
- *Multiple modes of behavior.* It is not unusual for two or more modes of behavior to be exhibited by the same nonlinear system. An unforced system may have more than one limit cycle. A forced system with periodic excitation may exhibit harmonic, subharmonic, or more complicated steady-state behavior, depending upon the amplitude and frequency of the input. It may even exhibit a discontinuous jump in the mode of behavior as the amplitude or frequency of the excitation is smoothly changed.

To develop non-linear systems analysis tools, it is useful to first consider representative cases of the most commonly applied non-linearities. They are exhibited in the sequence below of simple systems with one non-linear element.

- *Plain pendulum*

Consider first the simple pendulum shown in Fig. 2.3, where l denotes the length of a thin non-extensible suspension and m denotes the mass of the metal weight. Assume the suspension as rigid and having zero mass. Let θ denote the angle between the suspension and the vertical axis through the pivot point. The pendulum is free to swing in the vertical plane. Applying Newton's second law of motion gives

$$ml\ddot{\theta} + mg\sin\theta + kl\dot{\theta} = 0 \tag{2.1}$$

where g is the acceleration due to gravity and k denotes the coefficient of friction. Taking $x_1 = \theta$ and $x_2 = \dot{\theta}$ as the state variables yields the state equation

Fig. 2.3 A plain pendulum

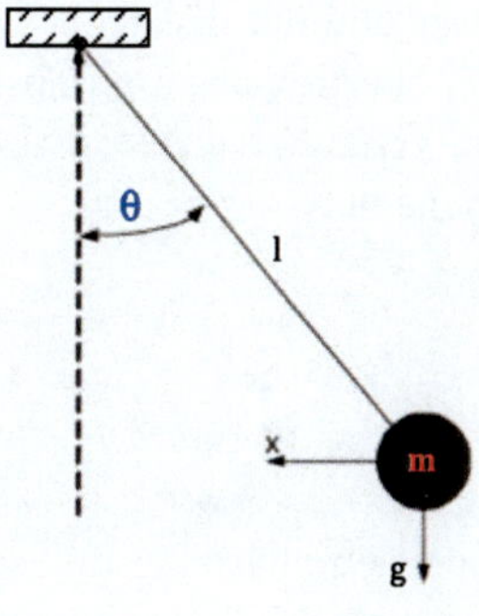

$$\begin{aligned} \dot{x_1} &= x_2 \\ \dot{x_2} &= -\frac{g}{l}\sin x_1 - \frac{k}{m}x_2 \end{aligned} \tag{2.2}$$

Setting $\dot{x}_1 = \dot{x}_2 = 0$ and then solving for x_1 and x_2 yields

$$\begin{aligned} x_2 &= 0 \\ -\frac{g}{l}\sin x_1 - \frac{k}{m}x_2 &= 0 \end{aligned} \tag{2.3}$$

i.e. the equilibrium points $(n\pi, 0)$, for $n = 0, \pm 1, \pm 2, \ldots$. From the physical point of view it is clear that the pendulum has only two equilibrium positions which are described as (0, 0) and (π, 0). Observe that these two equilibrium positions are quite distinct from each other importantly. The pendulum can rest easy at the (0, 0) equilibrium point, but it can hardly hold the (π, 0) equilibrium point because infinitesimally small disturbance from that equilibrium returns the pendulum to the first equilibrium point. Moreover, the analysis of the origin of such phenomenon clearly points out at the necessity of the existence of positive and negative feedback loops that make up the course of a given process. That case of oscillatory proceeding in time can be considered as the interaction of two feedback loops: a positive and a negative. The positive feedback loop is initiated at the lowest position of the pendulum (then the movement's velocity is maximal). However, at the highest position of the pendulum appears the negative feedback loop which firstly annihilates the movement and next changes the velocity direction to the lowest position.

At times the consideration is the pendulum equation when the coefficient of friction k is small. In this case, a version of the pendulum equation where the friction resistance is neglected by the k = 0 setting is frequently examined. The resulting system's state variables equation takes the form

$$\begin{aligned} \dot{x}_1 &= x_2 \\ \dot{x}_2 &= -\frac{g}{l}\sin x_1 \end{aligned} \tag{2.4}$$

The system described by (2.4) is conservative and after an initial pulse it oscillates forever without dissipating energy into the environment. This, of course, is unrealistic, but provides a glimpse of how the pendulum oscillates. An interesting form of the pendulum state variables equation appears if a torque T is applied to the pendulum, namely

$$\begin{aligned} \dot{x}_1 &= x_2 \\ \dot{x}_2 &= -\frac{g}{l}\sin x_1 - \frac{k}{m}x_2 + \frac{1}{ml^2}T \end{aligned} \tag{2.5}$$

It is worth emphasizing that several practically important physical systems are modeled by state variables equations similar to the pendulum equation. Models such as synchronous generator connected to an infinite bus, Josephson junction circuit, and a phase-locked loop take the forms similar to the pendulum equation. It should therefore be stressed that the pendulum equation is of major practical importance.

- *Negative-Resistance Oscillator*

An important class of nonlinear circuits is represented by the basic structure shown in Fig. 2.4a. It is assumed that the inductor and capacitor are linear, time invariant and passive, that is, $L > 0$ and $C > 0$. The third element is a nonlinear active resistor (with a tunnel diode) represented by the characteristic $i = h(v)$, shown in the Fig. 2.4b. The active resistor exhibits the cubic nonlinear property, that is, $i = h(v) = \gamma v^3 - \alpha v$ where i and v are current and voltage, respectively. Parameters α and γ are assumed to be constant. Applying the Kirchhoff current and voltage laws yields the equations.

The tunnel diode has the cubic current–voltage relationship

$$i = h_t(v) = h(V - E_0) + I_0, \tag{2.6}$$

with $h(V) = \gamma V^3 - \alpha V$. The Kirchhoff current equation for the circuit in Fig. 2.4 can then be written as

$$\frac{dV}{dt} = \frac{1}{C}(-h(V) - i_L) \tag{2.7}$$

with $\frac{d}{dt} i_L = V/L$. Thus differentiating (2.7) with respect to time yields

$$\frac{d^2}{dt^2}V - 1/C(\alpha - 3\gamma V^2)\frac{d}{dt}V + 1/LCV = 0. \tag{2.8}$$

Introducing new variables $x = \sqrt{3\gamma/\alpha}\,V, t' = t/\sqrt{LC}$, and $\varepsilon = \sqrt{L/C}\,\alpha$, Eq. (2.8) can be transformed as follows

$$\ddot{x} - \varepsilon(1 - x^2)\dot{x} + x = 0 \tag{2.9}$$

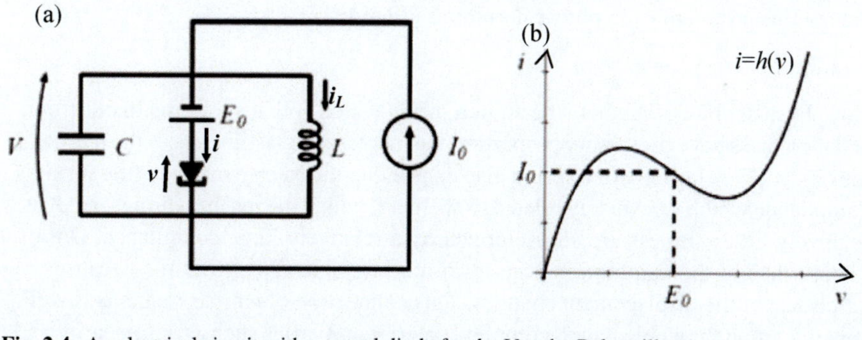

Fig. 2.4 An electrical circuit with a tunnel diode for the Van der Pol oscillator

where x is the dynamical variable and $\varepsilon > 0$ denotes a parameter.

The circuit that was considered by Van der Pol [27], includes an active resistor instead of a classic passive resistor. This way, when the current x is small, the quadratic term x^2 is negligible and the system becomes a linear differential equation with a negative damping $-\varepsilon\dot{x}$; the resistor acts as if it is pumping energy in the system. On the other hand, when the current x is large, the term x^2 becomes dominant and the damping becomes positive. Then the damping of the energy of the system takes place (unlike a usual resistor which simply dissipates energy). The interplay between energy injection and energy absorption results in a periodic oscillation in voltages and currents in the circuit.

Since its introduction, the Van der Pol equation has been used as a basic model for oscillatory processes in physics, electronics, biology, neurology, sociology and economics. The model was soon generalized to a forced system

$$\ddot{x} - \varepsilon\left(1 - x^2\right)\dot{x} + x = F(\omega t) \tag{2.10}$$

where $F(\omega t)$ is an external force, possibly depending on some parameter $\omega \in \boldsymbol{R}$. In particular, much attention has been devoted to the study of the VdP equation under an external periodic (sinusoidal) force $F(\omega t) = A\sin(\omega t)$, with $A, \omega \in \boldsymbol{R}$. Van der Pol himself built a number of electronic circuit models of the human heart to study the range of stability of heart dynamics [21, 27]. Such investigations with the addition of an external driving signal were analogous to the situation where a real heart is driven by a pacemaker. This allows interested cardiologists to find ways to stabilize an irregular heart rhythm (arrhythmias).

To develop a state model for the VdP circuit, it suffice to take $x_1 = x$ and $x_2 = \dot{x}$ in Eq. (2.10) to obtain

$$\begin{aligned} \dot{x}_1 &= x_2 \\ \dot{x}_2 &= \varepsilon\left(1 - x_1^2\right)x_2 + x_1 + F(\omega t) \end{aligned} \tag{2.11}$$

The electrical circuit elements with the nonlinear property can also be realized using operational amplifiers. By this method, much research has been done to study the nonlinear dynamics in physical systems [12–14, 22].

- *Mass-Spring System*

Any physical object, whether it be air or a spring which both have spring like stiffness and inert mass have oscillatory properties. An increase in mass increases the inertia of the object. The elasticity of a spring also can be quantified by a measure of a spring's compliance which is inversely related to stiffness, which means that springs that have relatively little stiffness are characterized by a relatively large compliance. During the oscillation, the amplitude of the restoring force changes over time because the amplitude of the displacement changes. This connection of simple elements is well-suited for modeling object with complex material properties such as nonlinearity and viscoelasticity. The mass-spring connections have proven very useful in numerous

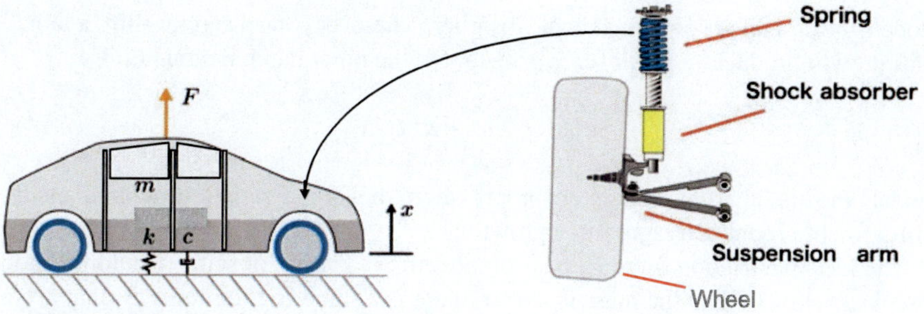

Fig. 2.5 Illustration of an application of a mass-spring system: **a** scheme of a car, **b** a wheel with shock absorber

industrial and utility processes. One of them is used in cars to dampen their vibrations when driving on a bumpy road [4, 5].

A spring-mass system is used to great effect in cars in the form of shock absorbers, which are placed above the wheel (Fig. 2.5). They are designed for preventing the car from being damaged when it passes over bumps and other obstacles. As a car goes over a bump, the spring above the shock absorber will compress and these springs must be designed to oscillate at the right amplitude and frequency to make the ride of the car comfortable.

The mass-spring connections have proven very useful in numerous industrial and utility processes and many applications of such systems have been demonstrated in the past decade.

In this case mass-spring system is in a vertical position; the equilibrium position is displaced downwards due to the gravitational force acting on the mass.

Denoting x as the displacement from a reference position and applying Newton's law of motion yields

$$m\ddot{x} + F_f + F_{sp} = F \tag{2.12}$$

where F_f is a resistive force due to friction and F_{sp} is the total restoring force of the shock absorber. It is assumed that F_{sp} is a function only of the displacement x and write it as $F_{sp} = g(x)$. Moreover, the reference position has been chosen such that $g(0) = 0$. The mass of the car, passengers and luggage per one wheel is denoted as m. The external force F is determined by the condition of driving and state of the road. Depending on F, F_f, and g, several interesting autonomous and nonautonomous second-order models occur.

For a relatively small jump, the restoring force of the absorber can be modeled as a linear function $g(x) = kx$, where k is the shock absorber constant. For a large jump, however, the restoring force may depend nonlinearly on x. For example, the function

$$g(x) = k\left(1 - a^2x^2\right)x, \quad |ax| < 1 \tag{2.13}$$

models the so-called softening shock absorber, where, beyond a certain shift, a large shift growth produces a small force growth. On the other hand, the function

$$g(x) = k\left(1 + a^2x^2\right)x \tag{2.14}$$

models the so-called hardening absorber, where, beyond a certain growth, a small shift growth produces a large force growth.

The resistive friction force F_f has components as a result of static, Coulomb and viscous friction. When the mass is at rest, there is a static friction force F_s that acts parallel to the absorber surface and is limited to $\pm\, \mu_s\, mg$, where $0 < \mu_s < 1$ is the static friction coefficient. This force is of some value, within its limits, to keep the mass at rest. The static friction force balance the restoring force of the shock absorber and maintain equilibrium for $|g(x)| \leq \mu_s\, mg$. Once motion has started, the resistive friction force F_f, which acts in the direction opposite to motion, is modeled as a function of the sliding velocity $v = \dot{x}$. The resistive force due to Coulomb friction F_c has a constant magnitude $\mu_k\, mg$, where μ_k is the kinetic friction coefficient, that is,

$$F_c = \begin{cases} -\mu_k mg & \text{for } v < 0 \\ \mu_k mg & \text{for } v > 0 \end{cases} \tag{2.15}$$

As the mass moves in a viscous medium, such as air, there will be a frictional force due to viscosity. This force is usually modeled as a nonlinear function of the shift speed; that is, $F_v = h(v)$, where $h(0) = 0$. For small speed, we can assume that $Fv = cv$.

The combination of a hardening shock absorber, linear viscous friction, and a periodic external force $F = A\cos\,\omega\, t$ results in the Duffing's equation

$$m\,\ddot{x} + c\dot{x} + kx + a^2x^3 = A\cos\omega t \tag{2.16}$$

which is a classic example in the study of periodic excitation of systems containing cubic nonlinear components.

The combination of a linear absorber, static friction, Coulomb friction, linear viscous friction, and zero external force results in

$$m\,\ddot{x} + k\dot{x} + cx + \sigma(x, \dot{x}) = 0 \tag{2.17}$$

where

$$\sigma(x, \dot{x}) = \begin{cases} \mu_k mg\, sign(\dot{x}), & \text{for } |\dot{x}| > 0 \\ -kx & \text{for } |\dot{x}| > 0 \quad \text{and} \quad |\mathrm{x}| \leq \frac{\mu_s mg}{k} \\ -\mu_k mg\, sign(\dot{x}) & \text{for} |\dot{x}| = 0 \quad \text{and} \quad |\mathrm{x}| > \mu_s \mathrm{mg/k} \end{cases} \tag{2.18}$$

The value of $\sigma\ (x, \dot{x})$ for $\dot{x} = 0$ and $|x| \leq \frac{\mu_s mg}{k}$ is obtained from the equilibrium condition $\ddot{x} = \dot{x} = 0$. Putting $x_1 = x$ and $x_2 = \dot{x}$, gives the state model, namely

$$\begin{aligned}\dot{x}_1 &= x_2 \\ \dot{x}_2 &= -\frac{c}{m}x_1 - \frac{k}{m}x_2 - \frac{1}{m}\sigma(x_1, x_2)\end{aligned} \tag{2.19}$$

It is worth emphasizing two features of this state model. First, it exhibits an equilibrium collection, rather than isolated equilibrium points. Second, the right-hand side function is a discontinuous function of the state variables. The discontinuity results from the idealization adopted in modeling friction. Note that such an idealization contradicts expectations that the physical friction changes from its static friction mode into its sliding friction mode in a smooth way, not suddenly and unexpectedly. However, the discontinuous idealization simplifies the analysis. For example, when $x_2 > 0$, we can model the system by the linear model

$$\begin{aligned}\dot{x}_1 &= x_2 \\ \dot{x}_2 &= -\frac{c}{m}x_1 - \frac{k}{m}x_2 - \mu_k g\end{aligned} \tag{2.20}$$

In a similar way, when $x_2 < 0$, the system (2.18) can be transformed to the linear model

$$\begin{aligned}\dot{x}_1 &= x_2 \\ \dot{x}_2 &= -\frac{c}{m}x_1 - \frac{k}{m}x_2 + \mu_k g\end{aligned} \tag{2.21}$$

Thus, in each parts of the state space, we can predict the behavior of the system via linear analysis. This is an example of the so-called piecewise linear analysis, where a system is represented by linear models in various parts of the state space, certain coefficients changing from part to part.

- *Oscillator with an active resistor*

The basic circuit structure representing a significant class of electronic oscillators is illustrated in Fig. 2.6a. The coil and capacitor are assumed to be linear, time invariant and passive, that is, $L > 0$ and $C > 0$. The active resistive element is determined by the characteristic $i = h(\mathrm{v})$ shown in Fig. 2.6b. The function $h(v)$ satisfies the conditions: $h(0) = 0$, $h'(0) < 0$, $h(v) \rightarrow +\infty$ as $v \rightarrow +\infty$, and $h(v) \rightarrow -\infty$ as $\mathrm{v} \rightarrow -\infty$, where $h'(v)$ is the first derivative of $h(v)$ with respect to v.

Applying Kirchhoff's current law yields

$$i + i_L + i_C = 0 \tag{2.22}$$

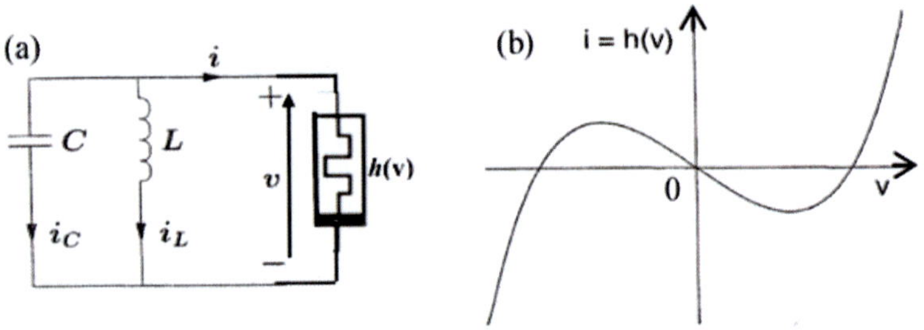

Fig. 2.6 Oscillator circuit with nonlinear resistor: **a** circuit scheme, **b** nonlinear resistor characteristic

Taking into account the dependencies determining the individual currents in the circuit, and then differentiating with respect to the time of the sides of the obtained equation, after a simple manipulation of the terms, one obtains

$$CL\frac{d^2v}{dt^2} + h'(\mathrm{v})L\frac{dv}{dt} + v = 0 \tag{2.23}$$

In order to locate the determined equation in the general theory of dynamical nonlinear systems, it is possible to transform the variables t into the form $\tau = t/\sqrt{\mathrm{CL}}$. Consequently, the derivatives of v with respect to t and τ are related by

$$\frac{dv}{d\tau} = \sqrt{CL}\frac{dv}{dt} \quad \text{and} \quad \frac{d^2v}{d\tau^2} = CL\frac{d^2v}{dt^2} \tag{2.24}$$

Denoting the derivative of v with respect to τ by v', we can rewrite the circuit equation as

$$v'' + \varepsilon h'(v)v' + v = 0 \tag{2.25}$$

where $\varepsilon = \sqrt{\mathrm{L/C}}$. This equation is a special case of *Lienard's equation*

$$\ddot{v} + f(v)\dot{v} + g(v) = 0 \tag{2.26}$$

where the symbol $\dot{}$ denotes differentiation with respect to independent variable. Equation (2.26) is known as the *Van der Pol equation.* To derive a state variables equation for (2.25), it suffices substitute $x_1 = v$ and $x_2 = v'$ to obtain

$$\begin{aligned} \mathrm{x}_1' &= \mathrm{x}_2 \\ \mathrm{x}_2' &= -\mathrm{x}_1 - \varepsilon \mathrm{h}'(\mathrm{x}_1)\mathrm{x}_2 \end{aligned} \tag{2.27}$$

This equation is an essential example in non-linear oscillation theory. It may exhibit a periodic solution that attracts every other solution except the zero solution at the unique equilibrium point $x_1 = x_2 = 0$.

- *Memristive devices and systems*

In recent decades, advances in science and engineering have focused on improving manufacturing and end-product applications at the nanometric scale that become a key factor in society's security, economic well-being and improved quality of life. Presently, nanotechnology is concentrated on the formation of purposeful materials, devices, and systems through the management of matter on the nanometer level and the utilization of new products and attributes at that scale. Nanometric structure materials have notably advantageous properties and constitute an effective alternative to conventional materials. The implementation of appropriate nanotechnology made it possible to produce a new component of the system in 2008, which was postulated by L.O. Chua in 1971. According to its properties of remembering the past state preceding a given state, it was called a memristor as the simplest element from a large class of nonlinear dynamical systems endowed with memristance, the so-called memristive systems. The memristor as a new element completes the symmetry in the set of passive elements used in electrical circuits for the tetrad of the following quantities: charge, current, voltage, magnetic flux, that is, q, i, u, φ, respectively. In the dyadic subsets, the interactions are expressed by the resistor R, the capacitor C and the inductor L, respectively, and the memristor M, missing until recently. Made in nanotechnology memristor exhibits a peculiar "fingerprint" characterized by a pinched hysteresis loop (Fig. 2.7b) confined to the first and the third quadrants of the v–i plane whose contour shape in general changes with both the amplitude and frequency of any periodic "sinewave-like" input voltage source, or current source. At the origin of the v–i plane both signals are identically zero crossing.

In its structure are coupled to one another, two thin layers of TiO_2 (having a thickness of about 10 nm) located between two platinum electrodes (Fig. 2.8a). One of the layers is doped with oxygen vacancies and constitutes a semiconductor. The second layer which is not doped is an insulator.

As a result of complex processes in the material layers the thickness $w(t)$ of the doped layer changes with the amount of electric charge $q(t)$ which flows through the memristor. Memristor has important advantages compared to other elements because being the two-terminal structure has a significant benefit over the three-terminal transistors. Since a fast flow of nanoscale ions takes place in the memristor, it is considered as a nanoionic electronic device.

The concise characteristic of its properties can be given as follows:

- the layer of titanium dioxide (TiO_2) reveals high resistance ROFF,
- layer of TiO_{2-x} (number of carbon impoverished of oxygen, O-Vacancies) shows low resistance RON,
- memristor's dimensions is of order of a few *nm*,
- the voltage of a few volts applied to memristor's terminals distant each other by a few *nm* is determined by

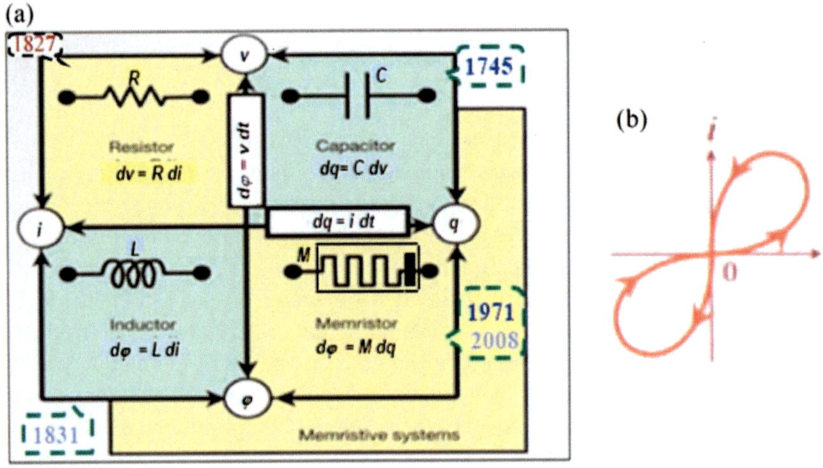

Fig. 2.7 Conceptual symmetries of resistor, capacitor, inductor, and memristor

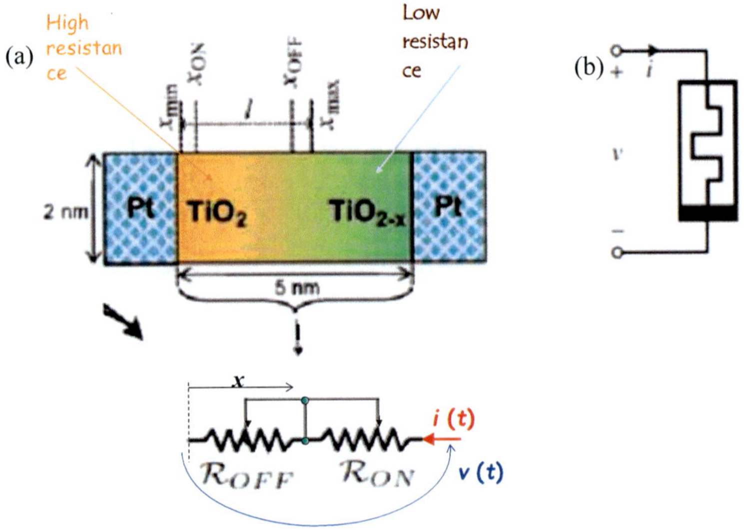

Fig. 2.8 Schetch of the memristor structure: **a** view of the real element, **b** circuit symbol

$$
\begin{aligned}
v(t) &= \left[R_{ON}\left(1 - \frac{x(t)}{D}\right) + R_{OFF}\frac{x(t)}{D} \right] i(t) \\
&= M(q)i(t), \\
M &= R_{ON}\left(1 - \frac{x(t)}{D}\right) + R_{OFF}\frac{x(t)}{D}
\end{aligned}
\tag{2.28}
$$

where $M(q)$ is the memristance and $v(t)$ and $i(t)$ denote the voltage and current, respectively, D is the total device length. The memory effect is due to the dependence of memristance M on $q = \int i(t)dt$. It should be emphasized that the magnetic flux φ is generalized from the circuit characteristic of an inductor. It does not represent a magnetic field here. The symbol φ may be regarded as the integral of voltage $v(t)$ over time. Physically considering $M(q)$ is the positive-constrained parameter for all q-values (assuming the element is passive). An extra thermodynamic peculiarity arises from the definition that memristors should energetically act like resistors. The instantaneous electrical power entering such an element is completely dissipated as Joule heat to the surrounding, so no extra energy remains in the system after it has been brought from one resistance state x_n to another one x_m.

Among the many applications of the memristor, let us consider in detail the use of its properties to build an oscillator. Typically, an electronic oscillator contains a passive linear inductor ($L > 0$) and a linear capacitor ($C > 0$) that interact with a non-linear resistor exhibiting a characteristic v-i having segments with a negative slope. In such a system, by selecting the appropriate parameters of the elements it is possible to obtain oscillations, e.g. sinusoidal waveforms, when the supply source of a constant voltage is applied. It turns out, however, that the structure of the oscillator can be simplified by using only one memristor and a DC voltage source (Fig. 2.9a).

Any 2-terminal electric element known as a memristor obeys the following voltage-controlled state-dependent Ohm's law

$$i = G(x, v)v \tag{2.29}$$

and the state-variables equation

$$\frac{\mathrm{dx}}{\mathrm{dt}} = g(\boldsymbol{x}, \boldsymbol{v}) \tag{2.30}$$

where v is the supplying voltage. The scalar G($\mathbf{x}$, v) in (2.29) is called the memductance (acronym for memory conductance) and $\boldsymbol{x}$ denotes the state vector.

Considering a passive memristor let its memductance be described as

$$G(x, v) = \alpha\left(x^2 - 1\right) \tag{2.31}$$

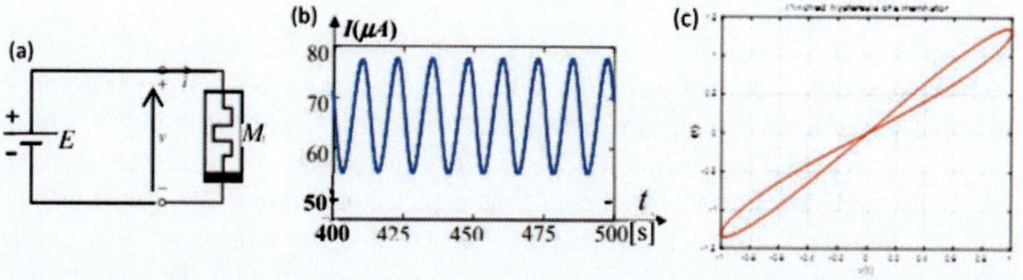

Fig. 2.9 One memristor oscillator: **a** scheme, **b** steady state current, **c** pinched hysteresis $v - i$

and

$$g(x, v) = v - \beta x - xv \tag{2.32}$$

With the appropriate selection of the memristor parameters and $E = 12$ V, oscillations of the current flowing to the memristor are obtained (Fig. 2.9b). It is worth noticing that when the memristor is driven by any periodic voltage source $v(t)$ with zero mean, the current response $i(t)$) is periodic of the same frequency. Then the memristor exhibits a unique fingerprint called a pinched hysteresis loop under excitation of any bipolar periodic signal with zero average. To illustrate the memristor in (2.29) exhibits this fingerprints, a sinusoidal voltage signal v(t) $= A \sin(2\pi ft)$ with amplitude $A = 1$ V, and frequency $f = 1$ Hz was applied across this memristor.

The right figure in Fig. 2.9 is a double-valued Lissajous figure plotted on the $i - v$ plane. Such a multi-valued figure of $v(t)$, $i(t)$, which passes through the origin is called a pinched hysteresis loop [28]. This unique feature is the characteristic property of a memristor that distinguishes it from non-memristive devices.

Numerous examples of pinched hysteresis loops of memristors can be found in [29–31]. In particular, it can be proved that beyond some critical frequency f_c, the area of each lobe of the pinched hysteresis loop of all memristors is a strictly monotone-decreasing function of the frequency f. Moreover, at sufficiently high frequencies, the pinched hysteresis loops must tend to straight lines (whose slope depends on the amplitude of the exciting periodic waveform) for all generic memristors [11].

It is worth noticing that memristors are omnipresent in real life constellations. Actually, many devices, including the "electric arc" dating back to 1801, have now been identified as memristors [32, 33]. Locally-passive memristors aside from serving as non-volatile memories [30] have been used for switching electromagnetic devices [31], for field programmable logic arrays [2, 34–37], for synaptic memories [3, 28, 38], for neuron in learning within artificial intelligence [39–41], and so on. Moreover, locally-active memristors have been found to exhibit many exotic dynamical phenomena, such as oscillations [42], chaos [43, 44], Hamiltonian vortices [45] and auto waves [46], etc.

2.2.3 *Nonlinear Oscillators Solvable in Elementary Functions*

Determination of a solution to an ordinary differential equation (ODE) may be important if that solution always appears in the mathematical model representing the given system, or is realized in special conditions. It is often desirable to determine the solution by using a traditional process of the direct integration of the given differential equation. This approach reveals much information about the continuity of the solution course and its long-term sustainability, which determines the equilibrium of the system. Thus, equilibrium solutions, which correspond to configurations, in which the physical system functions, do not occur in natural situations, unless they are stable. There are a number of approaches to realization of explicit integration of nonlinear ODEs. The integration techniques introduced so far for solving first-order ODEs share a common feature: each method transforms a differential equation of some special form into a differential equation that we can solve by performing integration. It should be stressed that there is no general method of determining nonlinear ODEs solutions, but those that can be directly integrated exhibit specific forms. Lots of differential equations that describe dynamic systems can be solved with power series methods and in cases where the solved equation is important those power series solutions tend to get names of their solution's authors like "Bessel functions", "Jacobi elliptic functions", "Sierpinski triangle", and so on.

Those and other special functions aren't really any less elementary than trigonometric functions *sin* and *cos*: they have recurrence relations, completeness/orthogonality relations, and reasonably simple transformation under derivatives, etc.

The motive for *sin* and *cos* is preferable that they form an enclosed basis under the translation of time:

$$\cos(t + \delta t) = A\cos(t) - B\sin(t) \tag{2.33}$$

where $A = \cos(\delta t)$ and $B = \sin(\delta t)$. This makes them attractive in problems where the equations of system dynamics themselves are time translation invariant because all time derivatives turn *cos* and *sin* into each other. Alternative approaches such as harmonic balance can also be used to determine the solution of an oscillator homogenous equation in terms of Fourier series.

A simple test that reveals whether a first-order ODE has a solution formula that arises by performing integration is given below. Consider first-order ODEs of the form

$$P(t, x)x'(t) + Q(t.x) = 0 \tag{2.34}$$

where P and Q are continuously differentiable functions in a rectangle R of the t-x plane formed by an interval I on the t-axis and an interval J on the x-axis.

It is very helpful to specify the exact ODE as follows: The ODE $P(t,x)x'(t) + Q(t,x) = 0$ is an exact ODE in a rectangle R of the t-x plane if P and Q are continuously differentiable, and if

$$\partial P/\partial t = \partial Q/\partial x \tag{2.35}$$

for all points (t, x) in R. For example, the equation

$$\mathrm{x}' = \frac{1 + 2tx^2}{2t - 2t^2x} \tag{2.36}$$

may be written as

$$\left(2t^2x - 2x\right)x' + 1 + 2tx^2 = 0 \tag{2.37}$$

with $P = 2\,t^2x - 2x$ and $Q = 1 + 2tx^2$. ODE (2.36) is exact in the $t - x$ plane because

$$\frac{P}{t} = 4tx \quad \text{and} \quad \frac{Q}{x} = 4tx$$

Note that if P is a function of x only and if Q is a function of t only, then $P(x)x' + Q(t) = 0$ is both separable and exact. In general, solutions of an exact equation are implicitly defined by following procedure.

Suppose that (2.34) is exact in a rectangle R. Then there is a twice continuously differentiable function $H(t,x)$ on R with the property that:

$$\partial H/\partial x = Q \text{ and } \partial H/\partial t = P, \tag{2.38}$$

for all (t,x) in R. If a differentiable function $x = x(t)$ with graph in R satisfies the equation

$$H(t, x) = C \tag{2.39}$$

for a constant C, then $x(t)$ is a solution of $Px' + Q = 0$. Conversely, any solution $x = x(t)$ of the ODE whose graph lies in R is also a solution of $H(t, x) = C$ for some constant C. In other words, $H(t, x)$ is an integral of ODE (2.34) and formula (2.37) gives the implicit general solution. This means that, if the ODE $Px' + Q = 0$ passes the exactness test (2.35) in some rectangle R, then we can solve it implicitly for $x = x(t)$ that defines also solution curves in R. Each solution curve $x = x(t)$ in R lies on a level curve (or contour) of the function H defined by Eq. (2.39). To construct an integral for (2.39) the function $H(t,x)$ should be determined taking into account the conditions given by (2.38). So we have

$$\frac{\partial H}{\partial t} = 1 + 2tx^2 \quad \text{and} \quad \frac{\partial H}{\partial x} = 2t^2 \tag{2.40}$$

Integrating of the first of these equations with respect to t, holding x fixed, yields

$$H(t, x) = t + t^2x^2 + q(x) \tag{2.41}$$

where $q(x)$ is a differentiable function to be determined. Using (2.41) and the second equation of (2.40), gives

$$\frac{\partial H}{\partial x} = 2t^2x + q'(x) = 2t^2x - 2x \tag{2.42}$$

Comparing the corresponding terms in the above equality, we get

$$q'(x) = -2x \tag{2.43}$$

Hence

$$q(x) = -x^2 \tag{2.44}$$

Substituting the above result into (2.41) yields

$$H(t, x) = t + t^2x^2 - x^2 = C \tag{2.45}$$

Solving (2.45) for x, we obtain the explicit solution formulas

$$x(t) = \left(\frac{C - t}{t^2 - 1}\right)^{\frac{1}{2}} \quad \text{or} \quad x(t) = -\left(\frac{C - t}{t^2 - 1}\right)^{\frac{1}{2}} \tag{2.46}$$

for values of t for which $(C - t)/(t^2 - 1)$ is positive.

Some observations are as follows: first, note that the test states that an exact ODE has an integral, and the verification of the test shows how to find one; second, note that the level set $H(t, x) = C$ may not be a single curve, but a collection of disjoint curves, or branches.

In cases of differential equations for which condition (2.35) is not met, good results are obtained by using the method of separation of variables, in which algebra allows one to rewrite an equation so that each of two variables occurs on a different side of the equation. The idea of the method will be presented on the example of a dynamical system described by the equation

$$\frac{dx}{dt} = x^2 - 1, \quad \text{with} \quad x(0) = x_0 \tag{2.47}$$

As long as $x^2 - 1 \neq 0$, we can rearrange terms to obtain

$$\frac{dx}{x^2 - 1} = dt \tag{2.48}$$

so that the two variables t and x have been separated. Integrating both sides of the equation with respect to t, we have

$$\int \frac{dx}{(x - 1)(x + 1)} = \int dt \tag{2.49}$$

Evaluating integrals on both sides of the above equality gives

$$\frac{1}{2}\ln|x - 1| - \frac{1}{2}\ln|x + 1| = t + C \tag{2.50}$$

Initial condition $x(0) = x_0$ allows to determine the integration constant C, namely

$$C = \frac{1}{2}\,\mathrm{In}\left|\frac{x_0 - 1}{x_0 + 1}\right| \tag{2.51}$$

Thus substituting (2.51) into (2.50) and then rearranging the terms yields the final form of the solution, that is

$$x(t) = \frac{e^{2t}(x_0 - 1) + x_0 + 1}{e^{2t}(1 - x_0) + x_0 + 1} \tag{2.52}$$

The dynamics of the system changes depending on the distance from fixed points. When a system has a stable point, it tends to reach this state. An unstable point is a point from which the trajectories of the system diverge. From Fig. 2.10 we can read that if the system starts from the point $x_0 = -1$, it stays there all the time. When $x_0 < 1$, the system always tends to a stable point, i.e. to the point $x^* = -1$. However, this happens with different dynamics that depend on the initial conditions. If x_0 is very close to the second fixed point, e.g. $x^* = 0.999$, the system will slowly move away from it. As it moves away from it, its velocity will increase, only to decrease again as the system approaches the stable point $x^* = -1$. The system behaves differently when it starts from the point $x_0 = -3$. At first, it will move at high speed, and then slow down as it approaches a stable point.

By a quite similar course of action, we can solve the problem of population growth of the human species described by the equation

$$\frac{dP}{dt} = kP\left(1 - \frac{P}{K}\right) \tag{2.53}$$

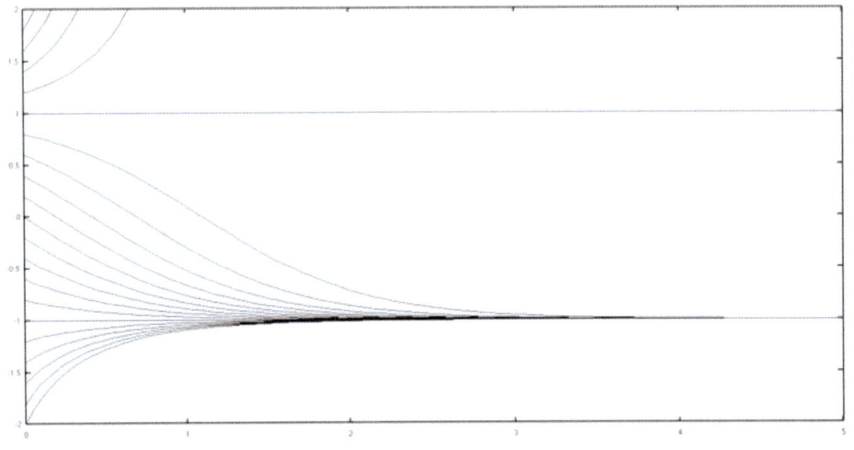

Fig. 2.10 Graphics of the solution (2.52) for different initial conditions

where P is the population with respect to time t, k is the rate of growth, and K is the carrying capacity of the environment. Separating variables and manipulating with the terms gives

$$\int \frac{K}{P(K-P)} dP = \int k dt \tag{2.54}$$

Decomposing the integrand expression in the left-hand side of (2.54) into simple fractions and then taking the integrals, we get

$$\ln|P| - \ln|K-P| = kt + C \tag{2.55}$$

Thus

$$\left|\frac{P}{K-P}\right| = e^{kt+C} \tag{2.56}$$

It follows that

$$\frac{P}{K-P} = e^{-C} e^{-kt} \tag{2.57}$$

Let

$$A = \pm e^{-C} \tag{2.58}$$

Therefore, the solution to the population growth equation is

$$P(t) = \frac{K}{1 + Ae^{-kt}} \tag{2.59}$$

To determine A, let $t = 0$ and $P(0) = P_0$. Then we have

$$A = \frac{K - P_0}{P_0} \tag{2.60}$$

Then we may rewrite the solution (2.59) as

$$P(t) = \frac{KP_0}{P_0 + (K - P_0)e^{-kt}} \tag{2.61}$$

This is the formula for solutions of the no war events population growth with $P_0 > 0$ and $P_0 \neq K$.

If $P_0 > 0$, then $P(t) \rightarrow K$ as $t \rightarrow \infty$ since $e^{-kt} \rightarrow 0$ and the denominator in formula (2.61) is always positive. To illustrate the above considerations let us take the society population growth with the rate coefficient $k = 0.8$ (per unit time, e.g.

one year) and the carrying capacity K = 40 millions, so that applying (2.61) yields

$$P(t) = \frac{40P_0}{P_0 + (40 - P_0)e^{-0.8t}} \tag{2.62}$$

Assuming various initial states, the population growth curves are obtained in the form indicated in Fig. 2.11. Since $e^{-0.8t} \to 0$ as $t \to \infty$, it is seen that $P(t) \to K =$ 40 *millions*. Diagram on the left of Fig. 2.11 shows the state line with attracting and repelling equilibrium points. Resources of the community can support a population size K, which is the asymptotic limit of the population curves.

Because the ODE (2.53) has a single state variable, the state space is the y-axis that forms a *state line*. Usually it is draw parallel to, but distinct from, the y-axis. An up-arrow or a down arrow is placing on the state line to indicate the direction of the course of the solution. An equilibrium solution corresponds to an equilibrium point on the state line. An equilibrium point is an attractor (A), or sink, if all orbits that originate near the equilibrium point on the state line tend to that point as $t \to \infty$. An equilibrium point is a repeller, or source, if all orbits that originate nearby on the state line move away from that equilibrium point as time increases. An equilibrium point on the state line is an attractor/repeller (AR) if it attracts on one side, but repels on the other.

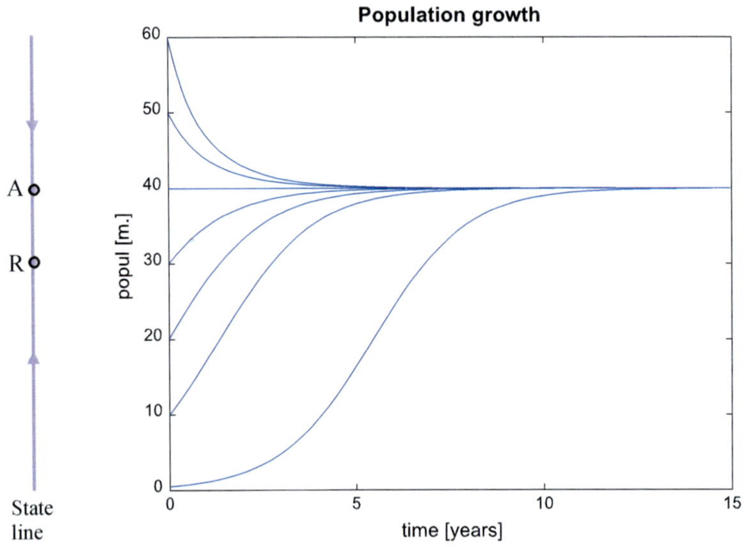

Fig. 2.11 Population growth changes: no war events

2.3 Models of Complex Nonlinear Dynamic Systems

2.3.1 *Basic Forms of Systems Oscillations*

Oscillations represent a very wide group of processes, which are generally characterised by their regular state repeating caused by the internal dynamics of a system. Such systems, whose internal couplings allow oscillations, are called oscillating systems. Oscillations occur not only in mechanical systems but also in dynamic systems in virtually every area of science: for example the beating of the human heart (for blood circulation), business cycles in economics, predator–prey population cycles in ecology, geothermal geysers in geology, vibration of strings in guitar and other string instruments, periodic firing of nerve cells in the brain, and the periodic swelling of Cepheid variable stars in astronomy. From the energy point of view, the oscillations are conditioned by the existence of two conservative forms of energy, which can reversibly exchange due to the internal dynamics of the system. There is, for example, potential energy ↔ kinetic energy (oscillations of mass on a spring) or electric field energy of capacitor ↔ magnetic field energy of inductor (an oscillating LC circuit). All real-world oscillator systems are thermodynamically irreversible. This means there are dissipative processes such as friction or electrical resistance which continually convert some of the energy stored in the oscillator into heat in the environment. This is called damping. Thus, oscillations tend to decay with time unless there is some net source of energy into the system.

If the oscillating system is isolated from external influences, it oscillates impulsively under influences of the initial energy stored in conservative components. Such systems are called autonomic. If the energy losses during the oscillations in the system are negligibly small the oscillations are maintained over time. The oscillations of the ideal system without loss are called unamortized auto-oscillations and represent only theoretical idealization because in every real system there are mechanisms of loss. For example, friction, heat losses due to internal friction, energy losses of the electrical system by radiation, etc. cause the irreversible transformation of the conservative form of the system energy into another non-conservative one. As a result of such losses the total energy of the conservative components decreases and the amplitude of oscillations gradually decreases over time too. Such type of oscillations is called damped self-oscillations. Due to the damping, these self-oscillations go out after some time, which depends on the rate of losses. This process generates a transient phenomenon in the system, for example, the vibrations of the string of the musical instrument fade; a swinging of pendulum stops after a certain time; oscillations of an *RLC* circuit gradually vanish, etc. If there is a means of covering energy losses with a device that gives energy, then the system can oscillate continuously without damping.

Analyzing the graphs presented in Fig. 2.12, it is easy to see that in an autonomous circuit (solutions for autonomous systems are often called free) the oscillations take different forms depending on whether damping is present that when it is not. In a no damped case free solutions satisfy the harmonic oscillator ODE

$$x'' + \omega_0^2 x = 0 \tag{2.63}$$

All solutions of (2.63) have the form

$$x(t) = A cos(\omega_0 t + \varphi) \tag{2.64}$$

where constants A and φ depends on the initial conditions. The quantity ω_0 is the *natural circular frequency* of the harmonic oscillator. Courses in time described by ODE (2.63) are periodic with period $= 2\pi/\omega_0$.

When damping is present in the system, then in a general case ODE takes the form

$$x'' + 2cx' + \omega_0^2 x = F(t) \tag{2.65}$$

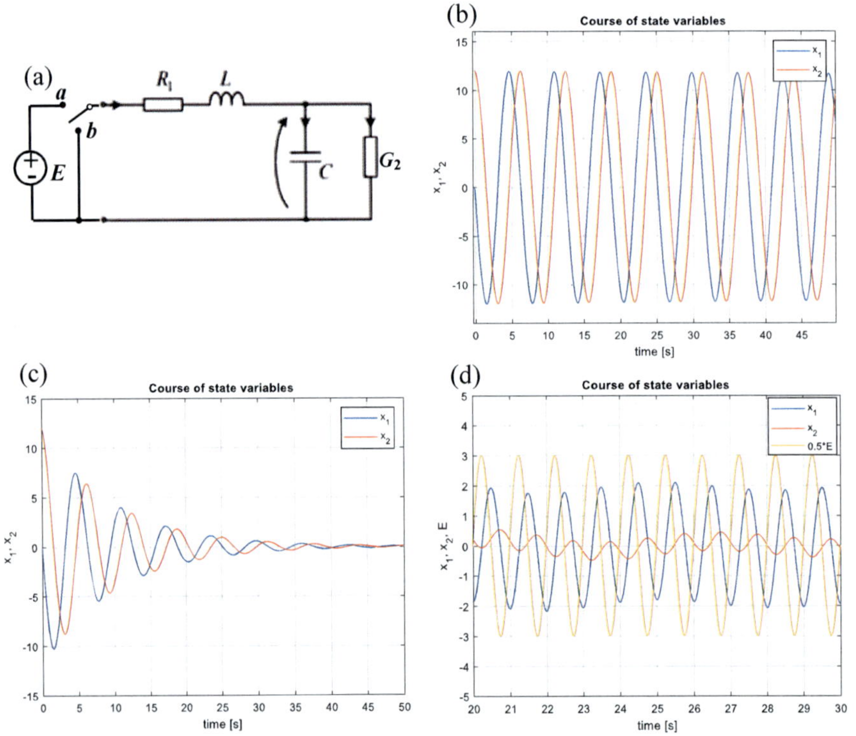

Fig. 2.12 Dynamics of a second order linear circuit: **a** circuit structure with a switching device, **b** course in time of state variables in an autonomous no damped circuit (switch in position **b**), **c** course in time of state variables in an autonomous damped circuit (switch in position **b**), **d** forced steady state oscillations (switch in position **a**) under sinusoidal excitation

where coefficient $2c$ represents the damping in the system and $F(t)$ is known as forcing term. If the system is autonomous, then $F(t) = 0$.

The characteristic polynomial $p(s) = s^2 + 2cs + \omega_0^2$ involves two characteristic roots

$$s_1,\ s_2 = -c \pm \sqrt{c^2 - \omega_0^2} \tag{2.66}$$

Since c and ω_0 are positive and real quantities, the real parts of s_1 and s_2 are always negative whatever the actual values of c and ω_0. Which is why the exact nature of the free solutions depends on the relative sizes of the constants c and ω_0. There are three cases:

- Overdamped when $c > \omega_0$ and s_1 and s_2 are real, negative and distinct;
- Critically damped when $c = \omega_0$ and $s_1 = s_2 = -c < 0$;
- Underdamped when $c < \omega_0$ and $s_1 = \alpha + i\omega,\ s_2 = \alpha - i\omega$ with $\alpha = -c$ and $\omega = \sqrt{\omega_0^2 - c^2}$, where $i = \sqrt{-1}$.

2.3.2 *Systems Under Periodic Pulsed Excitation*

The Dirac delta distribution (*δ distribution*), also known as the *unit impulse*[27] is a *generalized function* or *distribution* over the *real numbers*, whose value is zero everywhere except at zero, and whose integral over the entire real line is equal to one. In other words the Dirac distribution has large amplitude and is highly localized in time, what means that

$$\lim_{h\to 0} f_h(t - t_0) = \begin{cases} +\infty & if \quad t = t_0 \\ 0 & if \quad t \neq t_0 \end{cases} \tag{2.67}$$

The impulse I_h of the function $f_h(t)$ as h decreases ($h \to 0$) is

$$\lim_{h\to 0^+} I_h = \lim_{h\to 0^+} \int_0^\infty f_h(t - t_0)dt = 1$$

Some properties of the δ distribution are

$$\delta(t - t_0) = \lim_{h\to 0^+} f_h(t - t_0) = 0 \quad t \neq t_0$$

$$\int_0^\infty \delta(t - t_0)g(t)dt = \lim_{h\to 0^+} \int_0^\infty f_h(t - t_0)g(t)dt = g(t_0) \tag{2.68}$$

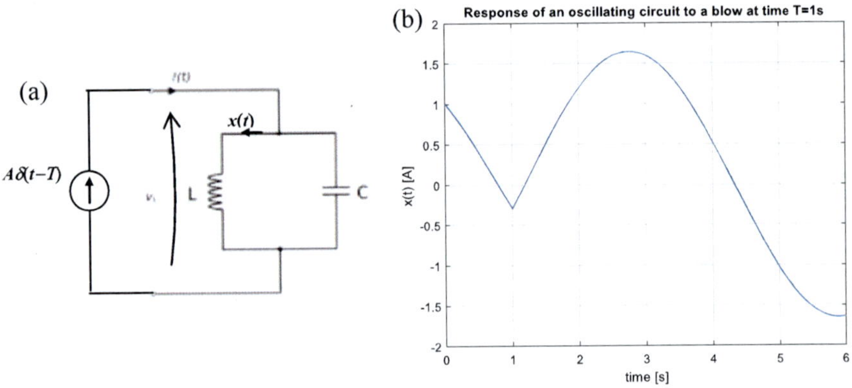

Fig. 2.13 A system with a δ distribution input: **a** lossless oscillating electric circuit, **b** response of the circuit to a blow at $\mathrm{T} = 1\mathrm{s}$

In physical terms the δ distribution represents a force of infinite amplitude that acts instantaneously at the time t_0 and generates a unit impulse.

To see how to determine the response of a system to a δ distribution input consider an electric circuit shown in Fig. 2.13 in which L is the inductance, C is the capacity and $I(t) = A\delta(t - T)$ denotes the driving current. If $x = x(t)$ denotes the current in the coil then the model of the circuit is

$$LCx'' + x = A\delta(t - T), \quad x(0) = a, \quad x'(0) = \beta \tag{2.69}$$

The instant of the time when current source begins supplying the circuit is denoted by $T > 0$. Dividing (2.69) by LC and setting $\omega^2 = 1/LC$ and then applying the Laplace transformation to both side of resulting equation yields

$$\left(s^2 + \omega^2\right)X(s) - \alpha s - \beta = \frac{Ae^{-sT}}{LC} \tag{2.70}$$

where $X(s) = \mathcal{L}[x(t)]$ is the Laplace transform of the coil current. Solving (2.70) with respect to $X(s)$ we obtain

$$X(s) = \frac{\alpha s}{s^2 + \omega^2} + \frac{\beta}{s^2 + \omega^2} + \frac{Ae^{-sT}}{\left(s^2 + \omega^2\right)LC} \tag{2.71}$$

After inverting $X(s)$ to the time domain we get

$$x(t) = \alpha \cos\omega t + \frac{\beta}{\omega}\sin\omega t + \frac{A}{LC\omega^2}\sin\omega(t - T)\mathrm{Step}(T) \tag{2.72}$$

The waveform of the current $x(t)$ in the coil is shown in Fig. 2.13b. The graph of $x(t)$ shows a jump discontinuity at $t = T$. The current in coil continues to oscillate after the impulsive blow.

Obtaining exact solutions of problems involved by periodic non-sinusoidal (often discontinuous) sources encounters serious difficulties, appearing in practical applications of the Fourier series. Overcoming these difficulties is possible by using the method of the unified representation of composite periodic non-sinusoidal waveforms by using a "*saw-tooth waveform*" and "*concatenation procedure*". For the construction of an appropriate algorithm we consider a general setting, i.e., waveforms $x(t)$ with discontinuities. However, they are subjected to two restrictions, namely: (*i*) a global condition requiring that $x(t)$ be absolutely integrable, (*ii*) a local condition constraining x(t) to have a finite number of maxima and minima and a finite number of discontinuities in every finite interval. Moreover, we use the results known in the general statement for the special case of the concatenation and show that this leads to an ingenious procedure in Fourier series-less analysis also in the situation of real jumps in the input as well as output waveforms [14, 15].

The method is carried out in the following manner: (*i*) for a given system and a non-sinusoidal periodic excitation f(t) $= f(t + T)$ the corresponding differential equation with respect to a suitably chosen response waveform $x(t) = x(t + T)$ is established, (*ii*) generalized eigenfrequencies s_1, s_2,..., s_q of the system are determined, (*iii*) a general form of the zero-excitation solution $x_{zin}(t)$ of the differential equation is established, (*iv*) the period carrying waveform $p(t)$ with the same period T is involved, (*v*) piecewise segments $f_1(t), f_2(t), \ldots, f_n(t)$ of the excitation waveform are selected for as many discontinuity instants as exhibited, (*vi*) for each piecewise segment of the excitation the corresponding forced response $x_{kf}(t)$, $k = 1, 2, \ldots, n$ is determined, (*vii*) for each piecewise segment of the excitation the general piecewise response $x_k(t) = x_{kzin}(t) + x_{kf}(t)$, $k = 1, 2, \ldots, n$ is established, (*viii*) the constant of the solution of step (*vii*) is now evaluated so as to satisfy the analytical continuity and periodicity conditions, (*ix*) for each discontinuity instant the corresponding switch-on waveform $r_k(t, t_k)$, $k = 1, 2, \ldots, n$ is established, (*x*) the period carrying waveform $p(t)$ is substituted instead of time variable t and to obtain the actual response $x(t) = x(t + T)$ the concatenation procedure for the piecewise responses $x_1(p(t))$, $x_2(p(t))$, ..., $x_n(p(t))$ is applied.

The so-called period carrying waveform $p(t)$ is defined as follows:

$$p(t) = p(t + T) = \frac{T}{2} - \frac{T}{\pi}\arctan\left(\cot\left(\frac{\pi}{T}t\right)\right) \tag{2.73}$$

The direct plot of (2.73) for $T = \pi$ is presented in Fig. 2.14. As is well known, the graphic representation of $p(t)$ is very often called the *saw-tooth function* or equivalently *saw-tooth waveform*.

Such time fragmentation into periods using the periodic function $p(t)$ is very useful in the study of dynamical systems with periodic excitations, both continuous and discontinuous. The specificity of such situations is well illustrated by the following problem.

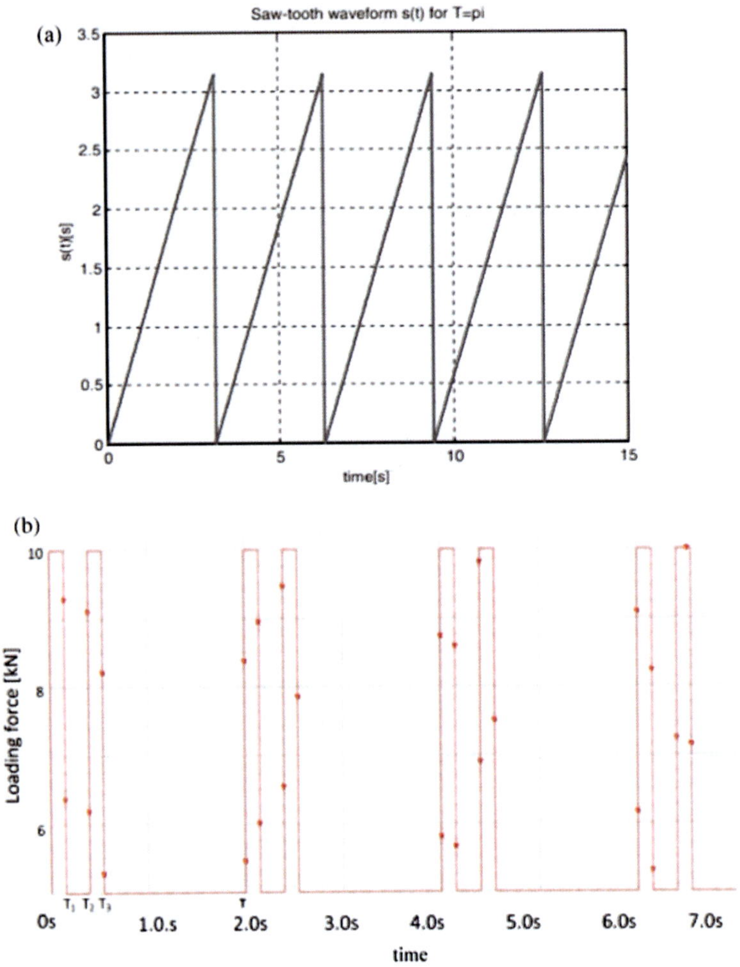

Fig. 2.14 Time-varying waveforms: **a** period carrying waveform for $T = \pi$ s, **b** periodic waveform of discontinuous loading force

The forcing T- periodic waveform $F(t) = F(t + T)$ with two pulses within each period can be represented as follows

$$F(t) = F(t+T) = \begin{cases} A \text{ for } & 0 \leq t \leq T_1 \\ 0 \text{ for } & T_1 \leq t \leq T_2 \\ A \text{ for } & T_2 \leq t \leq T_3 \\ 0 \text{ for } & \mathrm{T}_3 \leq t \leq T \end{cases} \tag{2.74}$$

where A and T_k, with $k = 1, 2, 3$, denote the magnitude and moments, respectively, describing the pulses in the forcing waveform. Using period carrying waveform $p(t)$ we can represent $F(t)$ as follows

$$F(t) = A + \sum_{k=1}^{3} H(p(t), T_k)[(-1)^k \mathrm{A}] \tag{2.75}$$

where $H = (p - T_k)/abs(p - T_k)$ is a *periodic jump function*, shifted at T_k.

The effectiveness of the presented dependencies in the calculations of dynamic systems with periodic non-sinusoidal excitations will be emphasized by examining the dynamics of the wheel-rail system while the train is running.

It is worth emphasizing that continuously increasing operational speeds and the demands for comfort require from the responsible services to adapting to the top level of knowledge on driving qualities of rail vehicles, safety and comfort of transporting passengers and other loads. In accord to challenges in the domain of railway vehicle dynamics [31, 36, 38–41], connected to higher speed and greater loads with a very high level of security, the use of more and more innovative engineering solutions, better understanding of technical issues and the use of new analytical tools are continuously required without interruption or gaps. A hard problem arises when the wheel-rail contact is subject to an action of time discontinuous forces. This corresponds to the system shown schematically in Fig. 2.15.

The contact problem translates to the normal action between the wheel and rail and the normal pressure at a point of the contact patch is proportionate to the interpenetration of the contacting bodies at the point.

A continuously vibrating system may be approximately modeled by an appropriate set of lumped masses properly interconnected using discrete spring and damper elements (Fig. 2.15). For each of segments as that shown in Fig. 2.15a a system of governing equations can be formulated as follows

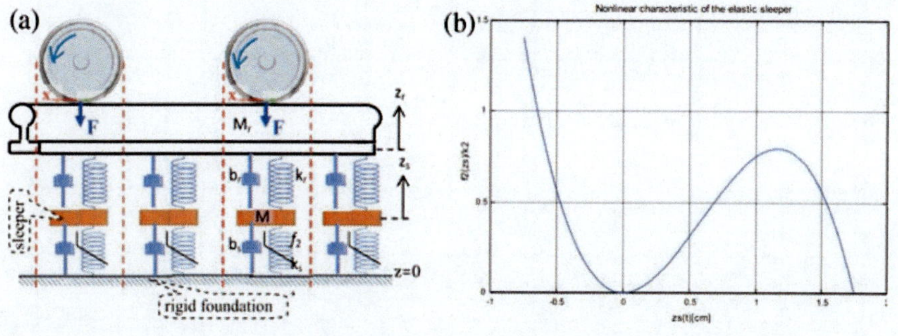

Fig. 2.15 Scheme of: **a** rigid rail with sleepers on the ground, **b** Nonlinear characteristics of the sleeper stiffness

$$\begin{bmatrix} M_r & 0 \\ 0 & M_s \end{bmatrix}\begin{bmatrix} \ddot{z}_r \\ \ddot{z}_s \end{bmatrix} + \begin{bmatrix} b_r & -b_r \\ -b_r & b_r + b_s \end{bmatrix}\begin{bmatrix} \dot{z}_r \\ \dot{z}_s \end{bmatrix} + \begin{bmatrix} k_r & -k_r \\ -k_r & k_r \end{bmatrix}\begin{bmatrix} z_r \\ z_s \end{bmatrix} = \begin{bmatrix} F(t) \\ -f_2(z_s) \end{bmatrix} \tag{2.76}$$

The meanings of particular quantities follow from the Fig. 2.15a directly. The nonlinear characteristic of the sleeper with respect to the ground is represented by the relation

$$f_2(z_s) = k_2\left(\alpha z_s^2 + \beta z_s^3\right) \tag{2.77}$$

where constant parameters k_2, α and β can be considered as bifurcating values. For $\alpha = 1.75$, $\beta = -1$ the plot of the relative value f_2/k_2 is shown in Fig. 2.15b. The set of system parameters and the forcing term shown in Fig. 2.14b and nonlinearity presented in Fig. 2.15b lead to the calculated time variations of the wheel-rail contact deformations which are depicted in Fig. 2.16.

Supplementing the data given above with the sleeper mass $M_s = 156$ kg and remaining parameters: $b_s = 520$ Ns/m and $k_s = 1.5$ kN/m and taking into account a number of excitation points equal to the number of wheels, i.e. two wheels in one side of the bogie, the rail deflection takes the form presented in Fig. 2.17a. The corresponding phase portrait $x_3(x_1)$ takes the form shown in Fig. 2.17b.

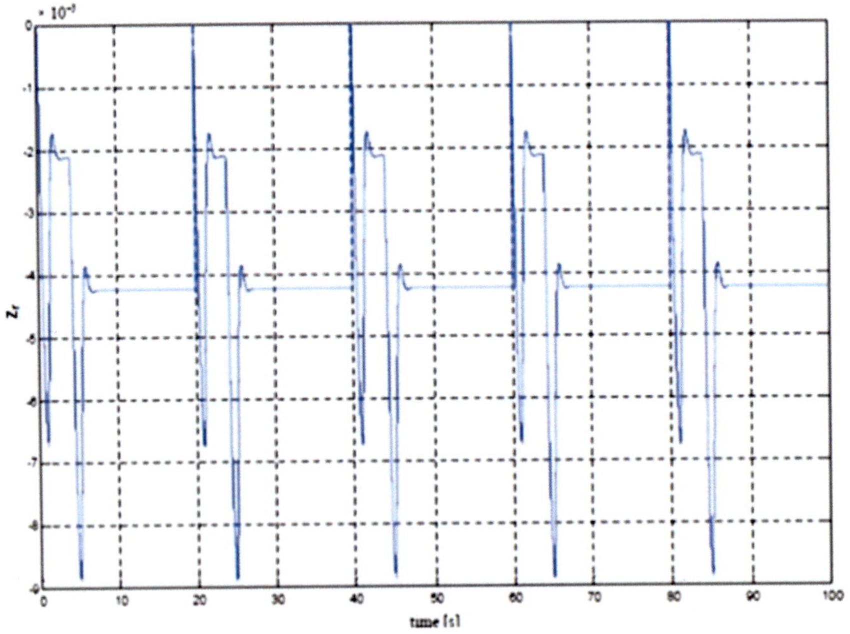

Fig. 2.16 Changes of wheel-rail contact deformations

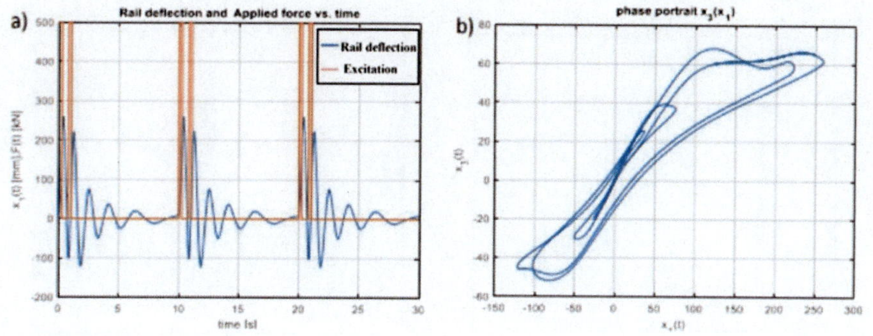

Fig. 2.17 Simulation results in the case $T = 10$ s: **a** input and output variables versus time, **b** phase portrait $x_3(x_1)$

A very interesting case is that when the wagon wheel presses onto a rail with the force with time-varying in the form of saw-tooth pulses (Fig. 2.18a). Then the response of the wheel-rail system takes a more complex form compared to the case with force in the form of square pulses. During the operation of the saw-tooth force, the response of the system is in the form of high-amplitude bursting vibrations, and after its extinction, small-amplitude vibrations appear (Fig. 2.18b, c).

The situation is different with regard to the sleeper. In the time intervals when the response shows large amplitude, the bursting vibration is less severe and after its decay, the small amplitude vibrations are relatively large (Fig. 2.18d).This situation leads to an increase in energy losses in the system and the absorption of more energy by the rail. The presented methodology can be applied to predict the durability of wheel rail systems subject to wear and crack growth. Under the action of the cyclic load obtained from the contact calculation the growth of the crack can be predicted.

2.3.3 *Regular Periodic Pulses in Linear Systems*

A *periodic pulse train* with period T_0 consists of *rectangular pulses* of duration T. The duty cycle of a periodic pulse train is defined by T/T_0. An application of the periodic pulse train is in the practical sampling process. An even periodic pulse train, as shown in Fig. 2.19a, can be analytically expressed as follows:

$$g(t) = \sum_{k=-\infty}^{\infty} r(t - kT_0) \tag{2.78}$$

where $r(t)$ is a *rectangular pulse*, as defined in (2.78). A **periodic impulse train** consists of impulses (delta functions) uniformly spaced T_0 seconds apart. An application of a periodic impulse train is in the ideal sampling process. Using (2.78), an even periodic impulse train, as shown in Fig. 2.19b, can be analytically expressed as follows:

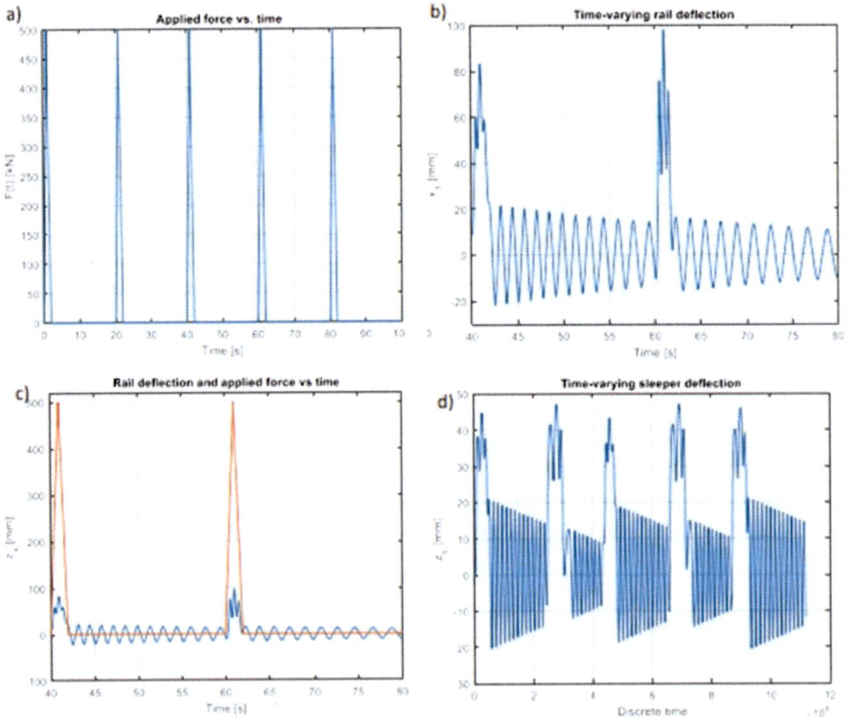

Fig. 2.18 Simulation results: **a** applied force of saw-tooth type, **b** rail deflection, **c** rail deflection and applied force, **d** sleeper deflection

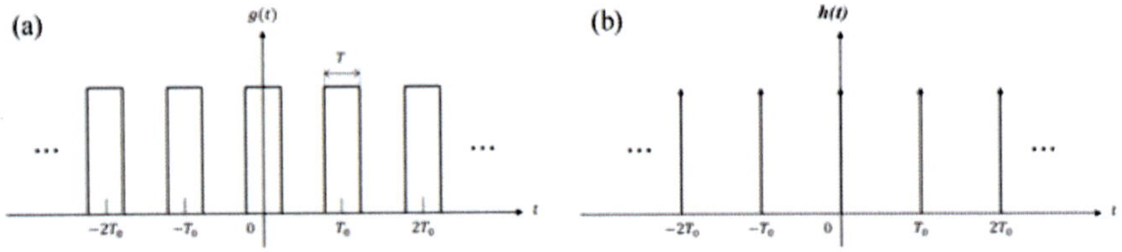

Fig. 2.19 Periodic pulse train: **a** rectangular pulses, **b** Dirac delta impulses

$$h(t) = \sum_{k=-\infty}^{\infty} \delta(t - kT_0) \tag{2.79}$$

The traditional tools for analysis of non-sinusoidal periodic waveforms are the Fourier series and the sampling theorem of Shannon, Whittaker and Kotel'nikov [47]. However, it is commonly known that a discontinuous waveform, like the square or saw-tooth waveforms, cannot be expressed as a sum, even an infinite one, of continuous signals. The extraneous peaks in the square wave's Fourier series never disappear; they occur whenever the waveform is discontinuous, and will always be present whenever the signal has jumps. Quite obviously, if the source waveform is subject to jump changes the linear smoothing procedure is not a good choice anymore, because all conservative system elements confuse and remove the high frequency components from the system output. For this reason, when a source waveform with jumps is applied to a linear system it causes a typical effect of 'edge blurring'. Taking into account the above requirements and insufficiencies of the methods based on Fourier series which have thus far been most commonly used for studies of periodic non-sinusoidal states of linear as well as nonlinear systems, it is proposed below a new method for obtaining, in analytic closed form, the response of any linear system exposed to piecewise-continuous periodic non-sinusoidal forcing terms. In this approach, the solution is exact, and by means of suitable unification of its piecewise representation, it is possible to obtain with ease the exact expressions for its time derivatives.

For the construction of an appropriate algorithm we consider a general setting, i.e., waveforms $f(t)$ with discontinuities. However, they are subjected to two restrictions, namely: (*i*) a global condition requiring that $f(t)$ be absolutely integrable, (*ii*) a local condition constraining $f(t)$ to have a finite number of maxima and minima and a finite number of discontinuities in every finite interval. To follow with these effects we will describe discontinuous waveforms by using the saw-tooth waveform and its relatives such as switch-on and switch-of waveforms. Thus the wave-form shown in Fig. 2.20a can be represented as

$$f(t) = f_1(t) + w(t, t_1)[f_2(t) - f_1(t)] + w(t, t_2)[f_3(t) - f_2(t)] \tag{2.80}$$

where the waveform $w(t, \tau_k)$, $k = 1, 2$ is determined by

$$w(t, \tau) = \begin{cases} 0 \text{ for t} \le \tau \\ 1 \text{ for t} \ge \tau \end{cases} \tag{2.81}$$

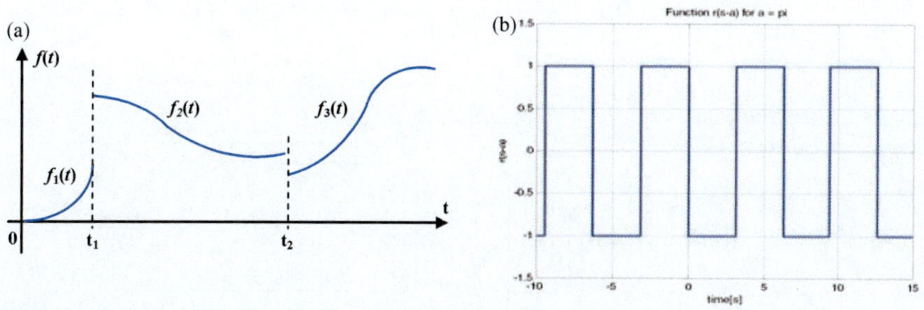

Fig. 2.20 Discontinuous waveforms: **a** a waveform with jumps at $t = t_1$ and $t = t_2$, **b** symmetric square waveform $f_T(t,\tau)$ with $T = 2\pi$ and jump at $\tau = \pi$

This intuitively appealing "switching rule" can be exploited in several ways. Using the switching approach (2.81) suggests the incorporation of a true smoothing element into the competition. For instance, in the case of the signal shown in Fig. 2.20b the period equals $T = 2\pi$, and the discontinuity appears at $t_I = \pi$. Prior to discontinuity the signal takes a constant value equal to − 1 and post discontinuity it equals 1. Thus applying (2.81) yields

$$r_T(t, \tau) = -1 + w(s, \pi)(1 - (-1)) = -1 + 2w(s, \pi) \tag{2.82}$$

These estimates demonstrate how the signal jumps can be taken into account to a governing equation of a given linear system under discontinuous periodic non-sinusoidal regime. In order to present a general algorithm let us consider steady state oscillations in the network shown in Fig. 2.21. It corresponds to a single-phase bridge voltage source inverter (*VSI*) supplying an *RLC* load.

The mathematical model of such a configuration takes the form

$$\ddot{z}(t) + 2\dot{z}(t) + 4z(t) = u(t) \tag{2.83}$$

where $z(t)$ denotes the voltage across the capacitor C, and the forcing term $u(t) = 0.8\, r_T(t)$ is presented in Fig. 2.24. The amplitude of this waveform was scaled by 50%.

In this case the self-frequencies of the network are

$$s_1 = -1 + j\sqrt{3}, \quad s_2 = -1 - j\sqrt{3}$$

The steady-state solutions for the periodic output waveform in successive semi-periods of the input wave take the forms.

- for $0 < t < \pi$

$$z_1(t) = 0.2 + e^{-t}[G_1 \cos(\sqrt{3}t) + H_1 \sin(\sqrt{3}t)] \tag{2.84}$$

- for $\pi < t < 2\pi$

$$z_2(t) = -0.2 + e^{-t}[G_2 \cos(\sqrt{3}t) + H_2 \sin(\sqrt{3}t)] \tag{2.85}$$

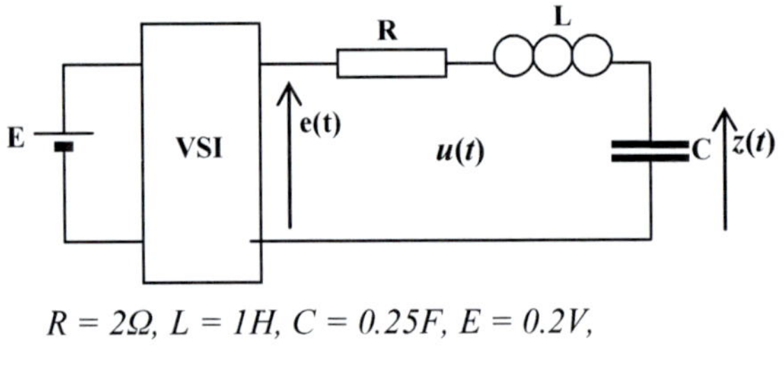

Fig. 2.21 A network with a voltage source inverter (*VSI*)

where the integration constants G_1, H_1, G_2 and H_2 are to be determined from the respective conditions for the periodicity and analytical continuity of the total solution $z(t)$, namely

$$z_1(\pi) = z_2(\pi), \quad \dot{z}_1(\pi) = \dot{z}_2(\pi)$$

$$z_1(0) = z_2(2\pi), \quad \dot{z}_1(0) = \dot{z}_2(2\pi)$$

Solving the above equalities with respect to the integration constants yields

$$G_1 = -0.3953, G_2 = 9.9876, H_1 = -0.2117, H_2 = -3.6316$$

Substituting these values into (2.84) and (2.85) and mapping the solution of (2.83) into the $p(t)$ domain gives

$$\begin{aligned} z(t) &= z_1(p) + [z_2(p) - z_1(p)]w(p, \pi) \\ &= -0.2r(p) + e^{-p}\Big\{[4.7988 + 5.1941r_T(p)]\cos(\sqrt{3}p) \\ &\quad + [-1.9177 - 1.706r(p)]\sin(\sqrt{3}p)\Big\} \end{aligned} \tag{2.86}$$

Figure 2.22 represents the forcing waveform $u(t)$, in twice reduced scale, and the output waveforms of the capacitive voltage $z(t)$ and load current $i(t)$. It is worth pointing out that the shapes of three plots shown in Fig. 2.22 differ importantly one to other but the period of the output waveforms is the same as that of the input waveform. Essentially, the peaks and troughs of the output current $i(t)$ correspond to maximum and minimum slopes in time of the output voltage $u(t)$, respectively.

2.3.4 Oscillators Under the Periodic Impulsive Loading

Many nonlinear systems exhibit far-reaching implications in their dynamic behavior such as limit cycle oscillations, jump phenomenon, quasi-periodic motions, bifurcations and chaos. Among the several types of nonlinearities, non-smoothness or discontinuities present special difficulties in identifications of their influences on course of processes in dynamical systems [48]. Non-smooth nonlinearities are mainly due to intermittent contacts and friction [49]. Based on the nature of discontinuity, these systems can be classified into friction systems, vibro-impact, piecewise smooth (PWS) continuous systems and systems with higher order nonlinearities [48, 49]. The dynamics of such systems are complex and may lead to rich behavior like discontinuity induced bifurcations, damaging bifurcations [49], torus bifurcation [50]. Dynamic systems with impulse inputs deserve special attention.

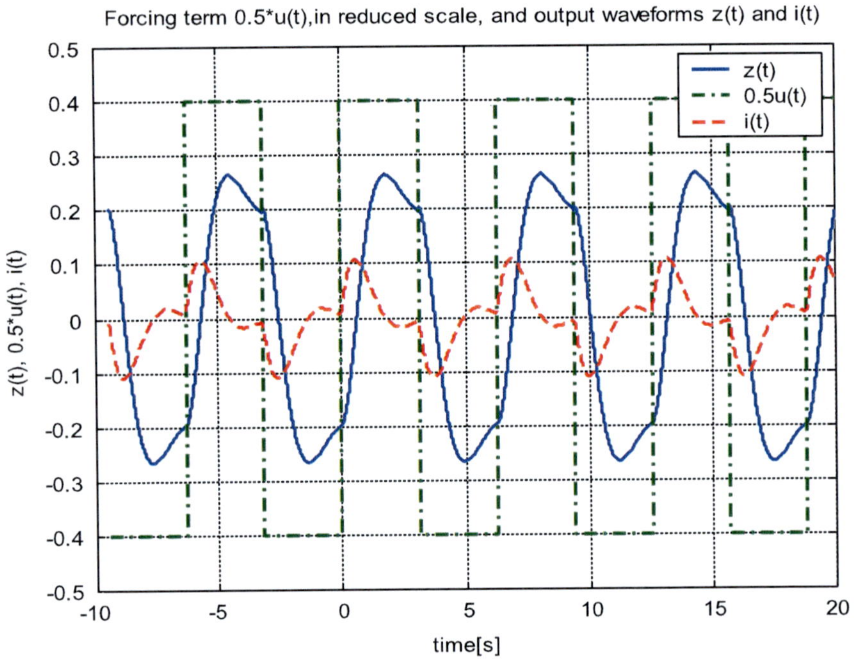

Fig. 2.22 The input and output waveforms in the network of Fig. 2.21

In the basic method of testing systems with impulse inputs, it is assumed that instantaneous impulses acting on a dynamical system can be modeled by imposing specific matching conditions on the system state vector at pulse times. For systems with periodic pulsed excitations the approach deals with the differential equations of a free system separately between the impulses, therefore a sequence of systems under the matching conditions are considered. The analytical tool, which is described below, on the one hand, eliminates the singular terms from the equations and, on the other hand, brings solutions to the unit-form of a single analytic expression for the whole time interval.

For instance, let us determine a particular solution of the first order differential equation

$$\dot{v} + bv = m \sum_{-\infty}^{\infty} [\delta(t + 1 - 4k) - \delta(t - 1 - 4k)] \tag{2.87}$$

where b and m are constant parameters. It describes the magnetic levitation of an electrically conductive small ball moving in a vacuum object (Fig. 2.23).

Next, the particular solution can be represented in the form

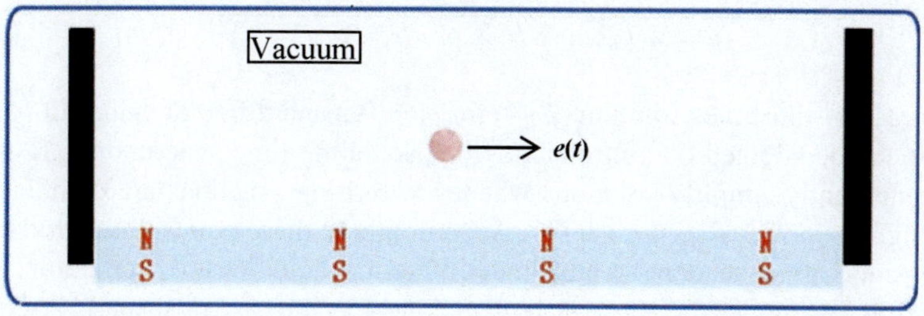

Fig. 2.23 Magnetic levitation in vacuum

$$v(t) = X(\tau(t)) + Y(\tau(t))e(t) \tag{2.88}$$

where τ is the so called 'oscillating time'. The basic functions $\tau(t)$ and $e(t)$ denote triangular and rectangular waves mapping the translation and reflection groups. They are expressed through the standard elementary functions in the closed form as

$$\tau(t) = 2\pi \cdot \arcsin(\sin(\pi t/2) \quad \text{and} \quad e(t) = \cos(\pi t/2)/|\cos(\pi t/2)| \tag{2.89}$$

In this case when $v(t)$ is an unknown periodic solution of a dynamical system, then equations for X and Y components are obtained by substituting (2.88) into the differential Eq. (2.87). Then either analytical or numerical procedures can be applied. For instance, one may seek solutions in the form of power series with respect to the 'oscillating time' τ. Therefore expressions (2.89) can be qualified as non-smooth time transformation, $t \to \tau$, on the manifold of periodic solutions.

Note that the right-hand side of Eq. (2.87) can be expressed through the generalized derivative of the rectangular wave function as follows

$$\dot{v} + bv = (m/2)\dot{e}(t) \tag{2.90}$$

Thus the particular solution can be represented in the form

$$v(t) = X(\tau(t)) + Y(\tau(t))e(t) \tag{2.91}$$

Substituting (2.91) in (2.90), gives

$$Y' + bX + \left(X' + bY\right)e(t) + (Y - m/2)\dot{e}(t) = 0 \tag{2.92}$$

It is clearly understood that the elements $\{1, \mathrm{e}\}$ and $\dot{\mathrm{e}}$ in combination (2.92) are linearly independent as functions of different classes of smoothness. Therefore

$$Y' + bX = 0, \quad X' + bY = 0, \quad Y\,|\tau = \pm 1 = m/2 \tag{2.93}$$

Solving the boundary value problem (2.93) and taking into account representation (2.91), gives periodic solution of Eq. (2.90) in the form

$$v = m/(2\cosh b)\exp[-b\tau(t)e(t)]e(t) \tag{2.94}$$

Figure 2.24 illustrates solution (2.94) for $m = 0.2$ and different values of b.

So it can be adopted that introducing the oscillating time as temporal argument may significantly simplify solutions whenever exciting functions are combined of the triangular wave and its derivatives. Accordingly to these points, the introduction of a nonsmooth independent variable into differential equations of oscillation can be purposeful. To amplify the above statement considers the case of strongly nonlinear exactly solvable oscillator excited by a train of periodic impulses. It is described as follows

$$\ddot{x} + \frac{\tan(x)}{cos^2(x)} = 2H\sum_{-\infty}^{\infty}(-1)^k\delta[t-(2k-1)a] \tag{2.95}$$

where $H > 0$ and $a = T/4 > 0$ characterize the amplitude and period of the impulsive excitation, respectively. The characteristic of a restoring force of stiff spring is presented in Fig. 2.25a. Taking into account properties of the rectangular cosine $e(t/a)$ and applying them to the right hand side of Eq. (2.95) yields

$$2H\sum_{-\infty}^{\infty}(-1)^k\delta[t-(2k-1)a] = \frac{H}{a}\frac{d}{d\left(\frac{t}{a}\right)}e\left(\frac{t}{a}\right) \tag{2.96}$$

Now, representing periodic solutions in the form of (2.91), that is $x(t) = X(p) + Y(p)e(t)$, and substituting this into Eq. (2.95), gives

$$\begin{aligned} &a^{-2}X''(p) + \tan(X(p)) + \tan^3(X(p)) = a^{-2}(aH - X')\dot{e}\left(\frac{t}{a}\right) \\ &X'_{|p=\pm1} = aH, \quad Y(p) \equiv 0 \end{aligned} \tag{2.97}$$

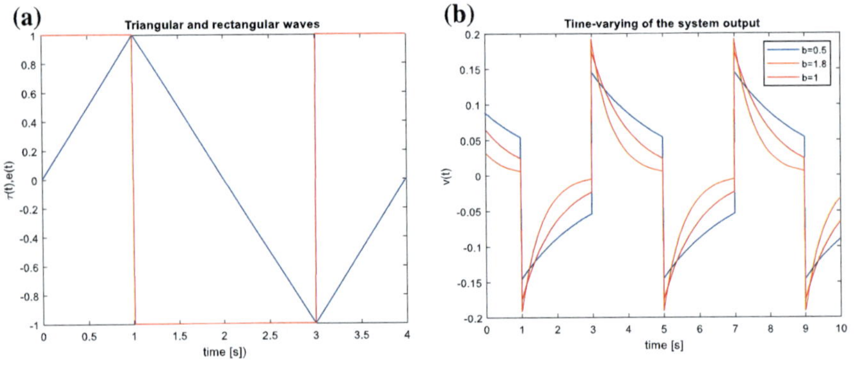

Fig. 2.24 Graphic representation: **a** triangular and rectangular waveforms, **b** the family of discontinuous periodic solutions

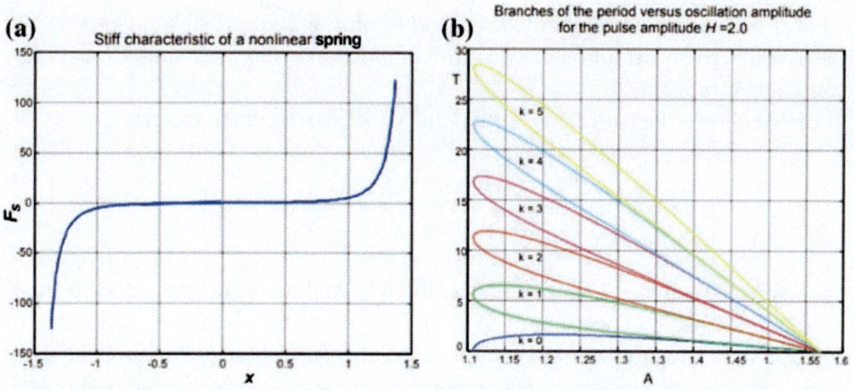

Fig. 2.25 Ingredients of a periodic impulsive oscillator: **a** stiff characteristic of a nonlinear spring, **b** branches of the period T versus oscillation amplitudes for the pulse amplitude $H = 2$

It should be emphasized that boundary value problem (2.97) describes the class of steady state periodic oscillations of the period $T = 4a$ of impulsively excited oscillations. With reference to Eqs. (2.91) and (2.97), we can present the solution exactly in the form.

$$X(p) = \arcsin\left(\sin(A) \cdot \sin\left(\frac{p}{\cos(A)}\right)\right) \tag{2.98}$$

where A is a constant, which can be computed by taking into account condition for $X'(p)$ in (2.97) and the symmetry of solution (2.98). Substituting the result for $p = \pm 1$ from (2.91) in (2.98) and performing analytical manipulations with elementary functions, yields

$$T = 4q \cdot \arccos\left(\frac{a^2 H}{\sqrt{(1-q^2)(1-q^2H^2)}}\right), \quad k = 0 \tag{2.99a}$$

and

$$T = 4q \cdot (k\pi \pm \arccos\left(\frac{a^2 H}{\sqrt{(1-q^2)(1-q^2H^2)}}\right), \quad k = 1, 2, \ldots, \tag{2.99b}$$

where $q = \cos(A)$, and $T = 4a$ is the period of impulsively excited oscillation.

The sequence of branches of solutions (2.99a) and (2.99b) at different numbers k, and the parameter $H = 2$ are presented in Fig. 2.25b. The diagram gives such combinations of the period and amplitude of the oscillation at which oscillator (2.95) can exhibit periodic oscillations with the period of external impulses $T = 4a$. The upper and lower branches of each loop correspond to plus and minus signs in expression (2.99b), respectively. Solution (2.99a) which corresponds to the number $k = 0$ has the only upper branch.

It can be shown that solutions (2.99a) and (2.99b) exist in the interval

$$\arccos\left(\frac{1}{\sqrt{1+\mathrm{H}^2}}\right) = \mathrm{A}_{\min} < A < \mathrm{A}_{\max} = \pi/2 \tag{2.100}$$

For the selected values of magnitude H, the minimal amplitude is found to be $A = 1.1071$ [rad], which corresponds to the left edges of the amplitude-period loops in Fig. 2.25. As a result, further slight increase of the amplitude is accompanied by bifurcation of the solutions as shown in Fig. 2.26a–d. The diagram includes only first three couples of new solutions ($k = 1, 2, 3$) from the infinite set of solutions. It follows from these diagrams that the influence of excitation pulses on the temporal shapes of state variables is decreasing as the amplitude grows. This is the result of dominating the restoring variable over the external pulses. When the amplitude becomes close to its maximum $A_{max} = \pi/2$, the oscillator itself generates high-frequency impacts.

It is worth mentioning that a slight increase of the amplitude is accompanied by bifurcation of oscillations. Moreover, it follows from the presented diagrams that the deformations of temporal shapes of oscillations are decreasing as the amplitude of external pulses grows and this influence is strong.

2.3.5 *Periodic Impulses with a Temporal 'Dipole' Shift*

In this Section, we will use a generalized version of representation (2.91) based on the saw-tooth function with an asymmetric slope, namely

$$\tau(t,\gamma) = \begin{cases} \frac{t}{1-\gamma}, & 1+\gamma \le t \le 1-\gamma, \\ \frac{-t+2}{1+\gamma}, & 1-\gamma \le t \le 3+\gamma \end{cases} \tag{2.101}$$

$$e(t,\gamma) = \frac{\partial \tau(t,\gamma)}{\partial t}, \quad \left(\tau(t,\gamma) \overset{\text{def}}{=} \tau(t+4,\gamma)\right) \tag{2.102}$$

where $\gamma \in (-1, 1)$ is a parameter characterizing the inclination of the 'saw's teeth' as Fig. 2.27 shows. The parameter of inclination γ is introduced in such a manner that the period $T = 4$ of the function and its amplitude value (= 1) do not depend on the inclination.

In this case, the algebraic structure becomes more intricate than in the previous section. The above multiplication rule, $e^2 = 1$, should be replaced with a more complicated one, that is

$$e^2 = \alpha(\gamma) + \beta(\gamma)e, \quad \alpha = \frac{1}{1-\gamma^2}, \quad \beta = 2\gamma\alpha \tag{2.103}$$

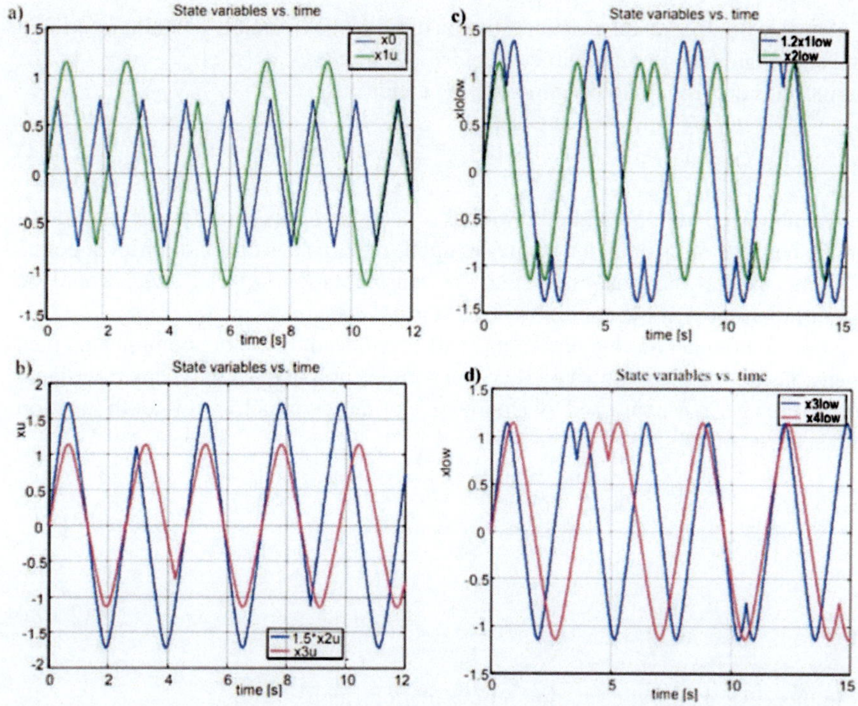

Fig. 2.26 The course of the state variable $x(t)$ in the oscillator with periodic impulse excitation at $A = 1.145$: **a** $k = 0$; 1, **b** $k = 2$; 3 with $x2$u rescaled by 1.5; (for upper branches of closed loops of the corresponding periods shown in Fig. 2.24b); **c** $k = 1$; 2, with $x2$low rescaled by 1.2; **b** $k = 3$; 4 (for lower branches of closed loops of the corresponding periods shown in Fig. 2.24b)

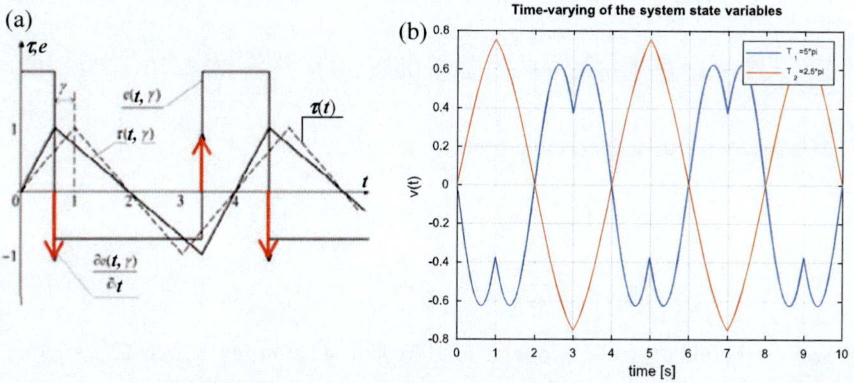

Fig. 2.27 System under temporal 'dipole' shift: **a** shifted triangular and rectangular waveforms, **b** time-varying system state variables

Note that now the result of differentiation of Eq. (2.91) formally contains a periodic series of singularities (delta Diraca impulses) at points $\Lambda = \{t\colon \tau(t,\gamma) = \pm 1\}$. The singularities are discarded by imposing the condition

$$Y|_{\tau=\pm 1} = 0 \tag{2.104}$$

The above condition can be viewed as a necessary condition for the continuity for $v(t)$ because of that that the stepwise function $e(t, \gamma)$ has discontinuities at points Λ. It means that the 'imaginary part' of representation (2.91), $Y(\tau)e$, would be discontinuous at Λ if the condition (2.104) is not met.

Now the boundary-value problem (2.90) is replaced by a more complicated one, when considering the impulsive excitation with a dipole-like shift, i.e. the right-hand side of Eq. (2.89) is expressed by a derivative of the asymmetric saw-tooth function with a slope γ [51]:

$$\begin{aligned}\ddot{v} + \omega_0 v &= \frac{2p}{1-\gamma^2}\sum_{k=-\infty}^{\infty}\left[\delta\left(\frac{2\pi t}{T} + 1 - \gamma - 4k\right) - \delta\left(\frac{2\pi t}{T} + 1 + \gamma - 4k\right)\right] \\ &= p\frac{\partial e\left(\frac{2\pi t}{T}, \gamma\right)}{\partial\left(\frac{2\pi t}{T}\right)}\end{aligned} \tag{2.105}$$

In this case, a periodic solution representation can be taken as

$$v = X(\tau(\omega t)) + Y(\tau(\omega t))e(\omega t), \quad \tau = \tau(\omega t, \gamma), \quad e = e(\omega t, \gamma) \tag{2.106}$$

Substituting expression (2.106) into differential Eq. (2.105) and recollecting separately the 'real' and the 'imaginary' ingredients and singular terms obtains

$$(\omega^2 X'' + \omega_0^2 X) + (\omega^2 Y'' + \omega_0^2 Y)e + (\omega^2 X' - p)\frac{\partial e(\omega t, \gamma)}{\partial(\omega t)} = 0 \tag{2.107}$$

As a result, the boundary-value problem takes the form

$$\begin{aligned}&X'' + \left(\frac{\omega_0}{\omega}\right)^2 \mathrm{X} = 0, \quad Y'' + \left(\frac{\omega_0}{\omega}\right)^2 Y = 0, \\ &X'_{|\tau=\pm 1} = \frac{p}{\omega^2}, \quad Y_{|\tau=\pm 1} = 0\end{aligned} \tag{2.108}$$

Solving boundary-value problem (2.108) and substituting it into (2.106) one obtains the periodic solution of the original equation having the form

$$v = \frac{p}{\omega\omega_0}\frac{\sin\left[\frac{\omega_0}{\omega}\tau(\omega t)\right]}{\cos\left(\frac{\omega_0}{\omega}\right)} \tag{2.109}$$

It is easy to check that the determined solution is continuous for all t and its graphic representation for $\omega_0 = 1$ rad/s and two values of ω is presented in Fig. 2.27b.

2.3.6 *Fractional Order Differential Models*

In a classical approach to study the dynamics of most natural and manmade systems are used laws based on exponential functions, but there are many systems where dynamics undergo faster or slower-than-exponential laws. Systems with anomalous changes in dynamics may be best described by suitable models using Mittag–Leffler functions [49]. Such systems are said to have *fractional dynamics* and are modeled by a fractional differential equation containing derivatives of non-integer order [52]. Derivatives and integrals of fractional orders are used to describe objects that can be characterized by power-law nonlocality, power-law long-range dependence or fractal properties. Fractional-order systems are useful in studying the anomalous behavior of dynamical systems in physics, electrochemistry, biology, viscoelasticity and chaotic systems. Anomalous diffusion is one of dynamic systems where fractional-order differential equations play significant role to describe the anomalous flow in the diffusion process. Viscoelasticity is the property of material in which the material exhibits its nature between purely elastic and pure fluid. In case of real materials, the relationships between stress and strain given by Hooke's law and Newton's law both have obvious disadvantages. In chaos theory, it has been stated, in the classical approach, that chaos occurs in dynamical systems of order three or more. With the introduction of fractional-order systems, many researchers study chaos in the system of total order less than three. Fractional order differential equations are accepted today as a new tool that extends the descriptive power of the conventional calculus, supporting mathematical models that, in many cases, describe more accurately the dynamic response of actual systems in various applications. While the theoretical and practical interest of these fractional order operators is nowadays well established, its applicability to science and engineering can be considered as an emerging topic. Among other, the need to numerically compute the fractional order derivatives and integrals arises frequently in many fields, especially in electronics, telecommunications, automatic control and digital signal processing. The purpose of this section is to introduce fundamentals of the fractional calculus and their applications to solutions of fractional order systems. In the literature, authors often use the term "fractional order calculus (FOC)", or "fractional order dynamic system (FODS)" where "fractional" actually means "noninteger" [49].

$$ {}_{a}D_t^{\alpha}[f(t)] = \frac{1}{\Gamma(n-\alpha)}\left(\frac{d^n}{dt^n}\int_a^t \frac{f(\tau)}{(t-\tau)^{1-(n-a)}}d\tau\right) \tag{2.110a} $$

$$ {}_{a}D_{t}^{\alpha}[f(t)] = \lim_{h\to 0} \frac{1}{\Gamma(\alpha)h^{\alpha}} \sum_{k=0}^{[t-a/h]} \frac{\Gamma(\alpha+k)}{\Gamma(k+1)} f(t-kh) \tag{2.110b} $$

$$ {}_{a}D_{t}^{\alpha}[f(t)] = \frac{1}{\Gamma(\alpha-n)} \int_{a}^{t} \frac{f^{(n)}(\tau)}{(t-\tau)^{\alpha-n+1}} d\tau \tag{2.110c} $$

for $n-1<\alpha<n$ and where $\Gamma(\cdot)$ is the Gamma function, and $\lceil z \rceil$ means the integer part of z. The "memory" effect of these operators is demonstrated by (2.110a) and (2.110b), where the convolution integral in (2.110a) and the infinite series in (2.110b) reveal the unlimited memory of these operators, ideal for modeling hereditary and memory properties in physical systems and materials. The initial conditions for the fractional order differential equations with the Caputo derivative (2.110c) are of the same form as the initial conditions for the integer-order differential equations [48]. In the sequel the approach is based on the fact that for a large class of functions, three above definitions are equivalent. Similarly to integer-order differentiation, fractional-order differentiation fulfils the relations

$$ {}_{a}D_{t}^{\alpha}\left({}_{a}D_{t}^{\beta}f(t)\right) = {}_{a}D_{t}^{\alpha+\beta}f(t) - \sum_{k=0}^{n} [{}_{a}D_{t}^{\beta-k}]_{t=a} \frac{(t-a)^{-\alpha-k}}{\Gamma(1-\alpha-k)} \tag{2.111} $$

$$ {}_{a}D_{t}^{\alpha}[\eta f(t) + \lambda g(t)] = \eta\left[{}_{a}D_{t}^{\alpha}f(t)\right] + \lambda\left[{}_{a}D_{t}^{\alpha}g(t)\right] \tag{2.112} $$

Moreover, fractional order systems may have other advantageous features that make them more suitable for the study of nonlinear dynamical systems when most of the desired depictions are not readily achieved by traditional models [10]. Note that by using the fractional differential equations, we get a total order of the system that is less than the number of differential equations.

For the analysis of a linear time-invariant complex fractional order system very useful is the following state space model

$$ \begin{aligned} {}_{0}D_{t}^{q}\boldsymbol{x}(t) &= \boldsymbol{A}\boldsymbol{u}(t), \\ \boldsymbol{y}(t) &= \boldsymbol{C}\boldsymbol{x}(t) + \boldsymbol{H}\boldsymbol{u}(t) \end{aligned} \tag{2.113} $$

where $\boldsymbol{x} \in \boldsymbol{R}^{n}$, $\boldsymbol{u} \in \boldsymbol{R}^{r}$ and $\boldsymbol{y} \in \boldsymbol{R}^{p}$ are the state, input and output vectors, and $\boldsymbol{A} \in \boldsymbol{R}^{n\times n}$, $\boldsymbol{B} \in \boldsymbol{R}^{n\times r}$, $\boldsymbol{C} \in \boldsymbol{R}^{p\times n}$, q is the fractional commensurate order. It has been shown that system is stable if the following condition is satisfied [53]:

$$ |\arg(\mathrm{eig}(\mathbf{A}))| > q\pi/2 \tag{2.114} $$

where $0<q<1$ and $\mathrm{eig}(\boldsymbol{A})$ represents the eigenvalues of matrix $\boldsymbol{A}$.

It has to be noted that in almost all cases the impulse responses of fractional order systems are related to the Mittag–Leffler function [54], which is effectively the fractional order analog of the exponential function, being common in the study of systems of integer orders. The knowledge of the possible dynamic behavior of fractional order systems (FOS) is fundamental as most properties and conclusions of integer order systems (IOSs) cannot be simply extended to that of the FOS. To demonstrate these facts we shall consider some selected cases of FOSs.

Let us begin with the fractional-order Volta's system [55], where integer-order derivatives are replaced by fractional-order ones and with $x(t)$, y(t), and z(t) standing as fractional state variables. Mathematical description of this system is expressed as

$$\begin{aligned} {}_0D_t^{q_1}x(\mathrm{t}) &= -x(t) - ay(t) - z(t)y(t), \\ {}_0D_t^{q_2}y(\mathrm{t}) &= -y(t) - bx(t) - x(t)z(t), \\ {}_0D_t^{q_3}z(t) &= cz(t) + x(t)y(t) + 1, \end{aligned} \tag{2.115}$$

where q_1, q_2, and q_3 are the derivative orders. The total order of the commensurate system is $q = (q_1, q_2, q_3)$. Parameters *a, b* and *c* are constant.

The vector representation of (2.115) is

$$D^q\boldsymbol{x} = \boldsymbol{f}(\boldsymbol{x}) \tag{2.116}$$

where $q = [q_1, q_2, q_3]^T$ for $0 < q_i < 1$, $(i = 1, 2, 3)$,$\boldsymbol{f} \in \boldsymbol{R}^3$ and $\boldsymbol{x} \in \boldsymbol{R}^3$. The stationary points of (2.116) are calculated via solving the following equation

$$\boldsymbol{f}(\boldsymbol{x}) = \boldsymbol{0} \tag{2.117}$$

and we suppose that $P^* = \left(x_1^*, x_2^*, x_3^*\right)$ is its stationary point.

Conditions for asymptotic stability of (2.115) result from the following statements. For a system of commensurate fractional orders the stationary points are asymptotically stable if all the eigenvalues λ_i $(i = 1, 2, 3)$ of the Jacobian matrix $\boldsymbol{J} = \partial \boldsymbol{f}/\partial \boldsymbol{x}$, evaluated at the stationary point P^* satisfy the condition (2.114) for all eigenvalues λ_i. Due to the complex form of the established system of nonlinear equations of fractional orders, the determination of their analytical solution is not an easy task and therefore, to examine the effect of various system parameters on the periodic dynamical process, we can perform numerical calculations applying a computer program MATLAB with using effective numerical integration procedures. Equation (2.116) was solved with the variable step ODE23 procedure from MATLAB with $RelEr = AbsEr = 10^{-8}$ and $0 \le t \le 50$ s and nonzero initial conditions.

Computed results for system parameter values (*a*, *b*, c) = (0.3 0.5 *c*) with *c* taken as bifurcation parameter and initial conditions fixed at (x(0) y(0) z(0)) = (18 5 2) are shown in Fig. 2.28a. They represent the time varying fractional system variables in the steady state. The corresponding phase space portrait of the given system is shown in Fig. 2.28b. Looking at the eigenvalues of the Jacobian it is easy to state that they correspond to a saddle node stationary point which involves instability of the system response for the wide range of possible values of the parameter *c*. Moreover, the above results can be used in many possible applications of FOSs and help assessing whether the latter are capable of addressing the industry's problems. As is well known, there exist some practical applications of FOSs in which the existence of non-overshooting step responses is critical, and consequently designing non-overshooting feedback control systems is of great importance. Therefore, these conditions are beneficial in predicting the existence of an overshoot in the step response of a fractional-order model whose numerical simulation is not simple.

The next problem provides a brief retrospective look on fractional calculus applications for studies of MMOs in nonlinear complex electric circuits. Let us consider the fractional order circuit shown in Fig. 2.29a which represents a modified Chua's circuit [56]. This circuit is different from the standard Chua's circuit in that the piecewise-linear nonlinearity is replaced by an appropriate cubic nonlinearity (2.29) which leads to a very particular behavior. In this case, the circuit forms a nonlinear oscillator comprising a nonlinear resistor with a cubic characteristic $I_n(x_1)$, three fractors represented by an inductor and two supercapacitors, a current controlled current source I and a biasing constant current source a. Its comportment depends on all six constant parameters involved. It can exhibit a large spectrum of dynamical behaviors such as the relaxation, mixed-mode oscillation, bifurcation and chaos. The mathematical description of dynamical components is based on general circuit laws and properties of the fractors and the studied circuit can be described as follows

$$\begin{aligned} C_1 \frac{d^{q_1}}{dt^{q_1}} x_1 &= -x_2 + \alpha x_1^2 + \beta x_1^3 \\ L \frac{d^{q_2}}{dt^{q_2}} x_2 &= x_1 - x_3 - R x_2 \\ C_3 \frac{d^{q_3}}{dt^{q_3}} x_3 &= a - b x_2 \end{aligned} \tag{2.118}$$

where q_1, q_3 and q_2 denote fractional orders of the supercapacitors C_1, C_3 and of the real coil L, respectively. Of particular interest here are the mixed-mode oscillations (MMOs). For instance, assuming $C_1 = 0.01\text{F}$, $L = 1\text{H}$, $C_3 = 1\text{F}$ and $a = 0.0005\text{A}$,

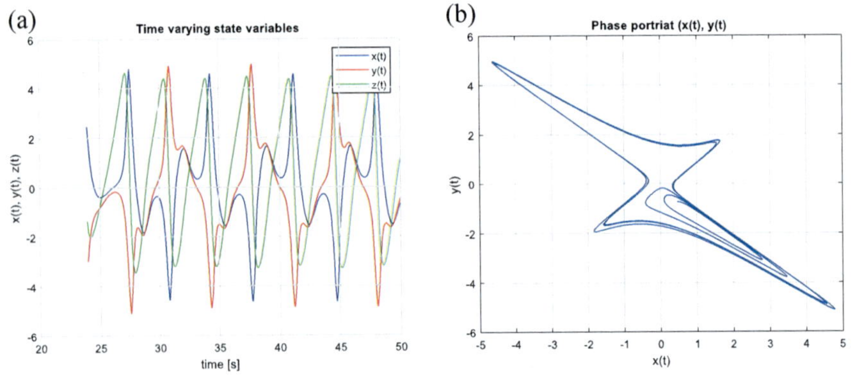

Fig. 2.28 Simulation results for Volta system of fractional orders $q_1 = q_2 = q_3 = 0.98$: **a** state variables: x(t)—in blue, $y(t)$—in red, $z(t)$—in green, **b** phase trajectory $y(t)$ versus $x(t)$

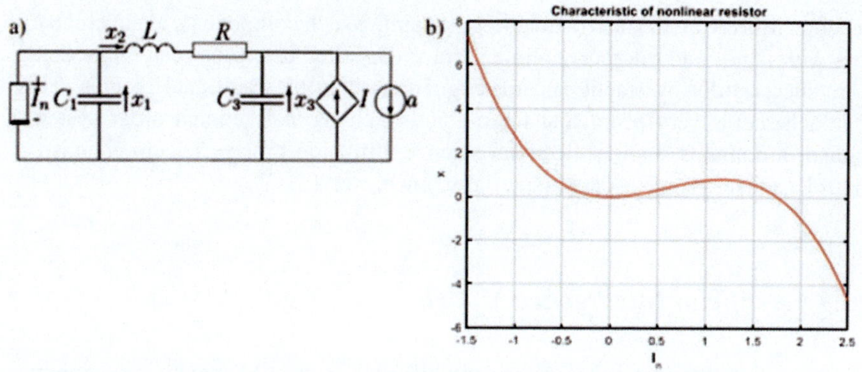

Fig. 2.29 Modified Chua's circuit: **a** scheme with fractors C_1, L, C_3 and $I = (1 + b)x_2$, $a = \text{const} > 0$, $b = \text{const} > 0$, $I_n = \alpha x_1^2 + \beta x_1^3$, $\alpha < 0$ and $\beta > 0$, **b** characteristic of a nonlinear resistor

$b = 0.0035$, $\alpha = 1.5$ and $\beta = -1$ with $q_1 = q_2 = q_3 = 0.9$ we get jacobian's eigenvalues $\lambda_1 = 0$ and $\lambda_{2,3} \approx 1.0107809162 \pm 0.0153011315j$. Such stationary point is the unstable focus node. The MMOs with respect to the state variable $x_1(t)$ are presented in Fig. 2.30a. The simulations were performed without applying the short-memory principle for time step $h = 0.001$ with the following initial conditions: x(0) = 0, y(0) = 0, and z(0) = 0. Computer simulations also show the limit cycle (Fig. 2.30b) exhibiting MMOs in the studied circuit.

The performed studies indicate significant differences between fractal circuits and integer order circuits. Most properties and conclusions relating to integer order circuits cannot be simply extended to that of the fractional order circuits. For instance, due to the lack of appropriate mathematical tools, chaos analysis in fractional order circuits is more complicated than that in integer order systems. The models of the fractional order circuits contain unlimited memory and they exhibit more degrees of

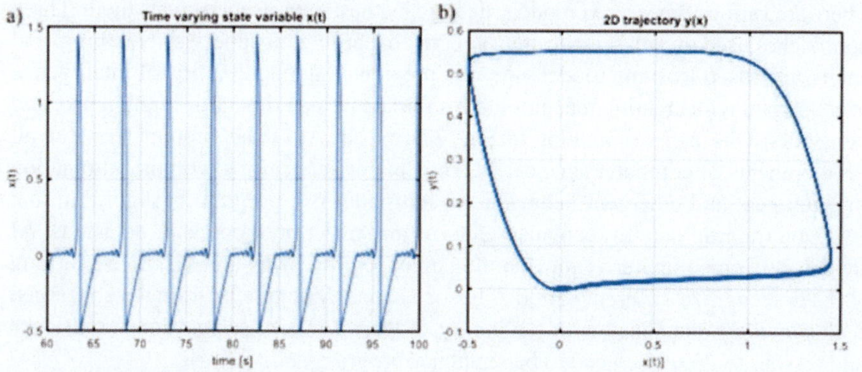

Fig. 2.30 Simulation results for a modified Chua's circuit of fractional orders $q_1 = q_2 = q_3 = 0.9$: **a** state variable $x_1(t)$, **b** trajectory $x_2(t)$ versus $x_1(t)$

freedom. Moreover, the microstructures containing such components as supercapacitors with nano- and microcrystalline surface deposited electrodes can be modeled more successfully by fractional order equations than by traditional models. This approach seems very promising for predicting chaos in fractional order systems studied in domains such as electrical science, diffusion process, electrochemistry, control science, viscoelasticity, material science, etc.

2.3.7 Artificial Intelligence Models

In concise characterization terms, an artificial intelligence (AI) model is a computer tool or algorithm which is based on a certain data set through which it can arrive at a decision—all without the need for human intervention in the decision-making process.

An AI model utilizes a huge set of data that enables it to recognize certain patterns. In simple words, it can reach a conclusion or make a prediction when provided with sufficient information, often a powerful amount of data. For these reasons, AI models are particularly suitable for solving complex problems while providing higher efficiency/cost rationalizations and suitable accuracy compared to simple methods. As AI models become more important and widespread in almost every sector of human activity, it is increasingly important for researchers and users to understand how these models work and the potential implications of using them. A more refined definition of artificial intelligence can be compressed as follows [54]: a computer program is said to learn from experience *E* with respect to some task *T* and some performance measure *P*, if its performance on *T*, as measured by *P*, improves with experience *E*.

Applying artificial neural networks freely designed according to the human brain, the models use those assessments to attempt to predict the future, forecasting the decisions as subject matter experts (SMEs) might make about the remaining documents. There are various types of AI models, defined by the means used to create them. Three approaches used in data science are: supervised learning, unsupervised learning, and semi-supervised learning models. As was previously defined, Artificial Intelligence is concerned with creating machines that simulate human thoughts, intelligence, and behaviors. One can see looking at Fig. 2.31a, that AI is the support term for all developments of computer science, which is mainly concerned with the automation of intelligent and emergent behavior. An AI model is a program or algorithm that relies on training data to recognize patterns and make predictions or decisions. AI models rely on computer vision, natural language processing and Machine Learning (ML) to recognize different patterns. In such scope, Machine Learning is concerned with providing machines with the ability to learn for themselves from experience and lessons without the need to be explicitly programmed.

Teaching AI models with few data are at the center of knowledge inference in the natural and engineering sciences, in contrast to the typical structure of AI in economics or computer science, where very large amounts of data are available for

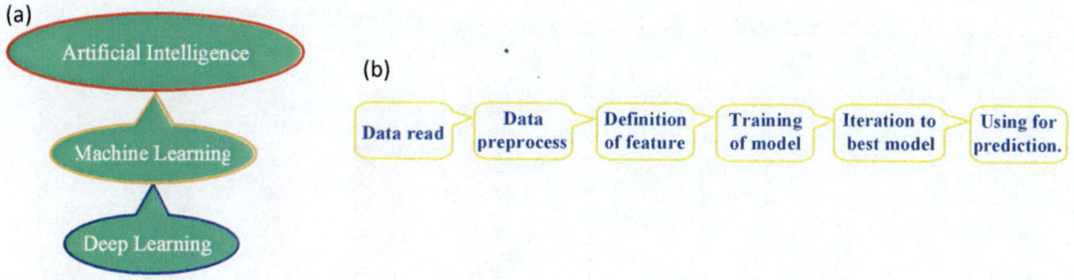

Fig. 2.31 Schematic of the AI Learning Process: **a** the hierarchy of artificial intelligence, machine learning and deep learning, **b** flowchart for the learning process with AI/ML

solution of the problem under consideration. Fewer input dimensions (i.e. number of excitations) induce fewer parameters or a simpler structure of the ML model, referred to degrees of freedom. A model with too many degrees of freedom is likely to over fit the training dataset and therefore may not perform well on new data. A generally valid scheme of steps involved in a successful ML project is presented in Fig. 2.31b.

The realization of every step is essential for construction of a sensible AI/ML application. The mathematical model in an AI yields an approximation of what the output will probably be, given a certain set of inputs. The model is accepted for its ability to predict outputs, with a certain level of probability.

2.4 Mixed-Mode Oscillations

All over nature occur oscillators producing rhythms and vibrations in the form of a strong, regular, repeating course of motion, lightning, or sound. Simple oscillators are based on a single mechanism, for instance in the case of an ordinary pendulum without damping that generates harmonic oscillations. But in more complex systems, different processes occur with different phases of oscillation. However, real-world oscillators seldom display the uniformity of the harmonic oscillator, but generally alternate between slow and fast course with large and small amplitudes, respectively. Such mixed mode oscillations abbreviated as MMOs form important issues in physics, chemistry, engineering, biology, medicine and economy [54]. Essentially speaking, MMOs are the time series oscillations of a dynamical system in which takes place an alternation between oscillations of distinct large and small amplitudes, abbreviated as LAOs and SAOs, respectively. There is no widely accepted strict criterion for this distinction between amplitudes, but usually the graduation between large and small is clear (e.g. in Fig. 2.32).

MMOs have been observed in, among others, the tunnel diode and Bonhoeffer-van der Pol oscillators, the Hodgkin-Huxley model of neuron dynamics, Taylor-Couette flow, surface oxidation and autocatalytic reactions, human heart arrhythmias and in epileptic brain neuron activity [29]. An important kinetic requirement for MMOs generation is that the differential equations describing dynamical systems must be nonlinear, must be in dimensional order at least 3, and span multiple time scales.

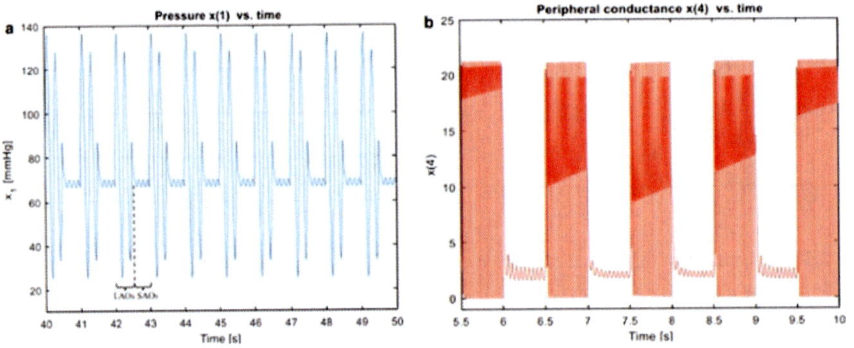

Fig. 2.32 MMOs at the load of a human left ventricle: **a** blood pressure variations at input of aorta, **b** variations of peripheral conductance in the cardiovascular system

However, it is not a unique sufficient condition for the occurrence of self-organization in the system, e.g., generation of spontaneous oscillations of the dynamic variables. Moreover, the analysis of the origin of such phenomena also clearly points out at the necessity of the existence of positive and negative feedback loops in the links that make up the entire mechanism of a given process. Therefore, the production of a given manifestation of selforganization requires, among others things, introducing appropriate feedback loops. In fact, even the simplest case of oscillatory proceeding in time can be considered as the interaction of minimum two feedback loops: a positive feedback loop and negative feedback loop. During the activity of the positive feedback loop, the increments of given system's variables growth in time at a great rate up to the moment when the negative feedback loop takes over the control and leads to a decrease in these changes until the instant at which positive feedback sets in again, and the process repeats in time continuously. Hence, the oscillatory process in the dynamics of the system requires the activity of fast positive and slow negative feedback loops. Quite simple conformity can be recognized in the case of oscillations of a plain pendulum when the positive feedback loop is initiated at the lowest position (then the movement's velocity is maximal). However, at the highest position appears the negative feedback loop which firstly annihilates the movement and next changes the velocity direction to the lowest position. The basic idea is that the small-amplitude oscillations (SAOs) are induced by a local mechanism while large-amplitude oscillations (LAOs) are produced by a global return mechanism. Typically, a basic MMOs cycle consists of s small amplitude oscillations (SAOs) followed by L spikes (LAOs) or vice versa. The notation L^s is conventionally used to describe these waveforms [48]. The global return mechanism, which is generic for LAOs frequently encounters a relaxation type oscillations induced by a cubic (or S-shaped) fast-variable nullcline, also called the critical manifold. The generation of MMO time series requires the coordinated action of various mechanisms: (*i*) a mechanism for the generation of the subthreshold regime (STOs); (*ii*) a spiking mechanism, including the onset of spikes and the description of the spiking dynamics; and (*iii*) a return mechanism from the spiking regime back to the STOs regime.

Singularly perturbed systems of ODEs can be used in the analysis of mixed-mode oscillations produced by the so-called modified Chua's circuits made up of the coupled *R, L, C* elements connected with controlled sources $(1 + \gamma)y - \mathrm{c}$ and nonlinear resistors with voltage-current characteristic $i = ax^2 + bx^3$ for circuit shown in Fig. 2.33. The circuit is described as follows

$$\begin{aligned} \varepsilon x' &= -y - \left(ax + bx^2\right)x \\ y' &= \alpha(x - Ry - z) \\ z' &= c - \beta y \end{aligned} \tag{2.119}$$

where the symbol $'$ means differentiation with respect to time $t \in \boldsymbol{R}, 0 < C_1 \equiv \varepsilon \ll 1, \alpha = L^{-1}$ and $\beta = \gamma / C_2$. The parameters of the controlled source depends on the variable y according to the expression $(1 + \gamma)y$-c with $\gamma > 0$ and $c = \text{const}$. Circuit described by (2.119) is a prototypical example for MMOs examined in [49]. Its behavior depends on all five constant parameters engaged. One can distinguish three different modes of oscillations exhibited by (2.119) within the time: only small amplitude oscillations (SAOs), large amplitude oscillations (LAOs), and a composition of both SAOs and LAOs producing the MMOs phenomenon. The three modes are illustrated in Fig. 2.34 which shows the solutions of (2.119) for $R = 0$ and $\varepsilon = 0.01$, $a = 1.5$, $b = -1$, $\beta = 0.005$ and three values of c. In the SAOs case only, the small amplitude oscillations around the equilibrium point at the origin (0, 0, 0) are due to Hopf bifurcation for $c = 0$. In the LAOs case only, a trajectory passing close to the origin bypasses the region of small amplitude oscillations. The MMOs case is in some sense a synthesis of the previous two cases. It is worth underlining that the mechanism in which small and large amplitude oscillations occur in a same circuit is quite complex [49]. In the case of MMOs, a series of SAOs considered as canard solutions appear around the equilibrium point at the origin (0, 0, 0) which undergoes a rapid canard explosion yielding an LAOs. But they through a special return mechanism bring back the circuit into the neighborhood of the origin. Depending on the circuit's parameters, the oscillation may continue with one (or more) LAOs, or may go through a new series of SAOs after which the trajectory leaves again the neighborhood of the origin and the phenomenon reduplicates. For fixed a and b in (2.119) the values of c, β and ε are responsible for various types of MMOs.

The two main ingredients of the dynamics interplay in generating the MMOs, the local flow close to the strong canard and the global return that underlies the basic

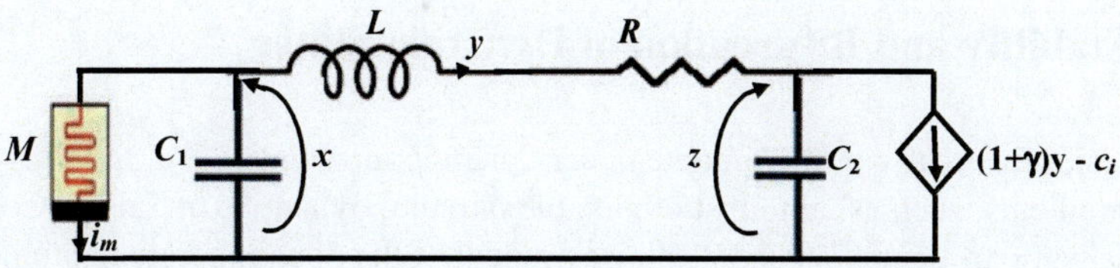

Fig. 2.33 Schemes of modified Chua's *R, L, C* circuit with nonlinear resistor $i = ax^2 + bx^3$

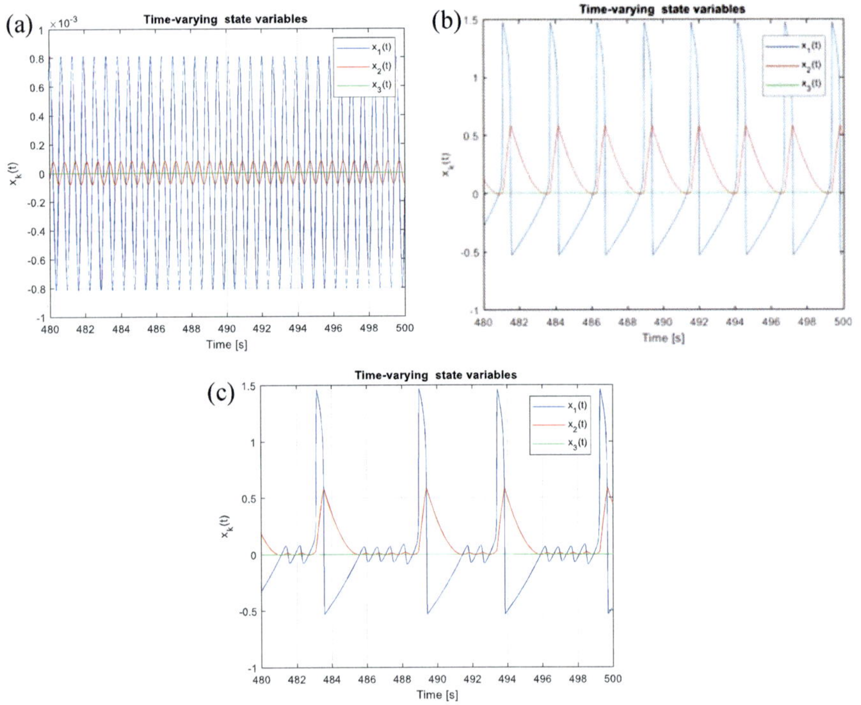

Fig. 2.34 Periodic oscillations produced by (2.119) for $\varepsilon = 0.01$, $a = 1.5$, $b = -1$ and $b = 0.005$ and different values of parameter c: **a** SAOs only for $c_1 = 0$, **b** LAOs only for $c_2 = 0.00105$, **c** MMOs for $c_3 = 0.00053$

canard mechanism for the emergence of MMOs in (2.119). Moreover, (2.119) is a normal form for the class of three time-scale systems, in the sense that the addition of higher-order terms in it will not fundamentally influence the resulting dynamics. Another aspect of the mixed-mode dynamics in (2.119), in addition to the return mechanism, is the family of so-called secondary canards viewed as a trajectory that undergoes k small (nonrelaxation) rotations, or "loops," during its passage "near" the lower fold $\mathcal{L}^-$ and that then remains $\mathcal{O}(\varepsilon)$—close to the critical manifold S_0 until it reaches the $\mathcal{O}(\varepsilon^{1/3})$—vicinity of the upper fold $\mathcal{L}^+$ [33, 48, 49].

2.5 Stability and Bifurcation of Dynamic States

In general, most real-world phenomena are dynamic, non-linear, and time-varying. Many problems, such as, amongst others, the bursting dynamics in Taylor–Couette flow, the compartmental Wilson–Callaway model of the dopaminergic neuron, stellate cell dynamics, the famous Hodgkin–Huxley model of neuron dynamics, the coupled calcium oscillator model, the complex plasma instability model, and the surface oxidation reaction and autocatalytic chemical reaction (for such examples see [27, 49]) are described by singularly disturbed systems of nonlinear ordinary differential equations (NODEs) with three (or more) state variables changing at different time scales (usually two or three).

Since each of these categories of processes has to deal with its specific expression, then the universality of dynamic comportment is clearly manifested entirely at the level of their mathematical description, which in turn is based on the run of the solutions of the respective NODEs. Therefore, the current achievements in the field of dynamics of nonlinear systems constitute a significant help in the analysis of processes occurring in real systems. In the modern approach to nonlinear complex systems, not having frequently explicit solution in terms of Newton's analytical methods, the emphasis is put on qualitative dynamics rather than strictly quantitative solutions. However, deterministic NODEs are an extremely complex example of dynamic phenomena whose solutions manifest themselves as periodic, continuous non-sinusoidal oscillatory changes in the state of a system. As a prototypical example, periodic oscillations in the form of pulsations of the heart of a healthy person can be used [52]. In general, spontaneous oscillations of the system state variables depend on the values of parameters that are included in the state variables equations creating equations in the state space of a given system. These parameters values we shall call control parameters. They may be, for instance, the flow rate through the reactor for the homogeneous chemical reactions, and the applied electrode potential or the imposed electric current for electrochemical processes and so on. It may happen that for some control parameters which we choose initially, the solutions of those equations will be quite trivial, i.e., they will predict the existence of only a single steady state. However, for another set of those parameters the same equations may generate, e.g., the variable spontaneous oscillations in time.

In the study of a non-linear system, it is important to find the critical value of one (or more) parameter at which such qualitative change in the system's behavior occurs. This is the bifurcation point (following the Latin word "bifurcus"—forked). This topological, essentially mathematical term, refers strictly to the theory of differential equations, as they describe the dynamics of any system. Then a typical sequence of waveforms is that even a relatively uncomplicated system, described in terms of deterministic differential equations, and may exhibit a sudden change to a completely new, qualitatively different behavior upon smooth variations of the control parameter. Therefore, a bifurcation corresponds to an abrupt or discontinuous change in system behavior as a system parameter is changed. At the bifurcation point the system may be particularly sensitive to the parameter fluctuations. In fact, it is the amplification of the small fluctuations to macroscopic scale which manifests itself as the qualitatively new system's behavior. This also means that in the trivial, linear range of the system's behavior the fluctuations are damped, but in a nonlinear state, due to the feedback loops, they are quickly (exponentially) enhanced. Furthermore, at the bifurcation point several possibilities for the further system's evolution may be opened and which of them will be chosen, depends on the actual random fluctuation; in this way in the essentially deterministic system, as defined by the respective differential equations, the stochastic element appears.

For instance, Fig. 2.35a shows a dynamic system corresponding to

$$\dot{z} = z^2 + \alpha$$

for which the fixed points are easily calculated:

$$f(z, \alpha) = 0 \quad \rightarrow \quad z = \begin{cases} \pm\sqrt{-\alpha} & \text{for} \quad \alpha < 0 \\ \text{No solutions for} & \alpha > 0 \end{cases}$$

So for $\alpha < 0$ the function f has two roots, and therefore two fixed points as plotted in the bifurcation diagram of Fig. 2.35b, whereas for all $\alpha > 0$ the function f has no roots, and therefore no fixed points. At $\alpha = 0$ we observe the spontaneous appearance/disappearance of a pair of fixed points, known as a *saddle bifurcation.* It is worth noting that in Fig. 2.35 is considered only the individual bifurcation; in practice a nonlinear system may have multiple bifurcations of different types, with a double saddle bifurcation, transcritical bifurcation, pitchfork bifurcation, etc., appearing quite commonly in the systems.

In Table 2.1 the interpretation of dynamics at fixed points of systems $\dot{z} = f(z)$ are presented in an abbreviate form.

All processes of the above type are strictly related to the stability of the states. When in the initially homogeneous system, a dynamical temporal or spatiotemporal order emerges, it means that the initial state lost its stability and a new behavior acquired it. If one type of self-organized phenomena turns into another one, it also means the exchange of their stability. It is very important to distinguish between the terms "steady state" and "stable state." The term "steady state" (or stationary state) means the state, the characteristics of which do not change as a function of time. In mathematics, the steady state is termed also the fixed point or the equilibrium point, although the latter term may seem a bit controversial for the physicists, chemists, biologists or mechanicians, if one considers the nonequilibrium steady state. It is

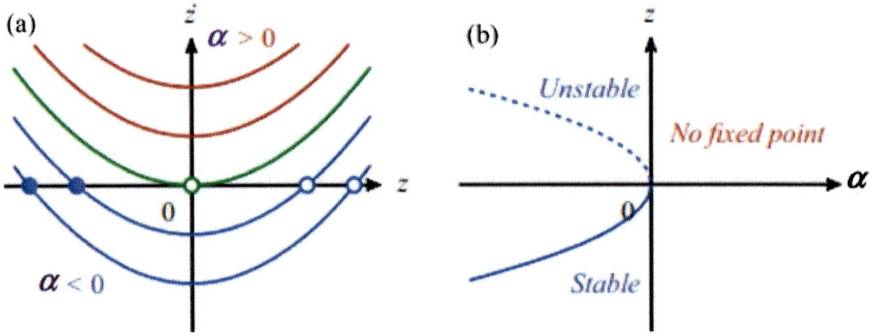

Fig. 2.35 A simple illustration for a bifurcation: **a** system dynamics, **b** a bifurcation is a sudden change in system behavior as a function of a system parameter α (blue corresponds to **α** < 0, green to the degenerate case right at **α** = 0, and red to **α**> 0)

Table 2.1 The interpretation of dynamics at fixed points

Dynamic vector	Fixed point interpretation
$f(z) \neq 0$	Not a fixed point
$f(z) = 0$	Fixed point at $z = z_1, z_2, \ldots, z_n$
$f'(z) > 0$	Stable fixed point
$f'(z) < 0$	Unstable fixed point
$f'(z) = 0$	Degenerate situation
$f''(z) \neq 0$	Bifurcation, Half Stable, Degenerate fixed point
$f''(z) = 0$	No bifurcation, regular fixed point
$f'''(z) > 0$	Unstable fixed point
$f'''(z) < 0$	Stable fixed point

further important to note that the steady state can be stable or unstable. We shall give here the simplest interpretation of these terms, which will be developed in more detail later. The stable steady state is resistant to external or internal perturbations in this way that if they happen, they are eventually damped and the system returns to its original state. The unstable steady states will not survive any fluctuation, as the system driven out of it goes then to the closest stable state. This is the reason for which the states which are observable must also be stable, since real systems are always a subject of fluctuations. Unstable steady states are not observable, unless they are stabilized by special procedures, but their existence can be indicated in theoretical models. One may additionally test the stability of the given state by introducing external excitations of increasing amplitude. If the system returns always strictly to the initial state, one calls its stability as the asymptotic one, as all the imbalances decay to zero asymptotically in a function of time. If the system characteristics survive in this way even relatively large fluctuations, covering the entire possible range of the dynamic variables considered, it is called "globally stable." Otherwise, if sufficiently large perturbations cause the switching of the system to another, concurrent stable state, the original one is called only "locally stable." More types of stability are formulated in terms of the system perturbation by relatively small fluctuations leaving the initial state, but remaining close to its characteristics. Then the system is called stable in a Lyapunov sense meaning that perturbations applied in the vicinity of the steady state cause that the system still remains in the neighborhood of this state and accepts certain tolerance for this deviation. On the other hand if this neighborhood reduces eventually to zero, then more general Lyapunov stability turns into its special case of asymptotic stability. Note that completely different, physically or biologically, processes can exhibit the same types of bifurcations, i.e., the same types of qualitative changes of dynamics upon variation of appropriate control parameters. In the example earlier, the bifurcation involved the loss of stability of the steady state and the birth of the stable oscillatory regime, instead. Such a change of stability of the states can be an example of the Hopf bifurcation, one of the most frequent mechanisms in which the oscillations are born from the trivial, nonoscillatory steady state.

At the beginning of in-depth discussion on stability properties of nonlinear systems it is valuable get acquainted with some preliminary concepts and definitions. First of all, in this chapter we shall only consider autonomous systems of ordinary differential equations in the form

$$\dot{x} = f(x) \tag{2.120}$$

where $x \in R^n$ denotes the state variable vector and $f \in R^n$ is the vector of system dynamics. Secondly, nonautonomous systems

$$\dot{x} = f(x, t) \tag{2.121}$$

where the function f can additionally be contingent on the independent variable t, may be written as an autonomous system (2.120) with $x \in R^{n+1}$ simply by letting $x_{n+1} = t$ and $\dot{x}_{n+1} = 1$. Next, considering f as differentiable function with respect to x gives

$$f(x) \cong Df(x_0)x \tag{2.122}$$

where

$$A = Df(x_0) = \left[\frac{\partial f_i}{\partial x_j}\right]_{x=x_0}, \quad i, j = 1, 2, \ldots, n \tag{2.123}$$

denotes the $n \times n$ Jacobian matrix at the equilibrium point x_0.

Multi-dimensional linear systems can have only a very limited number of possible behaviours, therefore most of our understanding of two-dimensional nonlinear systems will come via a linearization about every fixed point.

For a given two-dimensional nonlinear system, it is valuable to characterize it by determining its fixed points, linearized dynamics, and basins of attraction. The ***Fixed Points*** also called equilibrium points of a joint system are those points where all state elements are fixed, unchanging over time:

$$x = x_0 \quad \text{is a fixed point if} \quad \dot{x} = f(x_0) = 0 \tag{2.124}$$

At a stable fixed point the dynamic of system is inherently stable, requiring no control input of any kind to stabilize the system.

The ***Linearized Dynamics*** are determined by computing the Jacobian, (2.123), from which the linearized behaviour can be categorized on the basis of Table 2.2. As was discussed in Sect. 2.2.3 the linearization does, of course, introduce some degree of approximation to the original nonlinear system. Therefore if the linearized system type is degenerate (star, non-isolated, centre), there is more than one possible behavior of the corresponding nonlinear system.

At a ***Saddle Point*** the system dynamic has a combination of convergent (stable) and divergent (unstable) directions. A control strategy is therefore needed to stabilize

Table 2.2 Eigenfrequencies of a given matrix $\boldsymbol{A}$ showing one of the ten types of system dynamics

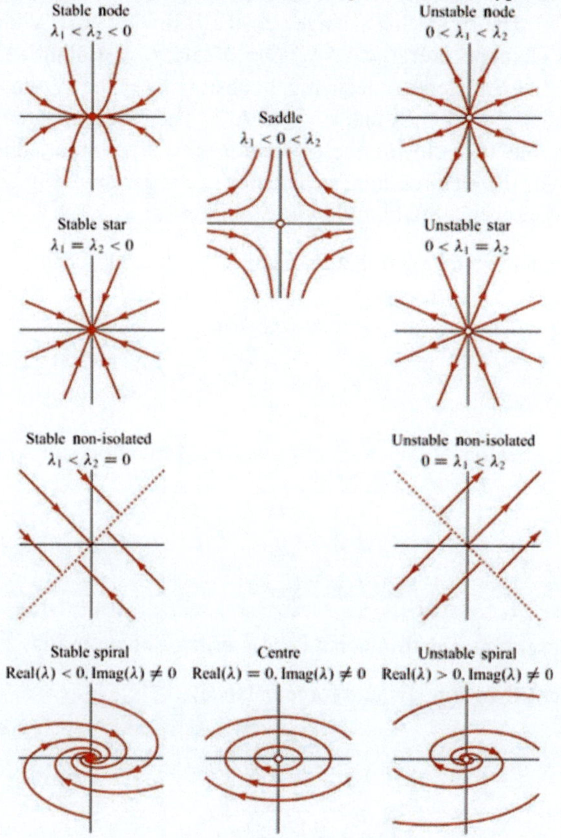

the divergent directions, which is a slightly simpler problem than at an unstable fixed point, for which all directions are divergent. At an ***Unstable Fixed Point*** all directions are divergent; therefore the feedback control needs to be sensitive to and acting upon all of these directions. Nevertheless the control problem remains relatively straightforward, since at the fixed point the divergent force is zero; as has been pointed out before, an unstable fixed point is not like a magnet whose repulsive force is increased as the pole is approached, rather the repulsive force shrinks to zero at the unstable fixed point, and grows away from the fixed point.

On a ***Stable Limit Cycle***, in contrast to the preceding cases, the desired point z_S is no longer a fixed point, therefore the control needs to produce a feedback signal opposing the dynamics. The response time of the feedback control is not so essential, since the limit cycle is stable and its dynamics are known. However, the control may need large inputs, since we now have to explicitly oppose the dynamics. The magnitude of the control input needed will be a function of the dynamics being opposed.

Consider a nonlinear system modeling a vibrating table of the shaker (vibrator) shown in Fig. 2.36. The supplying voltage $u = 0.141\sin(1.5915t)$ V and the parameters of the other elements are: $R = 1.5\ \Omega$; $L = 0.25$H; $C = 100$ mF; K = 500 N/m; $m = 10$ kg, $B = 2$ Ns/m; electro-mechanical constant $k_1 = 0.5$ Wb/m; mechanical–electrical constants $k_2 = 17$ m/A and $\alpha = 2.5\ \mathrm{A}^{-1}$. The problem is to determine the vibration of the table after closing the connector with zero initial conditions in the system. Diagram of the electromagnes characteristic is given on Fig. 2.37.

The system is described by the following equations.

- connecting feeder (voltage Kirchhoff law)

$$u = Ri_1 + u_C$$
$$u_C = L\frac{di_2}{dt} + e \tag{2.125}$$

- electromagnet (Faraday's law of electromagnetic induction and Ampere's law)

$$e = k_1\frac{dx}{dt} = k_1 v$$
$$f = g(i_2) \tag{2.126}$$

where $g(i_2)$ represent the electromagnet characteristic shown in Fig. 2.37.

– mechanical part (Newton law and force balance)

$$\frac{dx}{dt} = v$$
$$m\frac{dv}{dt} + Bv + Kx = f \tag{2.127}$$

By merging the above equations into one matrix equation for state variables gives

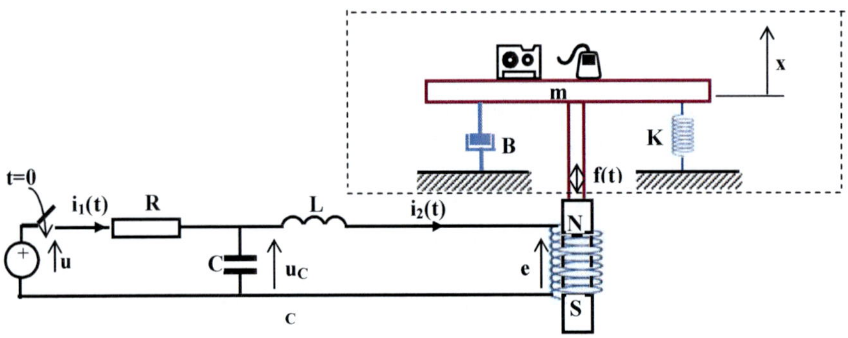

Fig. 2.36 Schema of the system with vibrating table

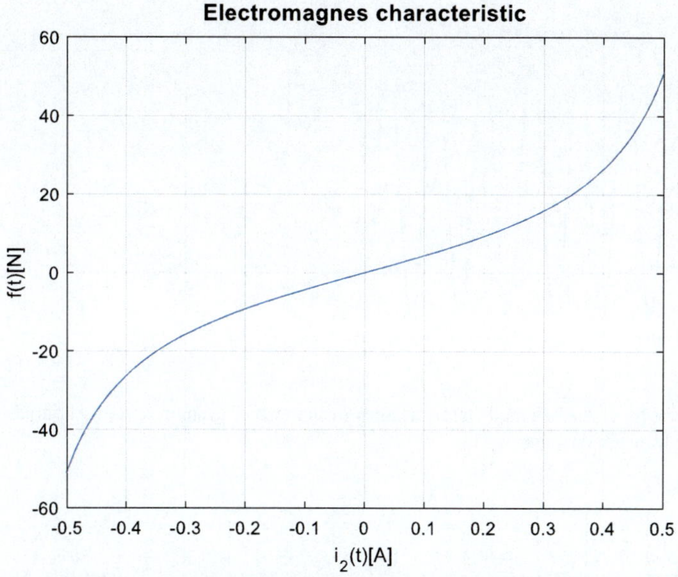

Fig. 2.37 Diagram of the electromagnes characteristic

$$\frac{d}{dt}\begin{bmatrix} i_2 \\ u_C \\ x \\ v \end{bmatrix} = \begin{bmatrix} \frac{u_C}{L} - \frac{k_1}{L}v \\ -\frac{i_2}{C} - \frac{u_C}{RC} \\ v \\ \frac{k_2 \tan(\alpha i_2)}{m} - \frac{Kx}{m} - \frac{Bv}{M} \end{bmatrix} + \begin{bmatrix} 0 \\ \frac{1}{RC} \\ 0 \\ 0 \end{bmatrix} u \tag{2.128}$$

This model allows for the study of the effects of changing the load on the table with the mass of the tested devices in terms of its resistance and stiffness of its suspension, which allows establishing a compromise between the effect and efficiency of operation. Design solutions are usually preceded by simulation tests. It is easy to see that the determined model is a system of four non-linear first-order differential equations. Its numerical solution based on the Matlab program is shown in Fig. 2.38a in relation to the state variables $i_2(t)$ and $x(t)$. The steady-state phase trajectory for $x(t)$ as a function of $i_2(t)$ is shown in Fig. 2.38b. The diagrams show that the state of the device evolves with time and after closing the switch a short transient occurs (about 8 s) and then steady sinusoidal oscillations are generated in the system.

The above nonlinear system of equations can be approximated by its Taylor series expansion neglecting the high order terms. In this case, the system (2.129) is replaced by the matrix of its first derivatives (Jacobian) evaluated at the origin and the following linear system is obtained.

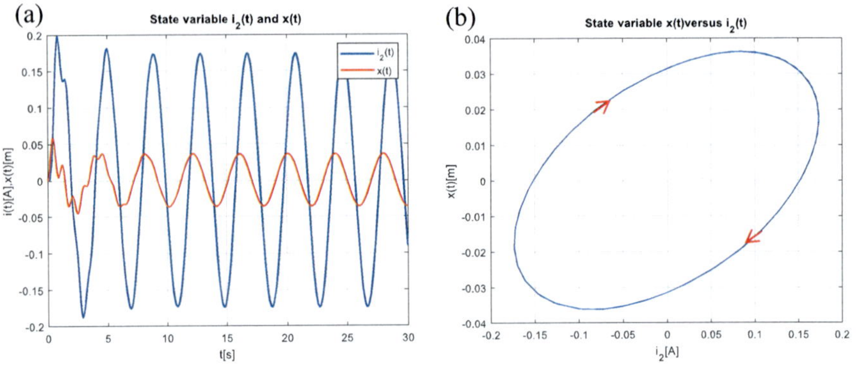

Fig. 2.38 Graphs of the selected state variables in the shaker system: **a** state variables $i_2(t)$ and $x(t)$, **b** phase trajectory $x(t)$ versus $i_2(t)$

$$\frac{d}{dt}\begin{bmatrix} i_2 \\ u_C \\ x \\ v \end{bmatrix} = \begin{bmatrix} 0 & \frac{1}{L} & 0 & -\frac{k_1}{L} \\ -\frac{1}{C} & -\frac{1}{RC} & 0 & 0 \\ 0 & 0 & 0 & 1 \\ \frac{k_2}{m} & 0 & -\frac{K}{m} & -\frac{B}{m} \end{bmatrix} \begin{bmatrix} i_2 \\ u_C \\ x \\ v \end{bmatrix} + \begin{bmatrix} 0 \\ \frac{1}{RC} \\ 0 \\ 0 \end{bmatrix} u \tag{2.129}$$

The solutions $i_2(t)$ and $x(t)$ can now be calculated from (2.129). These solutions have been plotted respectively in Fig. 2.39a and 2.29b.

Comparing the graphs of state variables determined for the non-linear model (2.128) and for the linearized model (2.129) presented in Figs. 2.38 and 2.39, respectively, it is easily seen that the applied linearization method very well converges to the true nonlinear solution of the system under study.

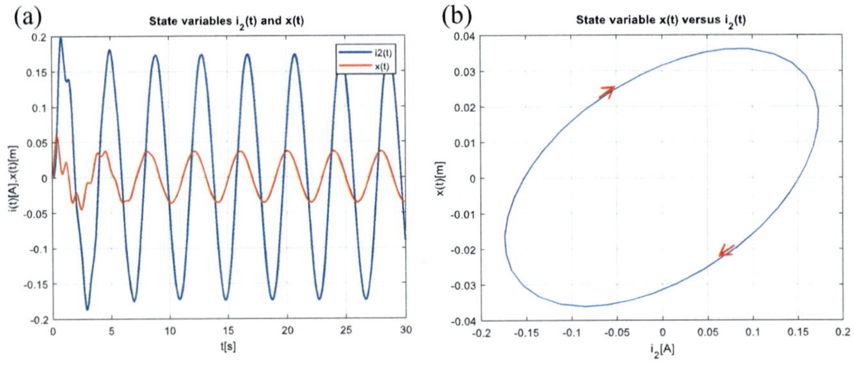

Fig. 2.39 Graphs of the selected state variables in the shaker system determined with linear model: **a** state variables $\mathrm{i}_2(\mathrm{t})$ and $x(t)$, **b** phase trajectory $x(t)$ versus $\mathrm{i}_2(t)$

2.6 Chaotic Oscillations

In general, a dynamical system is considered chaotic, irregular or complex if it does not correspond to any traditional model such as monotonic or periodic convergence or divergence, centre or limit cycle. Moreover, its time series appears to be disorderly as if it was a stochastic model although the system is completely deterministic and no random factors are present. Chaos theory perhaps started with E. Lorenz's [21] discovery of very complex dynamics arising from 3 nonlinear differential equations leading to turbulence in the weather system area for exploration and there always seems to be some new aspects that can be revealed. Chaotic behaviors have been observed in different areas of science and engineering and various mathematical definitions of chaos have been proposed, resulting in a lack of a commonly accepted standard definition for it. Thus, rather than selecting a definition of "chaos", it is much easier to list the properties of chaotic systems. It is therefore recognized that chaotic behavior is a recurrent, limited, non-periodic and long-term evolution of a system leading to a strange attractor in phase space. Also, chaotic systems present an extreme sensitivity to initial conditions i.e. small differences in the initial states can lead to extraordinary differences in the system states. It has led to many studies, both theoretical and applied, by scientists, engineers, and mathematicians trying to formulate a good definition and theory for chaotic but deterministic dynamics. The term ***chaotic*** highlights the difficulties of predicting future states of the system from information about the present state. The second term ***deterministic*** is in use because the fundamental theorem for systems guarantees that each set of initial values determines a unique solution. Despite these insecurities chaos is a desirable feature in many applications. For instance, in electronics and telecommunications chaos could be induced to desire spread modal energy at resonance or to achieve optimal spatial emission of electromagnetic waves. However, it all started with E. Lorenz who published a report [56] of the strangeness he had observed in computed solutions of a system of three autonomous, first order, nonlinear ODEs that model thermal variations in an air cell bellow a thunderhead.

The Lorenz system in dimensionless state variable is

$$
\begin{aligned}
\dot{x} &= -\sigma x + \sigma y \\
\dot{y} &= rx - y - xz \\
\dot{z} &= -bz + xy
\end{aligned}
\tag{2.130}
$$

where σ, r, and b are positive parameters that denote the physical characteristics for air flow, x is the amplitude of the convective currents in the air cell, y is the temperature difference between rising and falling currents, and z is the deviation from the normal temperature in the cell. For all positive values of the parameters σ, r, and b, the Lorenz system has equilibrium point at the origin. For $0 < r < 1$, there are no other equilibrium points. The eigenvalues of the Jacobian matrix of the system at the origin are

$$\lambda_{1,2} = \frac{1}{2}[-\sigma - 1 \pm ((\sigma - 1)^2 + 4\sigma r)^2], \quad \lambda_3 = -b \tag{2.131}$$

For values of r between 0 and 1 all of these eigenvalues are negative and the origin is an asymptotically stable global attractor. For $r > 1$, the first eigenvalue is positive, and the origin is unstable. Lorenz showed that for certain values of the parameters σ, r, and b the solutions of system (2.130) are very sensitive even to small changes in the initial data, a property now believed to be characteristic of chaotic systems.

Figure 2.40 shows the 3D phase portrait and time varying state variables which correspond to solutions that start at the points $(1, 1, 1)$ and $(-1, -1, 1)$ where $\sigma = 10$, $r = 11$, $b = 8/3$. It is easy seen that solution behavior is regular and uncomplicated. However it is worth emphasizing that one of the unsettling characteristic of the Lorenz model is that trajectory structure may change considerably with only small change in the parameter r or in the initial point. Assuming different values of the parameter r with unchanged parameters $\sigma = 10$ and $b = 8/3$, it is possible to determine the values of r at which chaos is generated in the system. The critical value of r is determined by the expression

$$r_c = \frac{\sigma(\sigma + b + 3)}{\sigma - b - 1} \tag{2.132}$$

Thus for $\sigma = 10$, $b = 8/3$ we have $r_c \approx 24.737$, so the above used value $r = 11$ is subcritical. For values of r greater than the critical value r_c trajectories head quickly toward the Lorenz attractor. Taking $r = 25$ and performing computer simulations leads to the 3D phase trajectory shown in Fig. 2.40.

In Fig. 2.41a–c is visible a part of the Lorenz attractor consisting of two almost planar pieces at an angle to one another, each piece containing oscillatory arcs of orbits near the equilibrium points. The attracting set in this case, although of zero volume, has a rather complicated structure and is called a ***strange attractor***. The orbit wanders from one piece to the other unpredictably as the component graphs in Fig. 2.41d–f show.

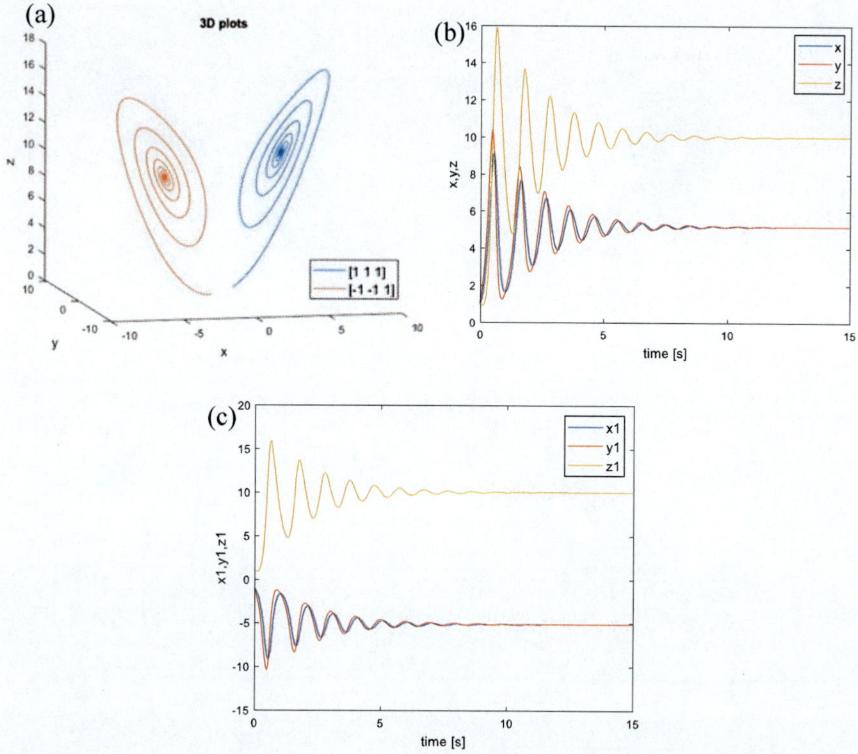

Fig. 2.40 Lorenz systems with $\sigma = 10$, $r = 11$, $b = 8/3$: **a** 3D phase trajectories for initial states: $[1\ 1\ 1]'$ and $[-\ 1 -\ 11]'$, **b** time-varying state variables for $[1\ 1\ 1]'$, **c** time-varying state variables for $[-1\ -1\ 1]'$

It is worth to emphasize that the Lorenz model is an excellent model for the general phenomenon of chaotic, deterministic behavior of particular dynamic nonlinear systems. Dynamic system practitioners, scientists, mathematicians, social scientists, economists, and even penmen use the Lorenz system as a paradigm for chaos [57].

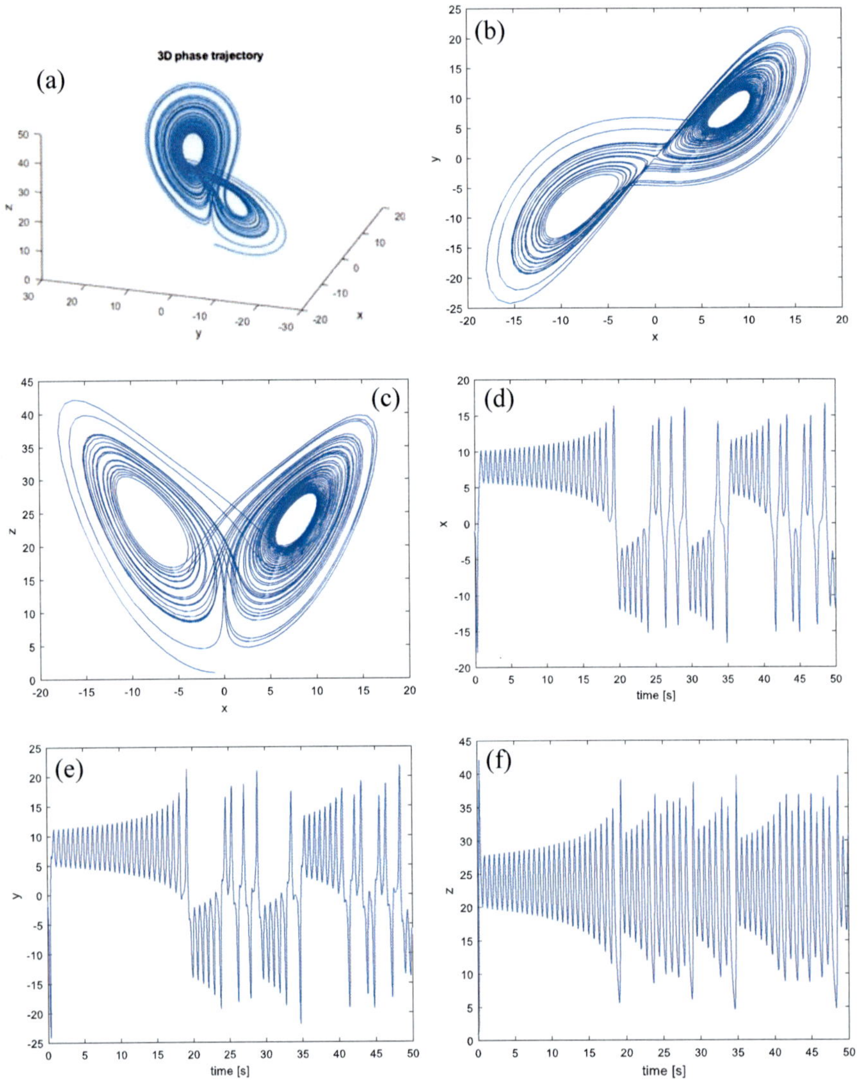

Fig. 2.41 Spatial and planar phase trajectories and time-course of state variables in Lorenz model with $r = 25$: **a–c** phase trajectories, **d–f** time-course of state variables

References

1. Trzaska, Z.: Dynamical processes in sequential-bipolar pulse sources supplying nonlinear loads. Electr. Rev. **90**(3), 147–152 (2007)
2. Trzaska, M., Trzaska, Z.: Energetic process modelling of thin-layer electrocrystallization. Electr. Rev. **87**(5), 173–175 (2011)

3. Trzaska, M., Trzaska, Z.: Straightforward energetic approach to studies of the corrosion performance of nanocopperthin-layers coatings. J. Appl. Electrochem. **37**, 1009–1014 (2007)
4. de Silva, C.W.: Vibrations. Fundamentals and Practice. CRC Press, London (2000)
5. Bachman, H., et al.: Vibration Problems in Structures. Birkhäuser-Verlag Basel (1995). https://doi.org/10.1007/978-3-0348-9231-5
6. Targosz, J.: Vibroisolation Systems in Railway and Automotive Transportation Applications. AGH University Science & Technology Press, Cracow (2007)
7. Benaroua, H.: Mechanical Vibrations. Analysis, Uncertain Control. Marcel Dekker, New York (2004)
8. Greenwood, D.T.: Advanced Dynamics. CUP, Cambridge (2003)
9. Zerzm, E.: Topics in Multidimensional Linear Systems Theory. Springer, London (2000)
10. Borrelli, R.R., Colleman, C.S.: Differential Equations: A Modeling Perspective. Wiley, New York (2004)
11. Trzaska, Z.: Analysis and Design of Electrical Circuits. Publishing Office, Warsaw University of Technology, Warsaw (2008)
12. Borrelli, R.L., Coleman, C.S.: Differential Equations. A Modeling Perspective. J.Wiley, Hoboken (2004)
13. Dimitrakopoulos, E.G.: Nonsmooth analysis of the impact between successive skew bridge-segments. Nonlinear Dyn. **74**, 911–928 (2013)
14. Heldt, T., Chang, J.L., Chen, J.J.S., Verghese, G.C., Mark, R.G.: Cycle-averaged dynamics of a periodically-driven, closed loop circulation model. Control Eng. Pract. **13**(9), 1163–1171 (2005)
15. Sagués, F., Epstein, I.R.: Nonlinear Chemical Dynamics. Dalton Trans. (2003)
16. Schupp, G.: Computational bifurcation analysis: an efficient method for studying stationary and periodic motions. SIMPACK User Meeting Eisenach **9–11**, 1–9 (2004)
17. Feynman, R.P.: There's plenty of room at the bottom: an invitation to enter a new world of physics. Caltech Eng. Sci. **23**(5), 22–36 (1960). http://www.zyvex.com/nanotech/feynman.htm
18. Trzaska, Z.: Straightforward method for studies of periodic non-harmonic states of linear systems. Arch. Electr. Eng. **53**(2), 191–215 (2004)
19. Trzaska, Z.: One-period energy in dynamical systems with periodic discontinuous excitations. Elektronika **48**(3), 26–32 (2007)
20. Orlova, A., Boronenko, Y., Scheffel, H., Fröhling, R., Kik, W.: Tuning von Üterwagen-drehgestellen Durch Radsatzkopplungen. ZEV-Glasers Annalen **126**(4), 270–282 (2002)
21. Kudrewicz, J.: Fractals and Chaos. Scientific and Technical Publishing House, Warsaw (1996)
22. Trzaska, Z.: Mixed Mode Oscillations (MMOs). Fundamentals and Applications. Springer, Cham, Switzerland (2021)
23. Schupp, G., Weidemann, C., Mauer, L.: Modeling the contact between wheel and rail within multibody system simulation. Veh. Syst. Dyn. **41**(5), 349–364 (2004). https://doi.org/10.1080/00423110412331300326
24. Trzaska, Z.: A new approach to shaping one-period energies of dynamical systems operating in non-sinusoidal states. Arch. Electr. Eng. **54**(3), 265–287 (2005)
25. Trzaska, Z., Marszalek, W.: Computing periodic solutions of linear differential-algebraic systems with nonsinusoidal excitations. Arch. Electr. Eng. **55**(3–4), 255–271 (2006)
26. Biernacki, M., Pidek, H., Ryll-Nardzewski, C.: Sur une inégalité entre des integrales definies, vol. A4, pp. 1–4. Ann. Univ. Mariae Curie-Sklodowska (1950)
27. van der Pol, B., van der Mark, J.: The heartbeat considered as a relaxation oscillation, and an electrical model of the heart. Lond. Edinb. Dublin Philos. Mag. J. Sci. Ser. **7**(6), 763–775 (1928)
28. Tavazoei, M.S., Haeri, M.: A note on the stability of fractional order systems. Math. Comput. Simul. **79**(5), 1566–1576 (2009)
29. Pilipchuk, V.N.: Analytical study of vibrating systems with strong non-linearities by employing saw-tooth time transformations. J. Sound Vib. **192**(1), 43–64 (1996)
30. Green, B.J., Wang, W., Hudson, J.L.: Chaos and spatiotemporal pattern formation in electrochemical reactions. Forma **15**, 257–265 (2000)

31. Kulka, A., Bode, M., Purwins, H.G.: On the influence of inhomogeneities in a reaction-diffusion system. Phys. Lett. A **203**, 33–39 (1995)
32. Marszalek, W., Trzaska, Z.W.: Mixed-mode oscillations in a modified Chua's circuit. Circ. Syst. Signal Proc. **29**, 1075–1087 (2010)
33. Marszalek, W., Trzaska, Z.W.: Properties of memristive circuits with mixed-mode oscillations. Electron. Lett. **51**(2), 140–141 (2015)
34. Ermentrout, G.B., Rinzel, J.: Reflected waves in an inhomogeneous excitable medium. SIAM J. Appl. Math. **56**, 1107–1128 (1996)
35. Trzaska, Z.: Fractional-order systems: their properties and applications. Elektronika **10**, 137–144 (2008)
36. Trzaska, Z.: Fractional order model of Wien Bridge oscillators containing CPEs. In: Proceedings of MATHMOD'09 Conference, Vienna, pp. 357–361 (2009)
37. Trzaska, M., Trzaska, Z.: Electrochemical Impedance Spectroscopy in Materials Science. Publishing Office of the Warsaw University of Technology, Warsaw (2010)
38. Trzaska, Z.: Impact oscillators: fundamentals and applications. J. KONES Powertrain Transp. **22**(1), 307–314 (2015)
39. Litak, G., Margielewicz, J., Gaska, D., Yurchenko, D., Dabek, K.: Dynamic response of the spherical pendulum subjected to horizontal Lissajous excitation. Nonlinear Dyn. **102**, 2125–2142 (2020)
40. Liu, N., Ouyang, H.: Friction-induced vibration considering multiple types of nonlinearities. Nonlinear Dyn. **102**, 2057–2075 (2020)
41. Trzaska, M., Trzaska, Z.: Nanomaterials produced by electrocrystallization method. In: Aliofkhazraei, M., Makhlouf, A. (eds.) Handbook of Nanoelectrochemistry, chap. 6, pp.135–168. Springer, Cham (2016)
42. Trzaska, Z., Marszalek, W.: Periodic solutions of DAEs with applications to dissipative electric circuits. In: Proceedings of the IASTED Conference Modeling, Identification and Control, pp. 309–314. Lanzarote (2006)
43. Steenbergen, M.J.: Modeling of wheels and rail discontinuities in dynamic wheel-rail contact analysis. Veh. Syst. Dyn. **44**(10), 763–787 (2006). https://doi.org/10.1080/00423110600648535
44. Santamaria, J., Vadillo, E.G., Gomez, J.: Influence of creep forces on the risk of derailment of railway vehicles. Veh. Syst. Dyn. **47**(6), 721–752 (2009). https://doi.org/10.1080/00423110802368817
45. Iwnicki, S.: Handbook of Railway Vehicle Dynamics. Taylor & Francis, Boca Raton (2006)
46. Nwokah, O.D.I., Hurmuzu, Y.: The Mechanical Systems Design Handbook: Modeling, Measurements and Control. CRC Press, London (2002)
47. Olver, P.J.: Nonlinear Ordinary Differential Equations. University of Minnesota (2022)
48. Podlubny, I.: Fractional Differential Equations: An Introduction to Fractional Derivatives, Fractional Differential Equations, to Methods of Their Solution and Some of Their Applications. Elsevier (1998)
49. Trzaska, Z.: Matlab solutions of chaotic fractional order circuits. In: Assi, A.H. (ed.) Chapter 19 in Engineering Education and Research Using MATLAB. InTech, Rijeka (2011)
50. Kozlovskaya, I.B., Grigoriev, A.I.: Russian system of countermeasures on board of the international space station (ISS): the first results. Acta Astronaut. **55**, 233–237 (2004)
51. Giusti, A., Mainardi, F.: A dynamic viscoelastic analogy for fluid-filled elastic tubes. Meccanica **51**(10), 2321–2330 (2016)
52. Kaczorek, T.: Selected Problems of Fractional Systems Theory. Springer, Berlin, Heidelberg (2011)
53. Lorenz, E.N.: Deterministic nonperiodic flow. J. Atmos. Sci. **20**, 130–141 (1963)
54. McCrea, N.: An Introduction to Machine Learning Theory and Its Applications: A Visual Tutorial with Examples. Toptal Engineering Blog (2014)
55. Izhikevich, E.M.: Dynamical Systems in Neuroscience: The Geometry of Excitability and Bursting. MIT Press, Cambridge (2007)

56. Trzaska, Z.: Nonsmooth analysis of the pulse pressured infusion fluid flow. Nonlinear Dyn. **78**, 525–540 (2014)
57. Evans, J., Iwnicki, S.: Vehicle Dynamics and the Wheel Rail Interface (2002). http://www.railtechnologyunit.com

Chapter 3
Oscillations in Physical Systems

3.1 Lorenz System and Its Properties

In the previous chapter, the genesis of the intensification of research focused on phenomena characterized by chaotic courses in the state of systems, both natural and man-made, was presented. It has been established that an excellent model for the general phenomenon of chaos in deterministic behavior of a system can be deduced from the Lorenz system. It should be emphasized that the vibrating system (2.139) with nonlinearity represents the nonlinear oscillator whose investigation is of prior interest in science and engineering. The Lorenz equations became famous because of their irregular near-random behavior, despite being a deterministic differential equation, and their unusually fine dependence on initial conditions, giving rise to the Butterfly effect already discussed in Chap. 2. It means that near to a catastrophic transition it is certainly true that an infinitesimal perturbation could push a nonlinear system very far one way or the other, the so-called Butterfly Effect. However there is a significant misunderstanding, at least in popular culture, regarding this effect. Nonlinear systems are most definitely not necessarily sensitive; far from a catastrophic transition a nonlinear system can be highly robust and unwilling to change [1, 2].

Nowadays, the belief has become already widespread that in many natural and man-made systems, prediction of their state is not entirely possible. This is just an illustration of the sudden end of predictions, even in a completely deterministic system. Despite this there is a reluctance to abandon the predictability of the classical universe. However, this is contradicted by the huge achievements of the scientific community in many fields, which are the aftermath of the discoveries made by E. Lorenz.

At this point it must be underlined that all of the computations of the Lorenz equations must be done numerically, as analytical solutions are impossible, using known methods. There are far too many properties of the Lorenz equations to place them all in this book. Thus it must be enough only highlighting the behavior of a

Z. Trzaska, *Mathematical Modelling and Computing in Physics, Chemistry and Biology*, Studies in Systems, Decision and Control 495,
https://doi.org/10.1007/978-3-031-39985-5_3

few properties, namely the ones needed to be enough to enable the reader could get acquainted what is going on with the Lorenz equations. In this section it will be proved that the Lorenz equations do not tend to infinity, and will show that the system may exhibit strange attractor.

The source of the research done by Lorenz concerns the motion of a layer of fluid, such as the earth's atmosphere, that is warmer at the bottom and cooler at the top (Fig. 3.1). If the vertical temperature difference ΔT is high enough then the steady convective flow breaks up and a complex as well as turbulent motion emerges. Investigating this phenomenon, E. Lorenz has formulated the nonlinear autonomous third order system of equations

$$\begin{aligned} \dot{x} &= -\sigma x + \sigma y \\ \dot{y} &= rx - y - xz \\ \dot{z} &= -bz + xy \end{aligned} \tag{3.1}$$

which are now commonly referred to as the Lorenz equations [3]. To localize critical points, the algebraic system of equations must be resolved

$$\begin{aligned} -\sigma x + \sigma y &= 0 \\ rx - y - xz &= 0 \\ -bz + xy &= 0 \end{aligned} \tag{3.2}$$

Now it is easy to check that point $P_1(0, 0, 0)$ is only one critical point for $r < 1$ but for $r > 1$ there are another two critical points, namely $P_2(\sqrt{b(r-1)}, \sqrt{b(r-1)}, r-1)$ and $P_3(-\sqrt{b(r-1)}, -\sqrt{b(r-1)}, r-1)$. Note that when $r = 1$ all three critical points coincide. As r increases through the value 1, the critical point P_1 at the origin bifurcates and goes to the critical points P_2 and P_3. Next in view of the results presented in Sect. 2.7 it is useful to determine the local behavior of solutions in the neighborhood of each critical point. Near the critical point P_1 the linearized system generates the following characteristic equation

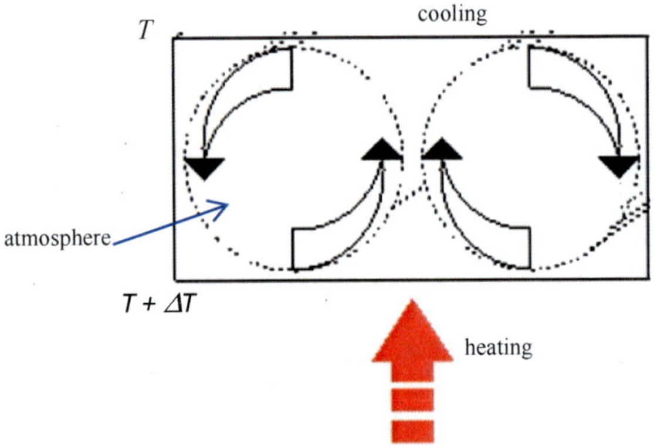

Fig. 3.1 A simplified model for weather forecasting

$$-\left(\lambda + \frac{8}{3}\right)\left[\lambda^2 + 11\lambda - 10(r-1)\right] = 0 \tag{3.3}$$

fulfilled by the values:

$$\lambda_1 = -\frac{8}{3}, \lambda_2 = \frac{-11 - \sqrt{81 + 40r}}{2}, \lambda_3 = \frac{-11 + \sqrt{81 + 40r}}{2}. \tag{3.4}$$

Note that for $r < 1$ all three eigenvalues are negative what means that the coordinate origin is asymptotically stable for this range of r both for the linear approximation and for the original system (3.1). However, when $r = 1$ then λ_3 changes sign and is positive for $r > 1$. The value $r = 1$ corresponds to the initiation of convective flow in the physical problem mentioned above. The coordinate origin is unstable for $r > 1$, and then all solutions starting near the origin tend to grow except for those lying precisely in the plane determined by the eigenvectors associated with λ_1 and λ_2.

To consider the neighborhood of the critical point $P_2(\sqrt{8(r-1)/3}, \sqrt{8(r-1)/3}, r-1)$ for $r > 1$ it is helpful to determine the eigenvalues that are the solution to the characteristic equation of the matrix of the linearized system at this point. So we have

$$3\lambda^3 + 41\lambda^2 + 8(r + 10)\lambda + 160(r - 1) = 0 \tag{3.5}$$

with solutions depending on r in the following way:

- for $1 < r < r_c \sim= 1.3456$ there are three negative real eigenvalues;
- for $r_c < r < r_g \sim= 24.737$ there are one negative real eigenvalue and two complex eigenvalues with negative real part;
- for $r_g < r$ there are one negative real eigenvalue and two complex eigenvalues with positive real part.

A plot of computed values of x versus t for a typical solution with $r > r_c$ is shown in Fig. 3.2a. Let's notice that the solution oscillates back and forth between positive and negative values in a rather irregular manner.

$$\lambda_1 = -\frac{8}{3}, \quad \lambda_2 = \frac{-11 - \sqrt{81 + 40r}}{2}, \quad \lambda_3 = \frac{-11 - \sqrt{81 + 40r}}{2}. \tag{3.6}$$

The solutions of the Lorenz equations (3.1) are also extremely sensitive to perturbations in the initial conditions. Figure 3.2b shows the graphs of computed values of $x(t)$ for the two solutions whose initial values are (5, 5, 5) and (5.01, 5, 5). Note that the two solutions remain close until $t \approx 10$, after which they are quite different and, actually, seem to have no relation to each other. It is this property that particularly attracted the attention of Lorenz in his original study of these equations, and caused him to conclude that detailed long-range weather predictions are probably not possible. In this case the attracting set, although of zero volume, has a rather complicated structure and is called a ***strange attractor***. The term *chaotic* has come into general use to describe solutions such as those shown in Fig. 3.2.

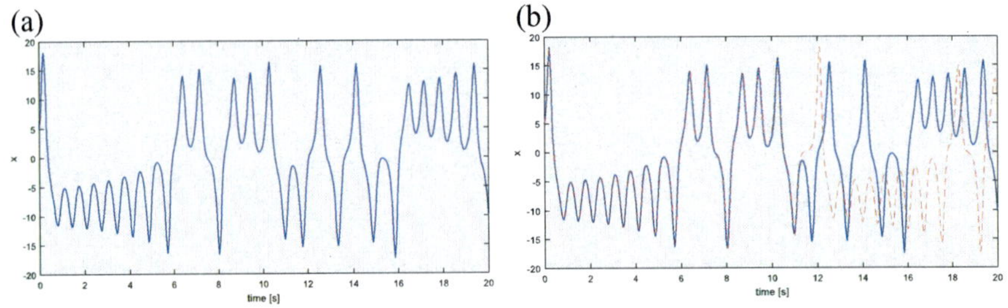

Fig. 3.2 A plot of $x(t)$ for the Lorenz equations (3.1) with $r = 28$: **a** initial condition (5, 5, 5), **b** initial condition (5, 5, 5) for dashed line and (5.01, 5, 5) for solid line

Solutions investigated for smaller values of r allow determining how and when the strange attractor is created. In Fig. 3.3 are shown solutions starting at three different initial points for $r = 21$. It is easy seen that for the initial values (3, 8, 0) the solution begins to converge to the point P_3 almost at once; see Fig. 3.3a. For the second initial values (5, 5, 5) there is a well short interval of transient behavior, after which the solution converges to P_2; see Fig. 3.3b. However, as shown in Fig. 3.3c, for the third initial values (5, 5, 10) there is a much longer interval of transient chaotic behavior before the solution eventually converges to P_2. It should be emphasized that as r increases, the duration of the chaotic transient behavior also increases. When $r = r_3 \sim= 24.06$, the chaotic transients appear to continue indefinitely and the strange attractor appears.

Solutions of the Lorenz equations for other parameter ranges exhibit further interesting types of dynamical system behavior. For example, for certain values of r greater than r_c, intermittent bursts of chaotic behavior separate long intervals of apparently steady periodic oscillation. It can also be shown that the trajectories of solutions of

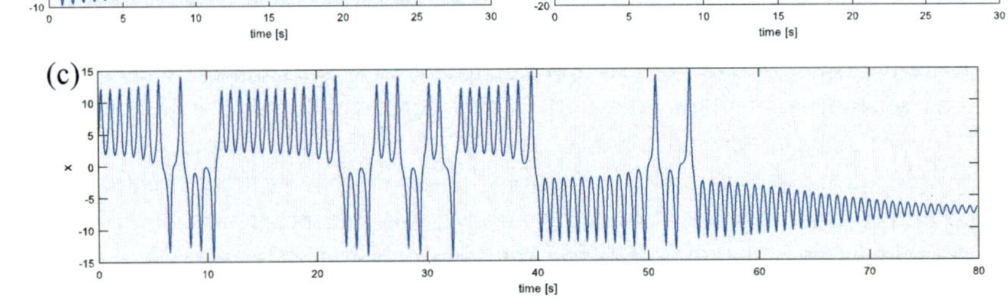

Fig. 3.3 Plots of x versus t for three solutions of Lorenz equations with $r = 21$: **a** initial values are (3, 8, 0), **b** initial values are (5, 5, 5), **c** initial values are (5, 5, 10)

the Lorenz equations in the three-dimensional phase space, or at least projections of them in various planes lead to strange attractors, a fractal, and a self-excited attractor. Its Hausdorff dimension is estimated from above by the Lyapunov dimension also known as Kaplan-Yorke dimension as 2.06 ± 0.01, and the correlation dimension is estimated to be 2.05 ± 0.01. In particular, the Lorenz attractor [4] is a set of chaotic solutions of the Lorenz system. The shape of the Lorenz attractor itself, when plotted in phase space, may also be seen to resemble a butterfly (see Fig. 3.4).

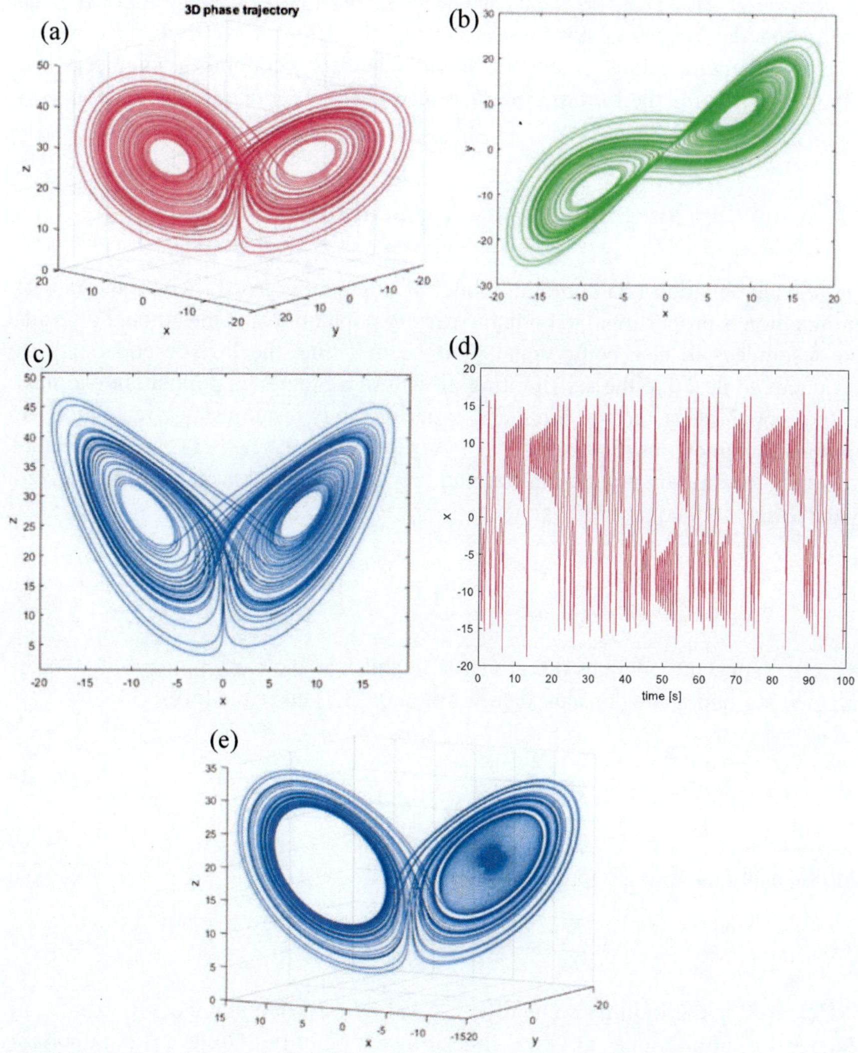

Fig. 3.4 Trajectories of the Lorenz equations with $r = 28$ in: **a** the xyz-space, **b** xy-plane, **c** xz-plane, **d** x versus t, **e** 3D phase portrait

Chaotic behavior of Lorenz system solutions appears to be much more common than was presumed at first and many problems remain unsolved. Some of these demonstrate natural mathematics, while others are related to applications in physics and correct interpretations of solutions.

It is to remember that in the field of integer order systems chaos requires autonomous systems with at least three state variables or a non autonomous system with at least two state variables in order to give the orbits enough dimensions to do their strange twisting. In the case of fractional order systems this principle is not confirmed. However, presently, it is recognized that Lorenz system [3] is an excellent model for general phenomenon of chaotic, deterministic dynamical system behavior. Mathematicians, scientists, social scientists, economists, engineers and even novelists apply the Lorenz system as an archetype for chaos.

3.2 Logistic Equation and Its Applications

The logistic equation is a common model of population growth where the rate of reproduction is proportional to both the existing population and the amount of available resources, all else being equal. For the first time the logistic equation was introduced to describe the self-limiting growth of a biological population and next for the growth of bacteria in broth and experimentally tested using a technique for nonlinear parameter estimation.

Letting P represent population size and t represent time, this model is formalized by the differential equation

$$\frac{dP}{dt} = rP\left(1 - \frac{P}{K}\right) \tag{3.7}$$

where the constant r defines the growth rate and K is the carrying capacity. When parameters r and K are constant then solution of (3.7) takes the form

$$P(t) = \frac{K P_0 e^{rt}}{K + P_0(e^{rt} - 1)} = \frac{K}{1 + \left(\frac{K - P_0}{P_0}\right)e^{-rt}} \tag{3.8}$$

with P_0 being the initial population. Thus

$$\lim_{t\to\infty} P(t) = K \tag{3.9}$$

Hence, K is the limiting value of $P(t)$, the highest value that the population can reach given infinite time (or come close to reaching in finite time). It is important to stress that the carrying capacity is asymptotically reached independently of the initial value $P_0 > 0$, and also in the case that $P_0 > K$. The expression $r(1 - P/K)$

is interpreted as an effective reproduction rate which, in expression (3.7), adjusts instantaneously to $P(t)$.

Usually the environmental conditions influence the carrying capacity, as a consequence it can be time-varying, with $K(t) > 0$, leading to the following mathematical model

$$\frac{dP}{dt} = rP\left(1 - \frac{P}{K(t)}\right) \tag{3.10}$$

A particularly important case is that of carrying capacity that varies periodically with period T, i.e. $K(t+T) = K(t)$. It can be shown that in such a case, independently from the initial value $P_0 > 0$, $P(t)$, will tend to a unique periodic solution $P^*(t)$ whose period is T with a typical value of one year. In such case $K(t)$ may reflect periodical variations of weather conditions.

Another interesting generalization is the case when the carrying capacity $K(t)$ is a function of the population at an earlier time, acquiring a delay in the manner by which population modifies its environment. This leads to a logistic delay equation [5] which exhibits a very rich behavior, such as bistability in some parameter package, as well as a monotonic decay to zero, smooth exponential growth, punctuated unlimited growth (i.e., multiple S-shapes), punctuated growth or alternation to a stationary level, oscillatory approach to a stationary level, sustainable oscillations, finite-time singularities as well as finite-time death. Let us remember that the term $r(1 - P/K)$ in expression (3.7) is interpreted as an effective reproduction rate which adjusts instantaneously to $P(t)$.

Taking into account that this effective reproduction rate lags with a delay time τ yields the following delayed logistic equation

$$dP(t)/dt = rP(t)[1 - P(t-\tau)/K] \tag{3.11}$$

The coexistence of both creative and destructive processes impacting the carrying capacity $K(t)$, it can be expressed as the sum of two different contributions, namely

$$K(t) = A + BP(t-\tau) \tag{3.12}$$

where the first term A is the pre-existing carrying capacity, e.g., provided by the natural environment. The second term is the capacity created (or destroyed) by the system.

In what follows particular considerations are reported only to one natural process, namely the evolution of human population of the planet Earth. Other cases can be found in the publications [6, 7]. First of all to make easier to write certain expressions it is reasonable to define the relative quantity

$$x(t) \equiv P(t)/P_{eff} \tag{3.13}$$

where P_{eff} denotes a quantity measured in some unit, could be millions of persons. In the same way are defined the dimensionless value a of the preexisting resources, the production (or destruction) factor b and the dimensionless carrying capacity

$$y(t) = a + bx(t - \tau) \tag{3.14}$$

Using a time measured in unit of 1/r, and keeping the same notation of time, Eq. (3.11) reduces to the evolution equation for $x = x(t)$ as follows

$$\frac{dx}{dt} = \sigma_1 x - \frac{\sigma_2 x^2}{y} \tag{3.15}$$

The above equation is complemented by an initial history condition for $t \leq 0$, namely

$$x(0) = x_0 \tag{3.16}$$

according to which

$$y_0 = a + bx_0 \tag{3.17}$$

Varying the system parameters yields the different solutions. Figure 3.5a, b demonstrate the behavior of $x = x(t)$ as a function of time for different values of the parameters a, b and the delay time τ. As shown in Fig. 3.5a the values of $x(t)$ tends to the stationary solution through a sequence of up and down alternations, the plot in blue corresponds to $b_1 = -0.6666$ while that in red report to $b_2 = -0.3333$. The solutions $x(t)$ non-monotonically grow by steps to their stationary points $x_2^* = a/(1 - b)$ as $t \to \infty$. The stationary points are: $x_2^* = 0.6000$ (for blue line) and $x_2^* = 0.7500$ (for red line). In the case of Fig. 3.5b it is seen that the solutions $x(t)$ converge by oscillating to their stationary point $x_2^* = -a/(b - 1) = 2/3$ as $t \to \infty$ and $x_2^* = 0.7500$, respectively.

The amplitude of the oscillations increases with increasing time lag τ. Note that the period of the oscillations is much lengthy than the time lag. Note also that the growth

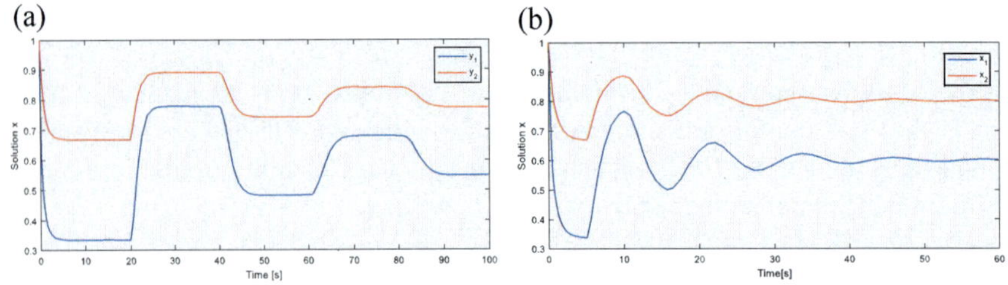

Fig. 3.5 Solutions $x(t)$ to (3.15) as functions of time for the parameters $b = -0.6666$ (blue line) and $b = -0.3333$ (red line). Other parameters are: $a = 1$, $x_0 = 1$, and: **a** $\tau = 20$, **b** $\tau = 5$

rate is positive, when gain (birth) prevails. The simulation results have been obtained by applying the Matlab delay differential equation procedure ***dde.m***. It allowed to find a rich and rather sensitive dependence of the structural properties of the solutions on the value of the delay time τ. For instance, in the regime where loss and competition are dominant, depending on the value of the initial carrying capacity and of τ, appears monotonic decay to zero, oscillatory approach to a stationary level, sustainable oscillations and moving finite-time singularities. It is not a big surprise, since delay equations are known to show much richer properties than ordinary differential equations. For instance population size in plants experiences significant oscillation due to the annual environmental oscillation. Plant dynamics experience a higher degree of this seasonality than do mammals, birds, or bovine insects. When combined with perturbations due to disease, this often results in chaotic oscillations.

3.3 Predator–Prey Systems

The simplest model of predator and prey association includes only natural growth or decay and the predator–prey interaction itself. It is assumed a priori that all other relationships to be neglected. Moreover, the prey population grows according to a first order rate law in the absence of predator, while the predator population declines according to a first order rate law if the prey population is extinct. To convert species interactions into terms of a mathematical model application of the ***Population Law of Mass Action*** is preferred, namely. *At time t the rate of change of one population due to interaction with another is proportional to the product of the two populations at that time t.*

Denoting by $x(t)$ the predator population at time t, by $y(t)$ the prey population at time t and applying the above formulated Population Law of Mass Action and the Balance Law "*Net rate of change a population = Rate in − Rate out*" arises the model of the autonomous predator–prey (or Lotka-Volterra) system [8]

$$\begin{aligned} \dot{x} &= -ax + bxy \\ \dot{y} &= cy - kxy \end{aligned} \tag{3.18}$$

where the rate constants a, b, c, k are positive. The linear rate terms $-ax$ and cy describe the natural decay and growth, respectively, of the predator and the prey population as if each were isolated from the other (so y would no longer be the food supply for x). The quadratic terms bxy and $-kxy$ describe the effects of mass action on the rates of change of the two species; food promotes the predator population's growth rate, while serving as food diminishes the prey population's growth rate. The predator–prey model makes a number of assumptions about the environment and biology of the predator and prey populations:

- The prey population finds ample food at all times and reproduces exponentially, unless subject to predation, this exponential growth is represented in the equation above by the term cx;
- The food supply of the predator population depends entirely on the size of the prey population and if either x or y is zero, then there can be no predation, the rate at which the predator population grows is not necessarily equal to the rate at which it consumes the prey;
- The rate of change of population is proportional to its size, the dynamics of predator and prey populations have a tendency to oscillate;
- During the process, the environment does not change in favour of one species, and genetic adaptation is inconsequential, but making the environment better for the prey benefits the predator, not the prey (this is related to the paradox of enrichment);
- Predators have limitless appetite;
- Both populations can be described by a single variable. This amounts to assuming that the populations do not have a spatial or age distribution that contributes to the dynamics.

Solving the equations $-ax + bxy = 0$ and $cy - kxy = 0$ yields the equilibrium points $(0, 0)$ and $(c/k, a/b)$ which are inside the population quadrant with $x \geq 0, y \geq 0$. It is worth underlying that the autonomous system (3.18) has a unique orbit through each point (x_0, y_0). Thus an orbit originating inside the population quadrant stays inside. For $a = 1$, $b = 0.1$, $c = 1$ and $k = 0.2$ the orbits and courses in time of the state variables are shown in Fig. 3.6.

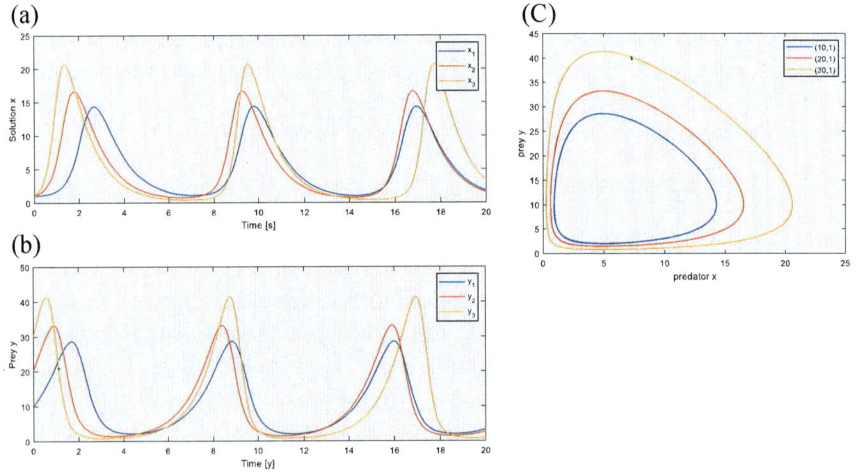

Fig. 3.6 Orbits and state variables of (3.18) for $a = 1$, $b = 0.1$, $c = 1$ and $k = 0.2$ with initial conditions: (10; 1)—blue lines, (20; 1)—red lines, (30; 1)—green lines: **a**, **b** state variables x(t) and y(t), **c** orbits

It is worth noting that the fluctuations of the populations of the predator and its prey are periodic. The period depends on the values of the rate coefficients of the system (3.18) and on the initial data. The period increases with the amplitude of the corresponding cycle. In system (3.18) the average populations of predators and preys over the period of every cycle are c/k and a/b, respectively.

The stability of the fixed points can be determined by performing a linearization of (3.18) using partial derivatives what leads to the Jacobian matrix of the predator–prey model

$$J = \begin{bmatrix} -a + by & bx \\ -ky & c - kx \end{bmatrix} \tag{3.19}$$

For the equilibrium point at the origin the above matrix becomes

$$J = \begin{bmatrix} -a & 0 \\ 0 & c \end{bmatrix} \tag{3.20}$$

The eigenvalues of this matrix are $\lambda_1 = -a$ and $\lambda_2 = c$. In the studied model a and c are always greater than zero, and as such the sign of the eigenvalues above will be always different. Hence the equilibrium point at the origin is a saddle point. The instability of this equilibrium point is of great significance because it indicates that the extinction of both species in the model is difficult. Indeed, this could only happen if the prey was artificially completely exterminated, causes the predators to starve to death and then the prey population would grow without bound. In reality, the populations of prey and predator can get infinitesimally close to zero and still recover.

Evaluating J at the second fixed point leads to

$$J = \begin{bmatrix} 0 & bc/k \\ -ka/b & 0 \end{bmatrix} \tag{3.21}$$

and the eigenvalues of this matrix are $\lambda_1 = j\sqrt{ac}$ and $\lambda_2 = -j\sqrt{ac}$ where $j = \sqrt{-1}$. As the eigenvalues are both purely imaginary and conjugate to each other, this fixed point must either be a center for closed orbits in the local vicinity or an attractive or repulsive spiral. In conservative systems, there must be closed orbits in the local vicinity of fixed points. Thus orbits about the fixed point are closed and elliptic, so the solutions are periodic, oscillating on a small ellipse around the fixed point, with a frequency $\omega = \sqrt{\lambda_1 \lambda_2} = \sqrt{ac}$ and period $T = 2\pi/\sqrt{\lambda_1 \lambda_2} = 2\pi/\sqrt{ac}$.

Empirical studies have shown that in systems with three or more species, behavioral indirect interactions can strongly impact the dynamics of predator–prey interactions and have a similar or stronger impact on prey demographics and atrophic cascades than direct effects. Such behavioral indirect interactions commonly lead to nonadditive (higher order) effects of multiple predators; when multiple predators are present, prey survival might be lower or higher than expected from their independent effects in the absence of the other predators. Such a problem is easy to explain by adding additional factors in the predator–prey system in the form of a harvester.

Quite accurate in this case is the simplest harvesting model as a constant effort harvest, in which the amount captured per unit time is proportional to the species population. Then the following model is adequate

$$\dot{x} = -ax + bxy - H_1x, \quad x(0) = 8;$$
$$\dot{y} = cy - kxy - H_2y, \quad y(0) = 16. \tag{3.22}$$

where the nonnegative numbers H_1 and H_2 are the harvesting coefficients. When harvesting takes place, the equilibrium point laying inside the population quadrant shifts to the left upward from $x = c/k$, $y = a/b$ to the point

$$x_0 = (c - H_2)/k, \quad y_0 = (a + H_1)/b \tag{3.23}$$

It is evident that must be $H_2 < c$ because if $H_2 > c$, then the heavy harvesting of the prey species y doesn't leave enough food for the predator x, so the predator species heads toward extinction.

By the law of averages, the population averages around any cycle are given by the coordinates of the equilibrium point. Thus harvesting raises the average of the prey population but lowers the average of the predator population. Note that if the harvesting coefficients in system (3.22) are too large, then the internal equilibrium point, $x_0 = (c - H_2)/k$, $y_0 = (a + H_1)/b$ crosses the y and one or both species becomes extinct, what is seen in follows.

The dynamics of the system (3.22) is represented by the phase trajectory graph shown in Fig. 3.7. It refers to four cases of harvesting intensity: $H = 0$—no harvesting, $H = 0.4$—light harvesting, $H = 1$—critical harvesting and $H = 5$—heavy harvesting. It is easy seen from the figure that in two cases (no harvesting and light harvesting) both species survive. On the other hand in the case of critical harvesting ($H = 1$) only the prey species survives. When H exceeds 1, then both predator and prey species become extinct. It follows that a decrease in the harvesting rate leads to an increase in the predator percentage, while an increase in the harvesting rate causes a decrease in the predator percentage. In this way, the predator–prey model continues to be the starting point for most substantial attempts to learn how predator–prey populations evolve with or without harvest.

3.4 Systems with the Double Scroll Attractors

In the studies of dynamical systems, the double-scroll attractor, also known as Chua's attractor, is a strange attractor observed from a physical electronic Chua's chaotic circuit with a single nonlinear resistor [9, 10]. The double-scroll system is often

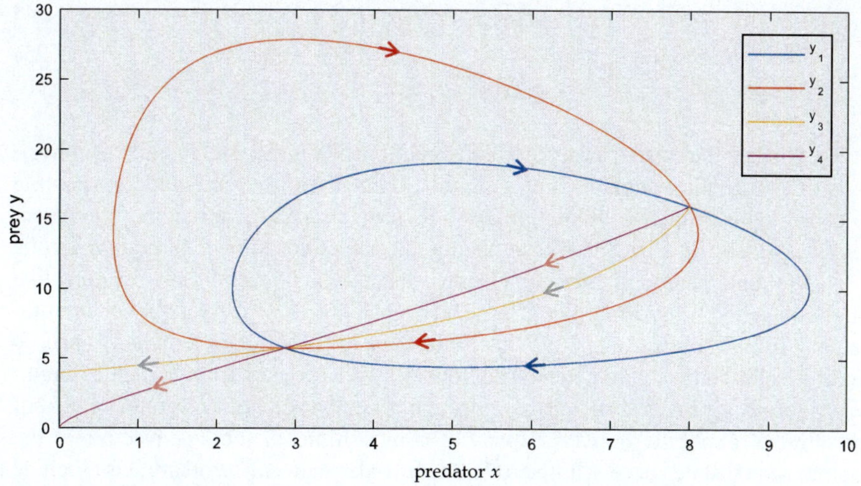

Fig. 3.7 The effects of harvestings with $H_1 = H_2 = H$: $y_1 - H = 0$; $y_2 - H = 0.4$; $y_3 - H = 1.0$; $y_4 - H = 5.0$

described by a system of three nonlinear ordinary differential equations and a 3-segment piecewise-linear equation

$$
\begin{aligned}
\frac{dx}{dt} &= \alpha[y - x - f(x)],\\
RC_2\frac{dy}{dt} &= x - y + Rz,\\
\frac{dz}{dt} &= -\beta y.
\end{aligned}
\tag{3.24}
$$

where α, β denote control parameters and R, C_2 are parameters of Chua's circuit (Fig. 3.8a, c).

One of the essential elements of the basic Chua circuit (Fig. 3.8a) is a nonlinear resistor whose current–voltage characteristics (Fig. 3.8b) is presented by the expressions

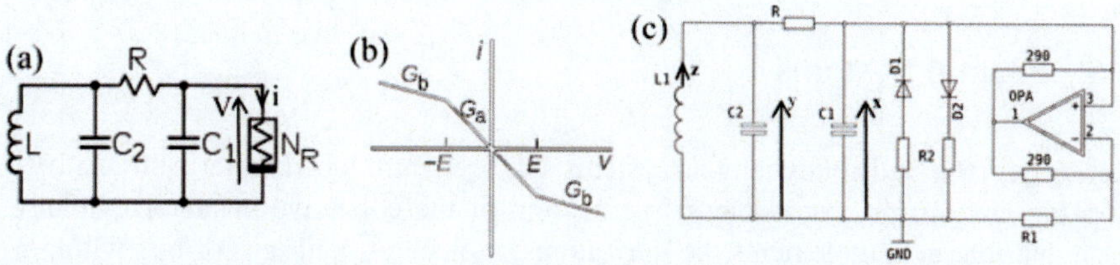

Fig. 3.8 Basic Chua circuit: **a** structure, **b** piecewise linear characteristic of nonlinear resistor, **c** circuit realization

$$i(v) = \begin{cases} G_1 v + (G_2 - G_1)v_1 & \text{for } v < v_1 \\ G_2 v & \text{for } v_1 < v < v_2 \\ G_1 v + (G_2 - G_{13})v_2 & \text{for } v > v_2 \end{cases} \tag{3.25}$$

This makes the system easily simulated numerically and easily manifested physically due to simple design of Chua's circuits. Using a Chua circuit model is possible to generate by computer simulation a set of strange scrolling attractors in different forms as shown in Fig. 3.9. These chaotic attractors are known as double scrolls because of their shapes in three-dimensional space, which resembles two Saturn-like rings connected by swirling lines, or also two balls of wool wound almost simultaneously from a single skein. Note that each shape of the attractor matches a specific set of system control parameters. The double-scroll attractor from the Chua circuit was rigorously proven to be chaotic through a number of Poincaré return maps of the attractor explicitly derived by way of compositions of the eigenvectors of the 3-dimensional state space. Multiscroll attractors also called n-scroll attractor include the Lu Chen attractor, the modified Chen chaotic attractor, PWL Duffing attractor, Rabinovich Fabrikant attractor, modified Chua chaotic attractor, that is, multiple scrolls in a single attractor. Multiscroll modified Chua chaotic attractor (Fig. 3.9b) arises in the system

$$\begin{aligned} \frac{dx}{dt} &= \alpha[y - h(x)], \\ \frac{dy}{dt} &= x - y + z, \\ \frac{dz}{dt} &= -\beta y. \end{aligned} \tag{3.26}$$

in which $h(x) = -b\sin\left(\frac{\pi x}{2a}\right)$ and $\alpha = 10.82$, $\beta = 14.286$, $a = 1.3$, $b = 0.11$, initial conditions are $x(0) = 1$, $y(0) = 1$, $z(0) = 0$.

The presented selected results of the calculation of the Chua'a circuit for various sets of parameters of its elements and initial conditions indicate the importance of even small changes in the values of both the parameters of linear elements and the initial conditions, which can lead to a significant change in the nature of vibrations and the dispersion of instantaneous values of individual phase trajectories.

3.5 Fractal Systems

Recently, in the study of dynamical systems, has been introduced the term "fractality" as a key notion for a new way of perception of the collective behavior of many basic but interacting elements, be they atoms, molecules, neurons, or bits within a computer [11, 12].

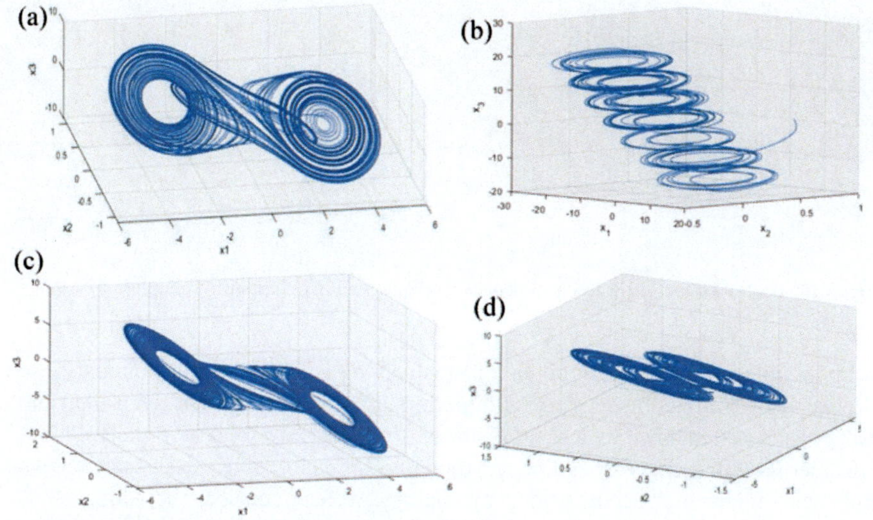

Fig. 3.9 Double scroll attractors of Chua's circuits with $x_1 = x(t)$, $x_2 = y(t)$, $x_3 = z(t)$: **a** $x_3(0) = -0.6$, **b** multiscroll with $h = 0.11\sin(\pi x_1/2a)$ and $x_3(0) = -0.6$; **c** $x_3(0) = -0.3$; **d** $x_3(0) = -1.2$

Chaotic dynamic systems and nonlinear processes, together with the resulting fractals and multifractals, are becoming increasingly fundamental for analyzing data and understanding processes in the physics, chemistry, biology, medicine and environmental systems. Many processes and phenomena, poorly known only a few years ago, can now be studied and understood with the help of conceptual models from the fields of fractals and dynamic systems. This represents a bold step towards the aim of understanding how the environment works. In the study of fractal properties of natural and manmade systems, we distinguish two basic approaches to their descriptions, namely those that are reported to the very system structure and those that are referred to varying in time of state variables. In the latter case, it is possible to extend the use of fractals to the absence of one characteristic time scale, i.e. the pattern. These issues were discussed in the previous Section. In this Section, the main attention will be devoted to fractals characterizing the system structure.

Fractal geometry is the geometry of rugged systems, that is objects having "non-smooth" boundaries and non-Euclidean shapes. From a study of rugged systems fractal geometers have derived many descriptive parameters for describing fractured systems such as broken rock and the random space filling structures of sedimentary systems such as sandstone and other porous systems. The term fractal geometry was coined by Mandelbrot from a Latin word "fractus" meaning fractured. In the 25 years since the publication of Mandelbrot's book [14], fractal geometry has found many applications in the various fields of practice.

If one looks at magnified portions of the boundary of the strange attractor it looks likes the larger image, a property which is described as self-similarity. Because of self-similarities, when analyzing a portion of the boundary of the Mandelbrot set it

Fig. 3.10 Fractal structures: **a** Koch snowflake, **b** Cantor beam, **c** Sierpiński triangle

is impossible to know the magnification at which one is inspecting the boundary. For ordinary self-similar objects, being n-dimensional means that when it is rep-tiled into pieces each scaled down by a scale-factor of $1/r$, there are a total of r^n pieces. Now, consider the ***Koch curve***. It can be rep-tiled into four sub-copies, each scaled down by a scale-factor of 1/3. So, strictly by analogy, we can consider the "dimension" of the Koch curve as being the unique real number D that satisfies $3^D = 4$. This number is called the *fractal dimension* of the Koch curve; it is not the conventionally perceived dimension of a curve. In general, a key property of fractals is that the fractal dimension differs from the *conventionally understood* dimension (formally called the topological dimension). This means that fractals cannot be measured in traditional ways. The fractal dimension of the Koch curve called the Koch snowflake (Fig. 3.10a) is of $D \approx 1.261859507$.

The original ***Cantor set*** also called Cantor beam (Fig. 3.10b) is constructed, taking an initial unit segment at stage 0 ($s = 0$) and splitting it in three identical parts. Then, the middle segment is erased and the first stage ($s = 1$) is obtained. This procedure is repeated infinitely to obtain the Cantor set. The number of self-copies in stage $s = 1$ is $N = 2$, while the scaling factor becomes $r = 1/3$, resulting in a fractal dimension

$$D = -\frac{\ln(N)}{\ln(r)} \tag{3.27}$$

The fractal dimension of the Cantor set called the Cantor beam (Fig. 3.10b) is of $D \approx 0.630929753$. The Cantor set has many applications in lens design, electrochemistry and acoustics [13, 14].

The ***Sierpiński triangle*** also called the Sierpiński gasket or Sierpiński sieve, is a fractal attractive fixed set with the overall shape of an equilateral triangle, subdivided recursively into smaller equilateral triangles. Originally constructed as a curve, this is one of the basic examples of self-similar sets—that is, it is a mathematically generated pattern that is reproducible at any magnification or reduction [11]. The Sierpiński triangle is created in an iterative process, which consists in successive transformations of an equilateral triangle with a unit side (it can also be any triangle) in such a way that in the next iteration step each side of the triangle is divided in half and the lines connecting the division points are the sides of new but smaller triangles. At the same time, interior triangles are removed from the object in each

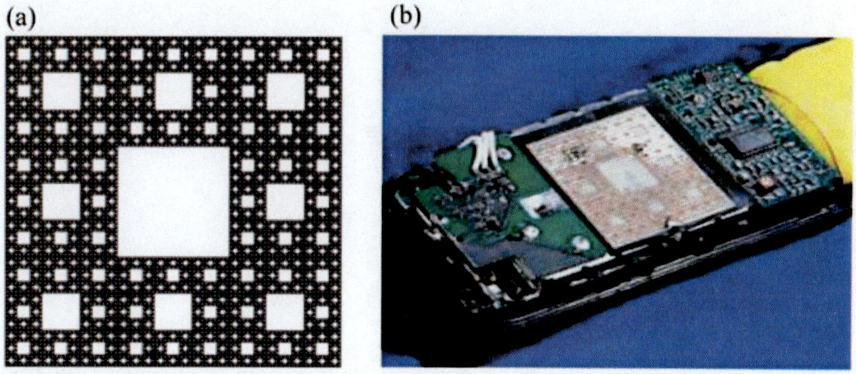

Fig. 3.11 Sierpiński carpet: **a** structure after n = 4 iterations of a square, **b** practical use in the implementation of an antenna for satellite communication

iteration step. The first few iterations in the construction of the Sierpinski triangle are shown in Fig. 3.10c. Note that in each iteration step the resulting structure is composed of $N = 3$ identical copies of the complete structure created in the previous step, but each of them is scaled with the factor $\delta = 0.5$. Hence the fractal dimension of similarity for a Sierpinski triangle is

$$D = \frac{\log(N)}{\log\left(\frac{1}{\delta}\right)} = \frac{\log(3)}{\log(2)} = 1.584962501\ldots \tag{3.28}$$

At this point, it is worth mentioning that another fractal called the Sierpinski carpet is also created in a similar iterative process (Fig. 3.11a).

The method of its construction is analogous to that used in the case of a triangle, except that the initiator is a square and in subsequent iterations the side is divided into three parts and the middle square is removed from the surface. Consequently, the fractal dimension of the Sierpiński carpet is defined by the expression

$$D = \frac{\log(N)}{\log\left(\frac{1}{\delta}\right)} = \frac{\log(8)}{\log(3)} = 1.892789261\ldots \tag{3.29}$$

This means that this object is located between the line and the plane. Such a facility has already been implemented in the design and practical implementation of multi-band radio-telecommunication antennas Fig. 3.11b.

Amazingly, the Sierpiński carpet is a curved line according to the currently accepted definition, which is equivalent to the definition of plane curves (on the plane) given by Cantor. Another property of the Sierpiński carpet is that its surface area is equal to 0. This is due to the following relations. In the next steps of the fractal construction, we remove each time $8n$ squares with side $(1/3)^{n+1}$ each, i.e. area $(1/9)^{n+1}$ each ($n = 0, 1, 2, \ldots,$). Thus, the area of the remaining figure after $n + 1$ iterations is

$$S_n = 1 - \frac{1}{9}\sum_{k=0}^{n}\left(\frac{8}{9}\right)^k \tag{3.30}$$

For $n \to \infty$ the sum of the geometric series in (3.30) is defined as

$$\sum_{k=0}^{\infty}\left(\frac{8}{9}\right)^k = \frac{1}{1-\frac{8}{9}} = 9 \tag{3.31}$$

So substituting (3.31) into (3.30) yields

$$S_\infty = 1 - \frac{1}{9}9 = 0 \tag{3.32}$$

This fact, among others, may provide additional evidence for qualifying the Sierpiński carpet as a curved line. The three-dimensional equivalent of the Cantor set and the Sierpiński carpet is a cube, also called the Menger sponge, given in 1927 by the Austrian mathematician Karl Menger. It is a fractal solid (e.g. a cube) whose similarity dimension is

$$D = \frac{\log(N)}{\log(\frac{1}{\delta})} = \frac{\log(20)}{\log(3)} = 2.726833\ldots \tag{3.33}$$

However, we will not devote more attention to such a structure due to the limited framework of this book and relatively few references to its practical applications in technology so far, and we recommend publications [13, 14]

In general, modeled fractals may be in sounds, digital images, electrochemical patterns, circadian rhythms, the immune system, the economy, corporations, societies, and so on. For instance, structures of trees, ferns, cells of the nervous system, blood and lung vasculatures, and other branching patterns in nature can be modeled on a computer by using recursive algorithms and L-systems techniques [3, 4, 14]. Not so long ago, scientists believed the universe and everything in it could be predicted and controlled. In spite of their efforts to find missing components, they have failed. Despite the use of the world's most powerful computers, weather has remained unpredictable, and despite intensive research and analysis, ecosystems and immune systems have not behaved as expected.

3.6 Applications of Van der Pol Equation

The Van der Pol oscillator is described by the nonlinear, second-order ordinary differential equation

$$\ddot{x} + \mu(x^2 - 1)\dot{x} + \omega^2 x = 0, \tag{3.34}$$

where μ and ω are positive parameters, and x (a function of time t) is the quantity of interest (e.g., position or voltage). Let's note that Eq. (3.34) contains a nonlinear damping term, namely $\mu(x^2 - 1)\,\dot{x}$ which, depending on the magnitude of x, introduces to the system positive or negative damping. Considering this in more details, then if $|x| > 1$, there is positive damping and energy is dissipated from the system to the environment what results in oscillations with decaying magnitude. Alternatively, if $|x| < 1$, there is negative damping and external energy supplies the system what results in magnitude amplification of oscillation. Therefore, in the steady-state appear sustained periodic oscillations and they form a limit cycle. It is worth to mention that this Van der Pol equation (3.34) is one of the simplest capable of compactly representing the various features of behavior of practical oscillators. When $\mu \to 0$, then it represents a purely basic oscillatory dynamics as a simply conservative harmonic oscillator [15].

The main properties of the Van der Pol oscillator can be revealed by particular analysis of Eq. (3.34). For this purpose, it is convenient to represent this equation in the form of an equation of state variables, namely: substituting in (3.34) $\dot{x} = y$ yields

$$\begin{aligned} \dot{x} &= y \\ \dot{y} &= -\omega^2 x + \mu y - \mu x^2 y \end{aligned} \tag{3.35}$$

Hence, the Jacobi matrix takes the form

$$\boldsymbol{J} = \begin{bmatrix} 0 & 1 \\ -\omega^2 - 2\mu xy & \mu - \mu x^2 \end{bmatrix} \tag{3.36}$$

Because the equilibrium point of the dynamic system is (0, 0), then the Jacobi matrix at this point is given by

$$\boldsymbol{J}_{(0,0)} = \begin{bmatrix} 0 & 1 \\ -\omega^2 & \mu \end{bmatrix} \tag{3.37}$$

Thus, the eigenvalues of the Jacobi matrix are

$$\lambda_{1,2} = \frac{\mu}{2} \pm \sqrt{\left(\frac{\mu}{2}\right)^2 - \omega^2} \tag{3.38}$$

Let us note that the real part of the eigenvalues is less than zero if relations $-2 < \mu < 0$, $\omega = 1$ are fulfilled. Then the oscillator (3.34) is stable. Therefore, when in this case the parameters and initial values are taken as $\mu = -1$, $\omega = 1$, and $(x(0), y(0)) = (1, 1)$, respectively, then the dynamic system (3.34) is asymptotically stable. However, if the initial values are fixed at $(x(0), y(0)) = (1, 2)$ and $\mu = -1$, $\omega = 1$, then magnitude of oscillations with over time will go to infinite, so the stability of dynamic system (3.34) is closely related to the initial values, i.e., it is locally stable in the region near equilibrium point (0, 0).

When the parameters $\mu > 0$, $\omega = 0.7746$, such as $\mu = 1$, $\omega = 0.7746$, and let $(x(0), y(0)) = (1, 2)$, then the dynamic system (3.34) has a locally stable limit cycle, and its phase portrait is shown in Fig. 3.12a. Time-varying state variables $x(t)$ and $y(t)$ are presented in Fig. 3.12b.

The Van der Pol oscillator with an external sinusoidal excitation component is described by the state variable equation as follows

$$\begin{aligned} \dot{x} &= y \\ \dot{y} &= -k\sin(\omega z)x + \mu y - \mu x^2 y \\ \dot{z} &= 1 \end{aligned} \tag{3.39}$$

where $z = t$ denotes the third state variable. For $\mu = 1.2$, $\omega = 5.9$, $k = 19.5$, $x(0) = 0.1$, $y(0) = 0.2$, the Lyapunov exponents of (3.39) are $\rho_1 = 0.57$, $\rho_2 = 0$, $\rho_3 = -4.01$. Thus, Eq. (3.39) produces a chaotic attractor, and its phase portrait is shown in Fig. 3.13a. Plots of the time-varying of state variables are presented in Fig. 3.13b.

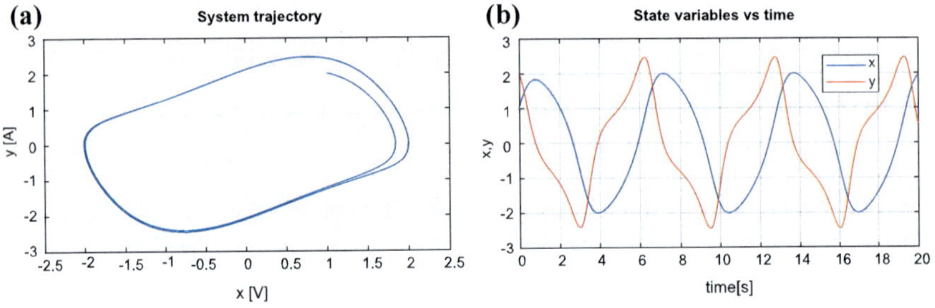

Fig. 3.12 Responses of the Van der Pol oscillator with initial values $(x(0), y(0)) = (1, 2)$: **a** phase portrait, **b** state variables versus time

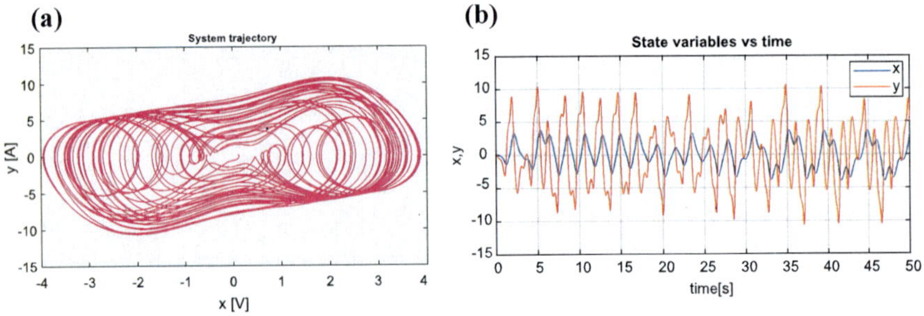

Fig. 3.13 Evolution of dynamic system (3.39) with initial values $(x(0), y(0)) = (0.1, 0.2)$: **a** strange attractor, **b** variables versus time

3.7 The Rössler Attractor

The Rössler attractor is significant for the Rössler system, as a system of three non-linear ordinary differential equations originally studied by Otto Rössler in the seventies of the twentieth century [6, 24]. These differential equations describe a continuous-time dynamical system that exhibits chaotic dynamics associated with the fractal properties of the attractor. Equations describing exactly the nature of the Rössler system are [16]

$$\begin{aligned} \dot{x} &= -y - z, \\ \dot{y} &= x + ay, \\ \dot{z} &= b + z(x - c) \end{aligned} \tag{3.40}$$

Here, $(x, y, z) \in R^3$ are dynamical variables defining the phase space and $(a, b, c) \in R^3$ are constant parameters. Rössler system has two fixed points called equilibriums: $F_\pm$ located at

$$(x_\pm, y_\pm, z_\pm) = \left(\frac{c \pm \sqrt{c^2 - 4ab}}{2}, \frac{c \pm \sqrt{c^2 - 4ab}}{2a}, \frac{c \pm \sqrt{c^2 - 4ab}}{2a} \right) \tag{3.41}$$

For values $a = 0.2$, $b = 0.2$ and $c = 5.7$ with initial conditions $[1\ 0\ 0]'$ the Rössler attractor takes the form presented in Fig. 3.14.

The stability of each of these fixed points can be analyzed by determining their respective eigenvalues and eigenvectors. The Jacobi matrix is

$$\boldsymbol{J} = \begin{bmatrix} 0 & -1 & 1 \\ 1 & a & 0 \\ z & 0 & x - c \end{bmatrix}.$$

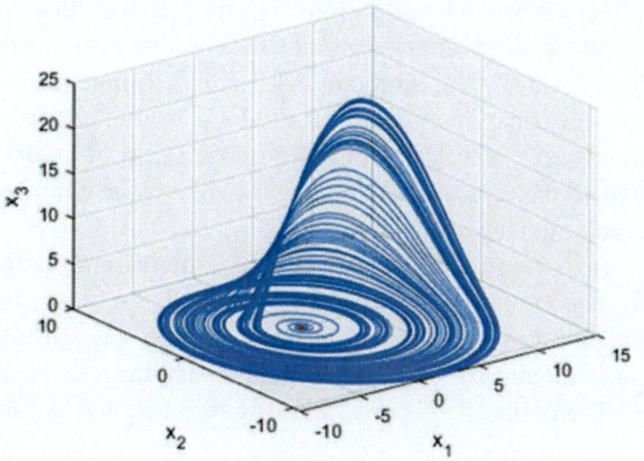

Fig. 3.14 Rössler attractor for $a = 0.2$, $b = 0.2$, $c = 5.7$ and initial conditions: $x(0) = 1$, $y(0) = 0$, and $z(0) = 0$

And the eigenvalues can be determined by solving the following cubic equation

$$-\lambda^3 + (a + x - c)\lambda^2 + (ac - ax - 1 - z)\lambda + x - c + az = 0,$$

For the centrally located fixed point, Rössler's original parameter values of $a = 0.2$, $b = 0.2$, and $c = 5.7$ yield eigenvalues

$$\lambda_1 = 0.0971028 + 0.995786i,$$
$$\lambda_2 = 0.0971028 - 0.995786i,$$
$$\lambda_3 = -5.68718.$$

The magnitude of a negative eigenvalue characterizes the level of attraction along the corresponding eigenvector. Similarly the magnitude of a positive eigenvalue characterizes the level of repulsion along the corresponding eigenvector.

The eigenvectors corresponding to these eigenvalues are

$$v_1 = \begin{bmatrix} 0.7073 \\ -0.07278 - 0.7032i \\ 0.0042 - 0.0007i \end{bmatrix},$$

$$v_2 = \begin{bmatrix} 0.7073 \\ 0.07278 + 0.7032i \\ 0.0042 + 0.0007i \end{bmatrix},$$

$$v_3 = \begin{bmatrix} 0.1682 \\ -0.0286 \\ 0.9853 \end{bmatrix}.$$

The above eigenvectors lead to several interesting conclusions that can be drawn from their components. First, the two eigenvalue/eigenvector pairs (v_1 and v_2) control the steady outward slide that occurs in the main disk of the attractor. The last eigenvalue/eigenvector pair is moving along an axis that runs through the center of the manifold and accounts for the z motion that occurs within the attractor. The red dot in the center of this attractor is FP_c. Let's us note that a sequence generated from the Rössler equations firstly begin to loop around FPc, next start being pulled upwards into the v_3 vector, creating the upward arm of a curve that bends slightly inward toward the vector before being pushed outward again as it is pulled back towards the repelling plane.

Let's us note that a sequence generated from the Rössler equations firstly begin to loop around FP_c, next start being pulled upwards into the v_3 vector, creating the upward arm of a curve that bends slightly inward toward the vector before being pushed outward again as it is pulled back towards the repelling plane.

The outlier fixed point, and Rössler's original parameter values yield eigenvalues

$$\lambda_1 = -0.0000046 + 5.4280259i,$$

$$\lambda_2 = -0.0000046 - 5.4280259i,$$
$$\lambda_3 = 0.1929830.$$

The eigenvectors corresponding to these eigenvalues are

$$v_1 = \begin{bmatrix} 0.0002422 + 0.1872055i \\ 0.0344403 - 0.0013136i \\ 0.9817159 \end{bmatrix},$$
$$v_2 = \begin{bmatrix} 0.0002422 - 0.1872055i \\ 0.0344403 + 0.0013136i \\ 0.9817159 \end{bmatrix},$$
$$v_3 = \begin{bmatrix} 0.0049651 \\ -0.7075770 \\ 0.7066188 \end{bmatrix}.$$

The influence of these eigenvalues and eigenvectors is confined to iterations of the Rössler system whose initial conditions are in the general vicinity of this outlier fixed point. However, if the initial conditions lie on the attracting plane generated by λ_1 and λ_2, this influence effectively involves pushing the resulting system towards the general Rössler attractor. When the resulting sequence approaches the central fixed point and the attractor itself, the influence of this distant fixed point (and its eigenvectors) it's getting weaker.

Because of the dissipativity of considered system, the sum of the Lyapunov exponents must be negative. Due to that condition, the dissipative Rössler attractor is an attracting set of measure zero in the phase space. All points on the phase trajectory, regardless of their initial position, approach the attractor exponentially fast and concentrate near it over time. For time series produced by Rössler systems (Fig. 3.15b), the presence of a positive characteristic exponent indicates chaos. The results of numerical calculations of the Lyapunov exponents for the Rössler attractor are the following

$$\rho_1 = 0.0714, \rho_2 = 0, \rho_3 = -5.3943$$

The Rössler strange attractor is fractal and its fractal dimension expressed by the Kaplan-Yorke formula

$$\mathrm{D_{KY}} = \frac{\rho_1}{\rho_1 + |\rho_3|}$$

gives the following approximate value $\mathrm{D_{KY}} \approx 2.0134$.

Embracing all, the Rössler system is a simple but powerful tool for studying the behavior of chaotic systems. Its power to exhibit strange attractors and its wide range of applications make it a valuable vehicle for researchers and practitioners in a

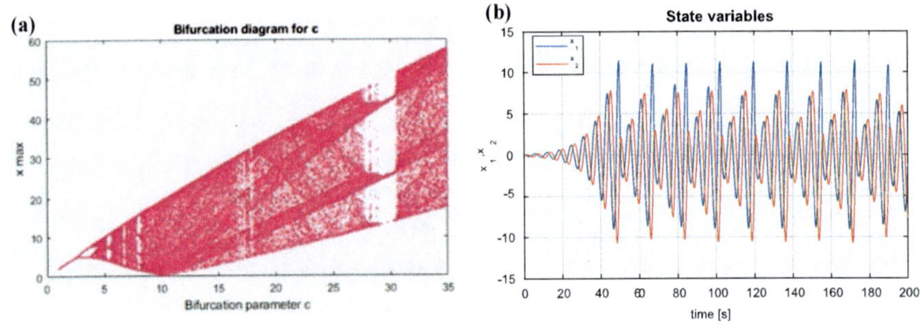

Fig. 3.15. Dynamical variables in Rössler system: **a** bifurcation diagram, **b** state variables versus time

variety of domains. To the unquestionable mathematical and visual advantages of the Rössler system, one should add its application in various fields, including physics, chemistry and engineering. It has been used to model the behavior of chemical reactions, electronic circuits, and even the human heart. Note that the Rössler system is not always dissipative as the divergence is given by $div(\boldsymbol{f}(x, y, z)) = a - c + x$. Thus, in a large region of parameters (especially when a grows) and for large values of the variable x, we will have a positive divergence and so escape orbits. Therefore, apart from regular and chaotic orbits, the Rössler system also has escape orbits with transient chaos or regular behavior before escaping. In Fig. 3.15a bifurcation diagrams on the (c, x) plane for constant values of the parameters a and b is presented. The plot shows a pattern structure that is repeated when the parameter c grows, that consists of interlacing bands in groups of three. When the parameter reach the value c for having escape behavior for almost all the orbits the funnel attractor experiments a boundary crisis and disappears, now we only have transient funnel chaos but finally the orbit escape through the direction of the fast unstable manifold of the equilibrium FP_c. Otto Rössler developed in 1979, a set of four equations that describe an autonomous "hyperchaotic" system. The term "hyperchaotic" means that the system of equations must be at least four-dimensional i.e. with four state variables and there must be two or more positive Lyapunov exponents. Moreover, the sum of all the Lyapunov exponents must be negative. The hyperchaotic set of equations that Rössler developed is

$$\begin{aligned}
\dot{x} &= -y - z, \\
\dot{y} &= x + 0.25y + w, \\
\dot{z} &= 3 + xz, \\
\dot{w} &= 0.05w - 0.5z
\end{aligned} \tag{3.42}$$

It is important to emphasize that when simulating or constructing Chua's circuit, we are working with an actual oscillator electronic circuit. Whereas, when dealing with the Rössler attractor, what we are simulating is an analog computer circuit which solves the Rössler equations.

3.8 Dynamic Analysis of the Nonlinear Energy Harvesting System

In the development of modern society, one of the key factors is to develop new technology which is able to save energy in different applications. More exactly, vibrations and other phenomena stem from either natural processes or man-made systems can be used as an important electric resource for low-power electronic devices such as transducers and wireless sensors. This Section initially gives to the reader an outline on the electromagnetic and piezoelectric energy-harvesting systems and then focuses on the theoretical and experimental approaches to the different harvester systems [17, 18].

Presently, this is the first key topic—energy harvesting—and becomes one of the most motivated fields of the multidisciplinary science due to the complicated features of the harvester materials, dependences on various mechanical, electrical, and magnetic parameters, rich responses on different external excitation frequencies and strength. Such factors as excitation frequency, electrical load, manufacturing tolerance, and ambient temperature play important roles in the practical realizations of the optimized energy harvesting from the environment. At the energy contributes the nonlinear nature of the vibration phenomena, and these nonlinear processes cannot be neglected for an optimized harvester system. It is fully justified an estimate that these "hot subjects" will be of interest for many decades and, at the same time, will be a challenge and hard task for the researchers all over the world, considering the new energy policies due to energy crisis. Today more than 50% of all energy consumption is in the form of electrical energy, which is expected to grow to 60% by 2040, and the power generation capacity worldwide is expected to increase by 44%, from 7282 GW in 2023 to about 10,700 GW in 2040. In spite of the efforts to decrease the energy input of the electronic devices, this ever increasing demand on energy surprisingly takes us to seek using existing power sources like human movement, means of transport, sunlight, wind and industrial vibrations but also specific natural processes. Complying with the first law of thermodynamics, principle of conservation of energy determines that the energy stored in existing across the environment, but dissipated, sources can be captured and filtered and then transfer into usable electrical energy. Individual harvester generator or energy harvesting source, also known as stand-alone EH, may produce low output energies to upload them to the system. To increase the low energy output of stand-alone energy harvestings the hybrid energy harvesting or multimodal energy harvesting takes place.

If the piezoelectric harvesting generator (PEH) is used, then the piezoelectric effect plays a decisive role. It is a phenomenon that occurs primarily in dielectric materials, such as quartz and tourmaline, on which surfaces an electric charge is produced owing to a voltage being generated when the material is compressed. On the other

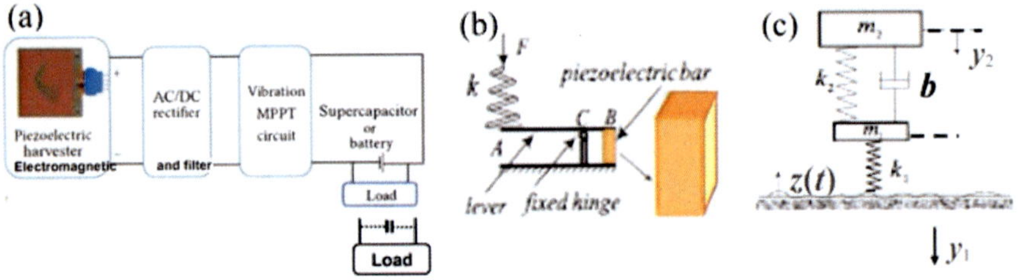

Fig. 3.16 The circuit diagram which enables to use of the harvested energy after rectification, MPPT and storage: **a** block scheme, **b** piezoelectric harvester, **c** vibrating mas harvester

hand in an electro-magnetic energy harvester (EMH) the energy is generated based on Faraday's law of induction, which determines the time derivative of the magnetic flux as an electromotive force. In order to get the optimized energy in the load an efficient maximal power point tracking (MPPT) technique should be applied. A general circuit model with rectification, storage, MPPT and electrical load is shown in Fig. 3.16a. In such circuit the harvested signal is initially rectified and filtered and then transferred to an MPPT arrangement just before the storage unit.

An effective way to improve the energy efficiency of an electric vehicle is the ability to rebuild the braking system so that the kinetic energy of braking on each wheel of the car can be converted into electricity that must be stored in a battery or/and in a supercapacitor. Full use of the possibilities of this process ensures the application of the disk Faraday generator, whose principle of operation is illustrated in Fig. 3.17a. It utilizes a strongly compacted device formed by two cylindrical permanent magnets separated by a thin good conductive disk. The disk homopolar generator distinguishes itself from other common generators in that no commutation or alternating of the magnetic poles is necessary for this machine in order to generate electric energy. When the car wheel rotates a voltage between brushes attached to the axle and the rim of the conducting disc is generated and directed appropriately from the driven compact and converts the kinetic energy to electrical energy that can be used for recharging the battery or supercapacitor [7–21].

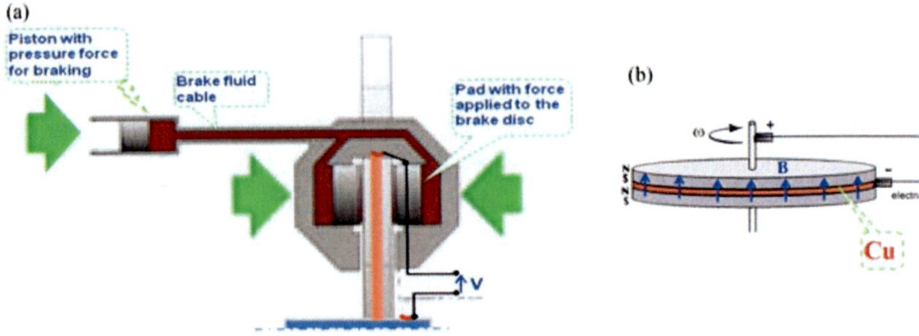

Fig. 3.17 Modified car brake: **a** with Faraday generator, **b** model for voltage evaluation

To examine the effect of electromagnetic characteristics on the output of the disk Faraday generator, and to investigate the conditions that result in the maximum efficiency of the generator a comprehensive model that takes into account most of the experimental variables is taken into account and validated by considering limited conditions (Fig. 3.17b). The value of the generated voltage can be derived by using Faraday's law of induction and the Lorentz force law, namely

$$V(t) = \int_r^R \overrightarrow{E_i} \cdot \overrightarrow{dr} = |\overrightarrow{B}|\omega \frac{R^2 - r^2}{2} \tag{3.43}$$

where $\overrightarrow{E_i}$, $\overrightarrow{r}$ and $\overrightarrow{B}$ denote the vectors of internal electric field intensity, of the radius from the axle and of the magnetic induction, respectively. The rotation velocity is denoted by ω and the external and internal radii of the conducting disk are denoted by R and r, respectively.

Assuming the radial direction of the current in the conducting disk, the resistance of a disk-shaped conductor can be expressed as:

$$R_t = \rho \int_r^R \frac{dl}{\int_0^{2\pi} hld\alpha} = \frac{\rho}{2\pi\pi} \int_r^R \frac{dl}{l} = \frac{\rho}{2h\pi} \ln\left(\frac{R}{r}\right) \tag{3.44}$$

where ρ and h denote the resistivity and thickness of the conducting disk.

The efficiency η of the generator can be defined as the total output energy divided by the input energy, namely

$$\eta = \frac{\int_0^T p(t)dt}{\frac{1}{2} J\omega_0^2} \tag{3.45}$$

where $p(t)$ and J denote the instantaneous power delivered by the generator to a load and the rotational inertia of the disk, respectively. The time of observation is denoted by T and ω_0 is the initial angular velocity of the disk.

In order to determine the power produced as a function of time, we can express $p(t)$ as

$$p(t) = V(t)I(t) \tag{3.46}$$

where $V(t)$ is determined by (3.43) and $I(t)$ denotes the current delivered by the generator. The instantaneous values of the current depend on the disk resistance (3.44) and parameters of dynamic elements representing the load and an intermediate connecting network.

Assuming that a supercapacitor with capacity C_s represents the load and the connecting network exhibits the structure of an R_c, L_c, C_c standard two-port network with constant parameters, the complete circuit is described by the following state space equation [22]

$$\begin{aligned}\dot{\boldsymbol{x}} &= \boldsymbol{A}\boldsymbol{x} + \boldsymbol{B}\boldsymbol{u} \\ \boldsymbol{y} &= \boldsymbol{C}\boldsymbol{x} + \boldsymbol{D}\boldsymbol{u}\end{aligned}, \tag{3.47}$$

where $x = [I(t) \quad V(t)]'$, $\boldsymbol{y} = V_s(t)$ and $\boldsymbol{u} = V(t)$ denote the state vector, the output variable and input variable, respectively, with $V_s(t)$ as the voltage at the load port. Matrices

$$\boldsymbol{A} = \begin{bmatrix} -\frac{R}{L_c} & -\frac{1}{L_c} \\ \frac{1}{C} & 0 \end{bmatrix}, \quad \boldsymbol{B} = \begin{bmatrix} \frac{1}{L_c} \\ 0 \end{bmatrix}, \quad \boldsymbol{C} = [0 \quad 1], \quad \boldsymbol{D} = [0] \tag{3.48}$$

depend on circuit element parameters with $R = R_t + R_c$ and $C = C_s + C_c$ as the equivalent resistance and capacitance of the complete circuit, respectively.

The state space Eqs. (3.47) with (3.48) can be useful to design a suitable control of regenerative braking of the car. A limitation of this model is that it does not accurately describe a situation when a current with a strong angular component is produced on the disk due to a high angular velocity, or with a high overall current. Thus the change in the magnetic flux along the path of the electrons creates eddy currents, which consequently resists the rotation of the disk.

To the above two governing equations must be added the description of the physical friction and the armature reaction which represent factors decelerating the rotation of the disk. The frictional torque $M(t)$ acting on the conducting disk can be modelled through the following equation [23]

$$M(t) = \alpha \cdot \mathrm{sign}(v(t)) + \beta v(t) \tag{3.49}$$

where $v(t)$ denotes the car velocity and α and β are coefficients of the dry kinetic friction and of the viscous friction, respectively.

The armature reaction commonly referred as a back torque $H(t)$ that resists the rotation of the generator disk is simply due to the Lorentz force by the induced current and is expressed by

$$H(t) = \int_r^R BI(t)x \cdot dx \tag{3.50}$$

where x denotes a point on the disk.

It has to be noted also that the resulting back torque affects the magnitude of the induced current. Therefore, Eqs. (3.47)–(3.50) constitute the mathematical model of the car braking energy regeneration with the Faraday disk generator which can be used to designed and consequently update the parameters and the variables every time the car is run.

Today, the problem of energy becomes so important that all the attention of modern societies is turning towards clean and renewable energies (solar energy, wind energy, etc.) and their efficient consumption. This is why energy regeneration systems of damping in electric cars become also into play [24, 25]. The even small amount of energy, that is produced when the engine and passengers compartments of a vehicle vibrate while in motion, is a possible source for energy harvesting. It is passable to convert vibration/kinetic energy to electric energy by using regenerative shock absorber effectively. The harvested energy from electric car shock absorbers will result in a much more economical vehicle performance and increased comfort of running [4].

Currently, the mostly available mechanisms suitable for vibrations-to-electric energy conversion are piezoelectric transducers. Among different types of energy transducers, the piezoelectric transducers (Fig. 3.16b) are preferred because of their efficiency which is much higher than the others. It was yet found out that the energy density of piezoelectric transducer is three times higher than the other transducers.

To achieve a new effective design with piezoelectric technology of damping energy harvesting for driving vehicles, a dual-mass piezoelectric bar harvester can be developed for absorbing energy from vibrations and motions of a suspension system under excitations of the vehicle from road roughness.

According to the Newton second law of dynamics the governing differential equations of the dual-mass piezoelectric bar harvester system (Fig. 3.16c) are expressed as follows

$$\begin{aligned} m_1\ddot{y}_1 + b(\dot{y}_1 - \dot{y}_2) + k_2(y_1 - y_2) + k_1(y_1 - z(t)) = 0 \\ m_2\ddot{y}_2 + b(\dot{y}_2 - \dot{y}_1) + k_2(y_2 - y_1) = 0 \end{aligned} \tag{3.51}$$

where k_1, k_2 and b denote elastances and damping coefficient of the springs and damper, respectively. The displacements of the unsprung mass and sprung mass with respect to their respective equilibrium positions are denoted by y_1 and y_2, respectively. The form of the road surface in the transverse motion of the car is denoted by $z(t)$.

Taking into account the principle that the dissipation energy of a damper is equal to the electric energy generated by the piezoelectric bar harvester, the damping coefficient b can be expressed as

$$b = n^2 d^2 k_2^2/(\pi^2 C f) \tag{3.52}$$

where n, d, C and f denote the ratio of the moment arms of the lever, the piezoelectric constant in the polling direction, the electrical capacity of the piezoelectric bar and the first natural vibration frequency of the car suspension, respectively.

Consequently, we can obtain the instantaneous displacements and velocities $\dot{y}_1$, $\dot{y}_2$ of the unsprung mass and sprung mass at their respective equilibrium positions. The relative displacements $y_{12} = y_1 - y_2$ and velocities $\dot{y}_{12} = \dot{y}_1 - \dot{y}_2$ of the unsprung mass and the sprung mass can also be determined. Then the generated charge $q(t)$, and voltage $V(t)$, from the piezoelectric bar at time t can be expressed by relations

$$\begin{aligned} q(t) &= (ndk_2)y_{21}(t) \\ V(t) &= \frac{q(t)}{C} \\ I(t) &= (ndk_2)\dot{y}_{21}(t) \end{aligned} \tag{3.53}$$

Thus, the instantaneous power $p(t)$ and energy $E(t)$ generated from 0 to t by the piezoelectric bar can be evaluated as

$$\begin{aligned} p(t) &= (1/C)(ndk_2)^2 y_{21}\dot{y}_{21} \\ E(t) &= \frac{(ndk_2)^2}{C}\int_0^t y_{21}\dot{y}_{21}d\tau \end{aligned} \tag{3.54}$$

It is easily seen that the energy $E(t)$ increases with an increase in the velocity of vehicles and the class of road surface, an increase in the ratio of the moment arms of the lever, and with a decrease in the width of the piezoelectric bar. It can be expected that in practice four or more of the novel piezoelectric bar energy harvesters could be installed on a vehicle and provide more efficient energy harvesting as an auxiliary energy of electric cars. This offers as much as possible the energy regeneration from the damped kinetic energy and store it to be used for a useful purpose such as an auxiliary on board electric energy source.

3.9 Duffing's Forced Oscillator

In physics and mathematics, in the area of dynamical systems, the Duffing equation is a non-linear second-order differential equation used to model certain damped and driven oscillators [26, 27]. Nowadays, the term "Duffing equation" is used for any equation that describes an oscillator that has a cubic stiffness term, regardless of the type of damping or excitation, namely

$$\ddot{x} + \delta\dot{x} + \alpha x + \beta x^3 = \gamma\cos(\omega t) \tag{3.55}$$

where $x = x(t)$ the unknown variable e.g., displacement at time t. The parameters δ, α, β, γ and ω are given constants. The equilibrium points, stable and unstable, are at $\alpha x + \beta x^3 = 0$. If $\alpha > 0$ the stable equilibrium is at x = 0. If $\alpha < 0$ and $\beta >$ 0 the stable equilibriums are at $x = \sqrt{-\alpha/\beta}$ and $x = -\sqrt{-\alpha/\beta}$. To analyze their stability, the application of the linearization procedure gives the Jacobian matrix

$$\boldsymbol{J} = \begin{bmatrix} 0 & 1 \\ -\alpha - 3\beta x^2 & -\delta \end{bmatrix}$$

The eigenvalues of $\mathbf{J}(0)$ are

$$\lambda_{1,2} = -\frac{1}{2}[\delta \pm \sqrt{\delta^2 - 4\alpha}],$$

and it is found that this equilibrium is stable for $\alpha \geq 0$. On the other hand, the eigenvalues of the equilibria $x = \pm\sqrt{-\alpha/\beta}$ are

$$\lambda_{1,2} = -\frac{1}{2}[\delta \pm \sqrt{\delta^2 + 8\alpha}],$$

which are unstable for positive values of δ and α.

When damping and forcing terms are negligibly small ($\delta \approx 0$, $\gamma \approx 0$) then the Duffing equation can be integrated upon multiplication by the velocity as follows

$$\dot{x}\left[\ddot{x} + \alpha x + \beta x^3\right] = \frac{d}{dt}\left(\frac{1}{2}\dot{x}^2 + \frac{1}{2}\alpha x^2 + \frac{1}{4}\beta x^4\right) = 0$$

Hence, after integration with respect to time of both sides of the above equation yields

$$H(t) = \frac{1}{2}\dot{x}^2 + \frac{1}{2}\alpha x^2 + \frac{1}{4}\beta x^4 = \text{constant} \tag{3.56}$$

The quantity $H(t)$ is called the **Hamiltonian** for the Duffing equation. The value of H is determined by the initial conditions $x(0)$ and $\dot{x}(0)$. Substituting $y = \dot{x}$ in H attests that the system is Hamiltonian, so that

$$\dot{x} = \frac{\partial H}{\partial y},\ \dot{y} = -\frac{\partial H}{\partial x},\ \text{ with } H = \frac{1}{2}y^2 + \frac{1}{2}\alpha x^2 + \frac{1}{4}\beta x^4$$

When both α and β are positive, the solution of the Duffing equation is bounded, i.e.

$$|x| < \sqrt{\frac{2H}{\alpha}} \text{ and } |\dot{x}| < \sqrt{2H}$$

with the Hamiltonian H being positive.

Denoting the time derivative of $x(t)$ as $y(t)$ gives the state variables form of the Duffing equation (3.55), namely

$$\begin{aligned} \dot{x} &= y \\ \dot{y} &= -\delta y - \alpha x - \beta x^3 + \gamma \cos(\omega t) \end{aligned} \tag{3.57}$$

The computer simulation results with parameters leading to chaotic state variables are shown in Fig. 3.18 where designation $c = \gamma$ has been entered. The diagrams look complicated, but in fact, most of the plot shows the initial period of time during which the motion is approaching its much simpler final behavior (Fig. 3.18a, b). The

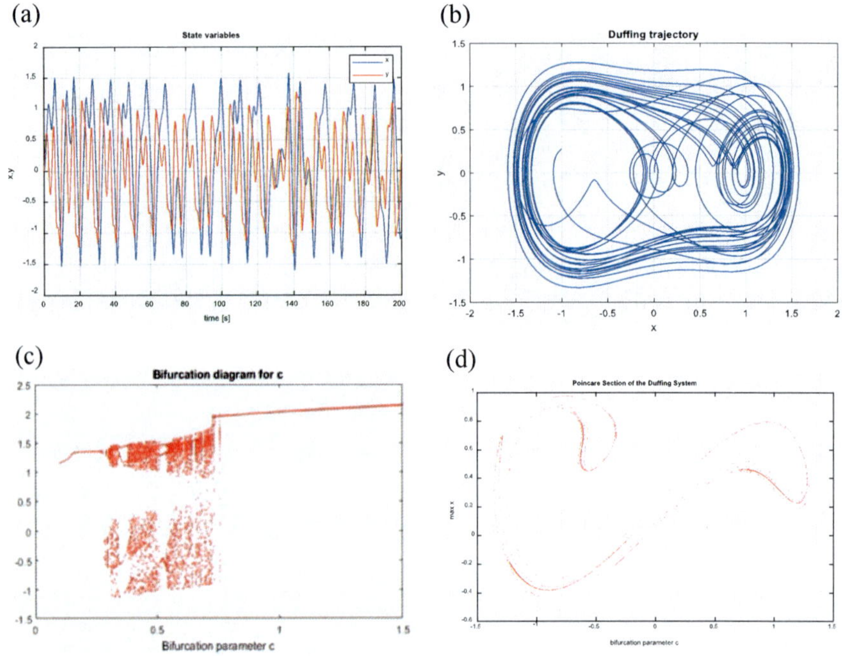

Fig. 3.18 Simulation results of Duffing equation: **a** state variables versus time, **b** strange attractor, **c** bifurcation diagram, **d** Poincaré section

initial course of individual quantities is called an "initial transient". It is interesting investigate how the phase space plots change when the strength of the driving force is changed. Figure 3.18c shows how the phase space plots change when the strength of the driving force is changed. It is seen that for the range $0.34 < \gamma < 0.65$ the system is prone to generate chaotic waveforms.

A useful way of analyzing chaotic motion is to look at what is called the **Poincaré section**. Rather than considering the phase space trajectory for all times, which gives a continuous curve, the Poincaré section is just the discrete set of phase space points of the trajectory at every period of the driving force, i.e. at $t = \pi/\omega$, $2\pi/\omega$, $3\pi/\omega$, …, etc. Clearly for a periodic orbit the Poincaré section is a single point, when the period has doubled it consists of two points, and so on. We define a function, **poincare**, which produces a Poincaré section for given values of A, γ, and ω, in which the first **ndrop** periods are assumed to be initial transient and so are not plotted, while the subsequent **ndrop** periods are plotted. The point size is given by the parameter **psize**. Note that the function **g[{xold, yold}]** maps a point in phase space **{xold, yold}** at time t to the point in phase space $\{x, y\}$ one period T later. This strange diagram is the **strange attractor.** It is the limiting set of points to which the trajectory tends (after the initial transient) every period of the driving force. The Poincaré section of the Duffing oscillator (3.57) is shown in Fig. 3.18d. It is the limiting set of points to which the trajectory tends (after the initial transient) every period of the source signal. Trying different values of the parameters gamma and omega one can see where the period doubling transition to chaos occurs.

References

1. Kolmanovskii, V., Myshkis, A.: Introduction to the Theory and Applications of Functional Differential Equations. Kluwer, Dordrecht (1999)
2. Arino, O., Hbid, M.L., Dads, E.A. (eds.): Delay Differential Equations and Applications. Springer, Dordrecht (2006)
3. Lorenz, W.E.: Fractals and fractal architecture. www.iemar.tuwien.ac.at
4. Peitgen, H.O., Jurgens, H., Saupe, D.: Chaos and Fractals: New Frontiers of Science, 2nd edn. Springer, New York (2004)
5. Rössler, O.E.: Different types of chaos in two simple differential equations. Zeitschrift für Naturforschung A **31**, 1664–1670 (1976)
6. Gopalsamy, K.: Stability and Oscillations in Delay Differential Equations of Population Dynamics. Kluwer, Dordrecht (1992)
7. Hoppensteadt, F.: Predator-prey model. Scholarpedia 1(10), 1563 (2006)
8. Leconte, M., Masson, P., Qi, L.: Limit cycle oscillations, response time, and the time-dependent solution to the Lotka-Volterra predator-prey model. Phys. Plasmas **29**(2), 022302 (2022)
9. Rössler, O.E., Letellie, R.C.: Chaos. The World of Nonperiodic Oscillations. Springer Nature, Cham, Switzerland AG (2020)
10. Cheng, C.-H., Chen, C.-Y., Chen, J.-D., Pan, D.-K., Ting, K.-T., Lin, F.-Y.: 3D pulsed chaos lidar system. Opt. Express **26**, 9 (2018)
11. Trzaska, Z.: Analysis and Design of Electric Circuits. Office of the Warsaw University of Technology, Warsaw (2008)
12. Liu, S.-T., Wang, P.: Fractal Control Theory. Springer, Singapore (2018)
13. Jaggard, D.L., Jaggard, A.D.: Cantor ring arrays. In: Digest of IEEE AP-S/URSI International Symposium, pp. 866–869 (1998)
14. Mandelbrot, B.B.: The Fractal Geometry of Nature. Times Book, New York (1977)
15. Van der Pol, B.: On relaxation-oscillations. Lond. Edinb. Dublin Philos. J. Sci. **2**(7), 978–992 (1927)
16. Tian, K., Ren, H.-P., Grebogi, C.: Rössler-network with time delay: univariate impulse pinning synchronization. Chaos **30**, 123101 (2020)
17. Ruiz-Oliveras, F.R., Pisarchik, A.N.: Synchronization of semiconductor lasers with coexisting attractors. Phys. Rev. E **79**, 0162022009 (2009)
18. Trzaska, Z.: Effective harvesting of braking energy in electric cars. KONES (2017)
19. Lozi, R., Pchelintsev, A.N.: A new reliable numerical method for computing chaotic solutions of dynamical systems: the Chen attractor case. Int. J. Bifurcat. Chaos **25**(13), 1550187–1550412 (2015)
20. Kuznetsov, N.V., Mokaev, T.N., Ponomarenko, V.I., Seleznev, E.P., Stankevich, N.V., Chua, L.: Hidden attractors in Chua circuit: mathematical theory meets physical experiments. Nonlinear Dyn. **111**(6), 5859–5887 (2023)
21. Peitgen, H.-O., Jürgens, H., Dietmar, S.: 12.3 The Rössler Attractor, Chaos and Fractals: New Frontiers of Science, pp. 636–646. Springer, Berlin (2004)
22. Rössler, O.E.: Chaotic behavior in simple reaction system. Zeitschrift für Naturforschung A. **31**(3–4), 259–264 (1976)
23. Rand, R.H.: Lecture Notes on Nonlinear Vibrations (PDF) (vol. 53, pp. 13–17). Cornell University (2012)
24. Griffiths, G., Schiesser W.: Linear and nonlinear waves. Scholarpedia **4**(7), 4308 (2009)
25. Li, M., Chenb, B., Yea, H.: A bioeconomic differential algebraic predator–prey model with nonlinear prey harvesting. Appl. Math. Model. **42**, 17–28 (2017)

26. Wawrzynski, W.: Duffing-type oscillator under harmonic excitation with a variable value of excitation amplitude and time-dependent external disturbances. Sci. Rep. **11**, 2889 (2021)
27. Chen, T., Cao, X., Niu, D.: Model modification and feature study of Duffing oscillator. J. Low Freq. Noise Vib. Active Control **41**, 1 (2022)

Chapter 4
Oscillatory Chemical Systems

4.1 Preliminary

In the last decades, despite considerable theoretical progress on the nature of chemical oscillations, the only known chemical oscillators show either biological origin, like the glycolysis [1–16] and glucose oxidase/peroxidase systems along with o-Dianisidine applied for the determination of blood and urinary glucose, especially the presence of fructose for both research and clinical purposes. It should also be replaced here reactions discovered accidentally, like the Bray-Liebhafsky reaction, being the first oscillating reaction in a stirred homogeneous solution. Certainly, the most widely known oscillatory chemical reaction is the Belousov–Zhabotinsky reaction [17]. Also many variants of those reactions are worthy of in-depth attention. This is why to specify necessary and sufficient conditions for chemical oscillations or to find new chemical oscillators had proved surprisingly frustrating. It is worth underlying that many of up-to date performed investigations of chemical oscillating reactions in a continuously stirred tank reactor (CSTR) [1–8] have revealed nonequilibrium phenomena such as oscillations, bistability, complex oscillations, mixed mode oscillations, and quasi-chaotic behavior of the reaction. For this reason one of the main challenges for researchers and practitioners it still is to predict and to control these phenomena in nonlinear chemical oscillations for potential applications. This is why in a present competitive world of globalization, activities of chemical industries specialists are striving to improve their processes and products for the dynamic and selective high expectations from consumers. However, the chemical businesses can move their attention to the cost challenges if they are able to fabricate innovative products that will carry a higher value to the customers, even if the cost of final products is high. The most common challenges in almost all chemical engineering fields are manufacturing high quality products with the lowest cost and generating abundant profit from market sales. Successful competitive chemical branches are not only able to decrease their product costs, but also capable to bring a high value profit through their product innovations. The innovations in this field are becoming more

Z. Trzaska, *Mathematical Modelling and Computing in Physics, Chemistry and Biology*, Studies in Systems, Decision and Control 495,
https://doi.org/10.1007/978-3-031-39985-5_4

challenging in developing products that have different applications. For this reason, chemical engineers need a systematic and reliable methodology to innovate their products, especially through current problem solving and forecasting. Therefore, it is critical for the investigations in chemical engineering to have a strong and reliable approach and method to support the product innovation and development. The challenge consists of problems that have various levels of complexity, constraints and limitations that represent various levels of difficulties. Problem-solving approach in chemical engineering requires the use of numerical language and software. However, the problem-solving scale only covers on optimization such as to generate the model equations for the problem at hand, appoint an appropriate numerical method to solve the model, write and debug a program to solve the problem using the fitted numerical algorithm, analyze and estimation of the obtained results for validity and precision. In the current condition, investigations in the chemical engineering show a higher constraint in innovation and creativity compared to research in other engineering fields. In this context, it is critical to have effective methodologies and tools by chemical researchers to develop interest in methods development leading to innovation for the competing applications. The main reason is that a typical sequence of events may be such that smooth changes of a control parameter even in a relatively simple chemical dynamical system, described by deterministic nonlinear differential equations, can cause a sudden transition to a completely new, qualitatively different state, what means establishing a non-equilibrium stationary state. When a control parameter such as temperature, flow rate, or influx concentration is under variation, bifurcation and transition of the system from one dynamic state to another occurs. Periodic or nearly periodic temporal changes of concentrations in the reacting system provide a systematic impulse to improve knowledge and develop new findings in the field of non-linear chemical dynamics. Oscillating chemical reactions can be realized as extremely sensitive to some components in result of modifying the system structure or determination in the non-equilibrium stationary state bifurcation, providing necessity of applying new methods for their investigations. When the differential equations, which present the model of a given system, are nonlinear, their analytical solution is often unknown and one has to use the computer to solve the problem numerically. In track of computer calculations not only numerical solutions of various problems are obtained, but also completely unexpected phenomena can be discovered [16]. It should be emphasized that mixed-mode oscillations, for which the oscillatory cycle consists of a number of large excursions combined with a number of small peaks, are frequently observed in both experiments and models appropriate to chemical and biological systems [1–11]. They have significant applications in nonlinear chemistry and medicinal chemistry. The progresses in the description of systems combining nonlinear chemical kinetics and produced oscillations are presented in the sequel.

4.2 Oscillations in Chemical Systems

Chemical oscillations belong to an interesting nonlinear dynamical phenomenon which arises due to complex stability condition of the steady state of a reaction running far away from the equilibrium state which is usually characterized by a periodic attractor or a limit cycle around an interior stationary point. For instance, the rate of chromium dissolution in acid periodically increases and decreases. Similarly, an electrochemical cell may produce an oscillating current between appropriately polarized electrodes submerged in a good solution. Generally speaking, **chemical oscillators** are open systems characterized by periodic variations of some reaction species concentration due to complex physico-chemical phenomena that may cause bistability, rise of limit cycle attractors, and birth of spiral waves, Turing patterns and finally deterministic chaos. In accord to the second law of thermodynamics chemical systems cannot oscillate about a position of final equilibrium. The thermodynamic requirement for a closed system, operating at constant temperature and pressure, is that the Gibbs free energy must decrease continuously and not oscillate. However it is possible that the concentrations of some *reaction intermediates* oscillate, and also that the *rate* of formation of products oscillates [18–23].

In an oscillating chemical system the energy-exciting reaction can follow at least two different ways, and the phenomenon periodically switches from one way to another. On one of these ways a specific intermediate is produced while on the second way it is consumed. The concentration of this intermediate product triggers the switching of ways. When the concentration of the intermediate is low, the reaction follows the producing way, leading then to a relatively high concentration of intermediate. When the concentration of the intermediate will reach a high level, then the reaction switches to the consuming way [24, 25].

One of several oscillating chemical subjects is a Belousov–Zhabotinsky (BZ) reaction, whose common element is the mixture of bromine and an acid. An essential aspect of the BZ reaction is its so-called "excitability"—under the influence of stimuli, for instance of light, color patterns (Fig. 4.1a) develop in what would otherwise be a perfectly colorless environment. Some clock reactions such as the Briggs–Rauscher reactions and the BZ using the chemical ruthenium bipyridyl as catalyst can be excited into self-organizing activity through the influence also of light. A system that demonstrates temporal order is the Brusselator [4]. It is characterized by the rate equations [26–31]

$$\frac{dX}{dt} = A + X^2Y - BX - X$$
$$\frac{dY}{dt} = BX - X^2Y \tag{4.1}$$

where, for convenience, the rate constants have been set to 1. The Brusselator has an equilibrium point at

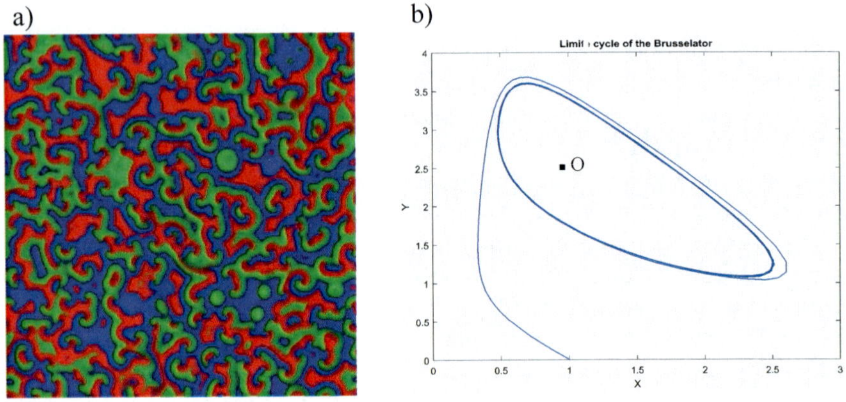

Fig. 4.1 Computer simulation pattern of: **a** the Belousov–Zhabotinsky reaction at an instant $t_k \gg 0$, **b** trajectory of the Brusselator in the unstable regime: A = 1, B = 2.5, X(0) = 1, Y(0) = 0. The system approaches a limit cycle

$$X_0 = A \quad \text{and} \quad Y_0 = \frac{B}{A} \tag{4.2}$$

The fixed point becomes unstable when

$$B > A^2 + 1 \tag{4.3}$$

leading to an oscillation of the system. Unlike the Lotka-Volterra equation, the oscillations of the Brusselator do not depend on the amount of reactant present initially. Instead, after sufficient time, the oscillations approach a limit cycle (Fig. 4.1b).

It is worth to emphasizing that the oscillatory proceeding is strictly transient and the effects of reactant expenditure cause an end of oscillation and a final monotonic entrance to the chemical equilibrium state.

Since chemical reactions have no momentum so thus they are unlike mechanical oscillators such as a mass on a spring or the drum membrane. However, in a chemical oscillator, the values of certain parameters change cyclically in time around the state of equilibrium. A simple formal example of an oscillatory mechanism including positive and negative feedback steps is the model presented by block scheme in Fig. 4.2.

The rate equation for reagent X is given by

$$\frac{dX}{dt} = k_i A X^2 + k_u A - k_2 X \tag{4.4}$$

The consumption of concentrate A in reaction (4.4) also limits the rate of increase of product X and can be considered as a negative feedback. However, this contribution to the rate term, directly linked to the production of X, cannot lead to an oscillatory instability in a homogenous system. In the absence of the autocatalytic

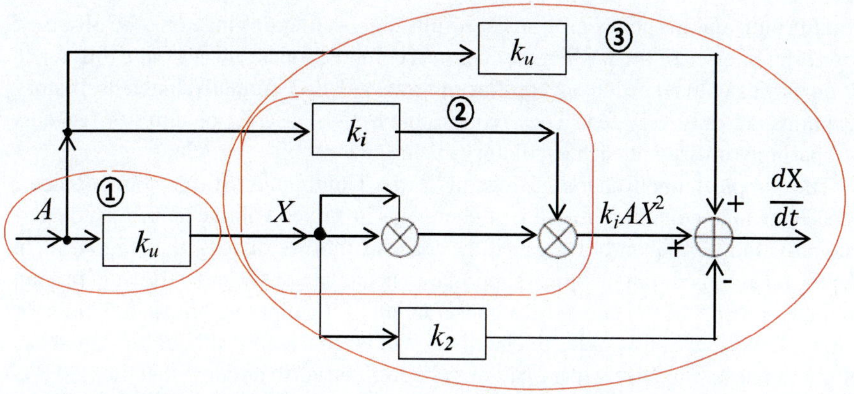

Fig. 4.2 Block scheme illustration of a chemical oscillatory mechanism: ①—initiation, ②—autocatalyst, ③—scavenging

term in reaction (4.4), this would be a simple linear cascade of reactions that could not produce oscillations. Note, that the autocatalytic reaction (4.4) exerts a positive feedback on component X while reaction $-\ k_2X$ which independently opposes to the increase of X is a negative feedback process. Positive feedback called briefly autocatalysis means an increase in the speed of a chemical reaction under the influence of one of the products of this reaction, which acts as a catalyst. The rate of the autocatalytic reaction increases as it progresses and the resulting increase in the concentration of the catalyst product, and then decreases due to the decrease in the concentration of the reactants. Positive feedbacks are always destabilizing processes, while stabilizing ones are the antagonist negative feedbacks. Anything stating on the complexity of the kinetic mechanism, oscillations develop only when the positive feedback evolves on a shorter time scale τ_p than the negative feedback τ_{n}. In a strict sense, there must be a "delay" between the positive and negative feedbacks actions in the given system. In general, in the chemical world, the primary source of nonlinearities causing instabilities is implicated in local reactive dissipative processes. Thus they may be exhibited even in the absence of spatial degrees of freedom, as for instance in a well-mixed reactor, what is leading to a wealth of nonlinear behaviors. This places nonlinear kinetics in chemical systems at the forefront for understanding the origin of endogenous oscillating and model phenomena [32].

In real chemical systems, autocatalysis usually occurs in a multistep process, like

$$\mathrm{A} + \mathrm{nX} \rightarrow \mathrm{mY} \quad \mathrm{with\,m} > \mathrm{n} \tag{4.5}$$

$$\mathrm{Y} \rightarrow \cdots \rightarrow \mathrm{X} \tag{4.6}$$

with constants m and n depending on the kind of chemical reaction and reactor structure. Usually, chemical reactors can present an open form or a closed one. In closed systems, all chemical reactions spontaneously evolve to their thermodynamic equilibrium, and in such conditions, it results that oscillations may only be observed for a brief interval of time, when the chemical composition is still at a finite difference of quantities with respect to an equilibrium state [33, 34]. Actually, batch oscillatory reactions are only very few. They require that the initial great quantity of reagents be sparingly utilized at each oscillatory cycle.

The simplest open chemical reactor is the Continuous Stirred Tank Reactor (CSTR), which contains a mixture of chemicals in a fixed volume tank, stirred energetically and permanently refreshed by constant inflows of reagents (Fig. 4.3). In the ideal case, the input flows are supposed to instantaneously and uniformly mixing into the contents of the tank with a conservation of the input volume at any time. In what follows the liquid phases of chemistry will be considered although the methodology can be applied to a wider class of systems. It is worth noticing that the CSTR is playing a crucial role in the investigations in the field of nonlinear chemistry when the acting between reaction and diffusion is out of the question (i.e. well mixed reactor). It should be emphasized that in the case of nonlinear chemical systems, even finely dispersed residual feed mixing inhomogeneities in the reacting solution can dramatically affect the overall dynamics [35]. In the case of the well mixed reactor, all spatial concentration gradients do not exist. Then the dynamics of reactants is excited only by the kinetic agent and input–output balance conditions. Hence, the design and control of new oscillators should take into account the appropriate condition of a constant feed of the CSTR. Normally, the oscillatory behavior of a CSTR is evident through time-periodic changes in the reaction temperature and/or the concentrations of the reacting substances. In order to model such behavior, let us consider now a chemical system involving species X and Y with respective concentrations x and y, and ruled by the following reaction mechanisms

$$
\begin{aligned}
&A \to \mathrm{X}\\
&B + X \to Y\\
&2X + Y \to 3X\\
&\mathrm{X} \to Q
\end{aligned}
\tag{4.7}
$$

where A, B are the initial reactants, Q the final product and X, Y the intermediate products. The initial reactants considered in large excess fulfill the conditions for the CSTR. Thus the system may be considered as open to the external environment and the corresponding kinetics takes the form

$$
\begin{aligned}
\frac{dX}{dt} &= AX - BY - \alpha X^2 Y\\
\frac{\mathrm{d}Y}{\mathrm{dt}} &= BX + AY - \alpha XY^2
\end{aligned}
\tag{4.8}
$$

where X and Y are an activator and inhibitor, respectively. From (4.8) it is easily seen that in the chemical world the first principal source of nonlinearities leading to instability arises from the local reactive dissipative processes. However, two caveats

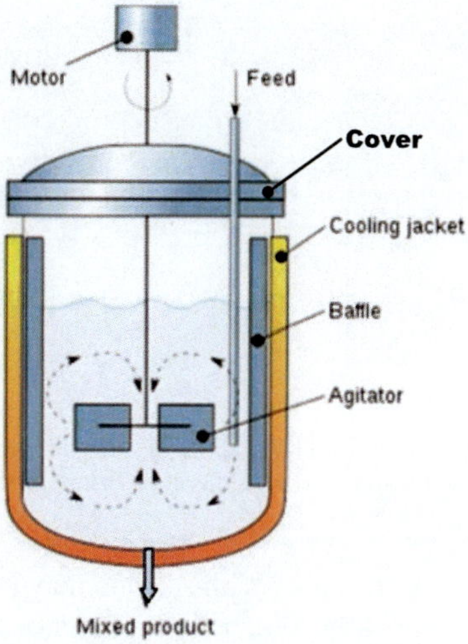

Fig. 4.3 An isothermal instantaneously stirred tank reactor

are to be explained. First, from an ideal point of view, it is fair to admit that this system cannot represent a real chemical system in the accurate sense as the "concentrations" X and Y may become negative what is unacceptable [36–39]. However allowing exact solutions enable the introduction, in not too formal a way, of some important concepts which explain in satisfactory means most of the natural chemical reactions. The other caveat pertains to the coefficients A and B, which can be thought of as representing the concentrations of other species that evolve on a much longer time scale and that may be considered as constant on the timescale considered in the given reaction. In this pool chemical approximation Eqs. (4.2 and 4.5) can be considered as representing an open system, the parameters of which may be varied to control the difference with respect to some reference state. The parameter α measures the non-linear coupling force that brings the cubic term into the expression describing concentrations. From Eq. (4.8) it follows that $X_0 = 0$ and $Y_0 = 0$ is the steady state solution regardless of the values of the parameters. By linearizing Eq. (4.8) around (X_0, Y_0) it is possible to test its stability. The characteristic equation from the determinant of the Jacobian matrix takes the form

$$\det\begin{bmatrix} \lambda - A & B \\ -B & \lambda - A \end{bmatrix} = \lambda^2 - 2\lambda A + A^2 + B^2 = 0 \tag{4.9}$$

Thus the complex conjugate pair $\lambda_{1,2} = A \pm jB$ form the generalized eigenfrequencies of the system. As long as A remains negative ($A < 0$) the state (X_0, Y_0) stays asymptotically stable. When A becomes positive, then the state of the system is unstable and for $t \to \infty$ all trajectories in the (X, Y) plane tend to the circle of radius $R_s = \sqrt{\frac{A}{\alpha}}$. In this case the new state $X(t) = R_s\cos Bt$ and $Y(t) = -R_s\sin Bt$ is stable and the concentrations evolve periodically in time. This corresponds to a limit cycle that coexists with the unstable state $A = B = 0$. The radius of the limit

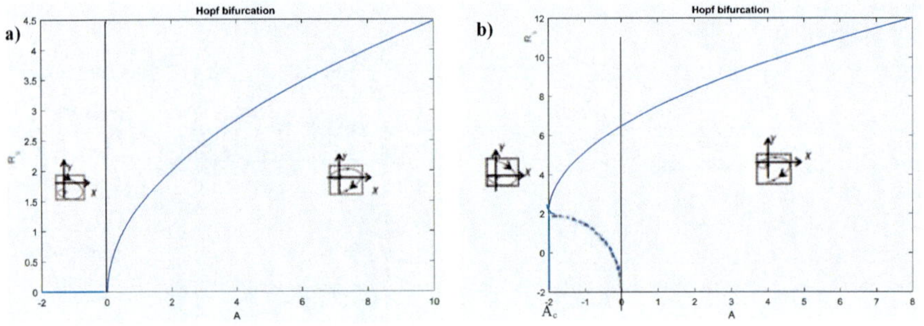

Fig. 4.4 Hopf bifurcation: **a** subcritical for the $\lambda - \omega$ system, **b** supercritical stabilized through the interaction with a saddle-node bifurcation

cycle increases with A. This may be represented by the bifurcation diagram shown in Fig. 4.4a. When the real part of the least stable single pair of complex conjugate eigenvalues changes sign, we say that the system undergoes a Hopf bifurcation [40]. The bifurcation is supercritical when the limit cycle is born with an increase of the control parameter above its marginal value $A = 0$. In a case, when the bifurcation parameter increases and the unstable limit cycle collapses into the steady state we have a subcritical Hopf bifurcation.

From the point of view of stability, it shows a limitation of the linear stability analysis. Often, to stabilize the system, another bifurcation scheme with A_c can be considered (Fig. 4.4b), which shows the essential elements of such a process. In the subcritical region, when $A_c < A < 0$, in addition to steady state $X_0 = Y_0 = 0$, there are two limit cycles with different amplitudes. The one with the smallest radius is unstable, and the other generation of mixed mode oscillations (MMOs). For A_c, the radii of the two cycles become equal and fall within the so-called saddle- node bifurcation, connected by their unstable states. It is worth noticing that in the vicinity of a supercritical Hopf bifurcation point it may be shown that the dynamics obeys the generic form that is the well known as the complex Stuart-Landau equation [12] which constitutes a universal model for oscillators with a weak nonlinearity. Note, that in the phase space spanned by all concentrations the limit cycle is not a circle anymore, although still it forms a closed curve. Moreover, it changes in time with a speed that now depends on the amplitude of oscillations as the equations for amplitude and phase angle can't longer be decoupled. In result, the oscillations lose their harmonic character [41–43] with the larger radius, which is stable, keeps the system stable (even in supercritical region). These two limit cycles of different amplitudes provide the basis for the generation of mixed mode oscillations (MMOs). For A_c, the radii of the two cycles become equal and fall within the so-called saddle-node bifurcation, connected by their unstable states.

4.3 Autocatalysis, Activation and Inhibition

In recent times nonlinear autocatalytic reactions have attracted a great deal of attention of chemical researchers and practitioners. In an autocatalytic reaction a chemical reactant promotes its own production. For instance, two units of Y react with one unit of X to produce three units of Y, i.e. a net gain is one unit of Y. In this reaction the rate of transformation from one reactant into another does not follow a first order rate law. Recall that a **first-order reaction** is a reaction that proceeds at a rate that depends *linearly* on only one reactant concentration. In order to model this kind of reaction it is beneficial to use the following empirically based principle also known as Chemical Law of Mass Action: *Suppose that* n *reactant agents* $X_1, X_2, \ldots, X_n$ *react to produce* m *product agents* $P_1, P_2, \ldots, P_m$ *in one step of reaction. Then the rate of decrease of the concentration of each reactant agent and the rate of increase of the concentration of each product agent is proportional to the product of the concentration of the* n *reactants.*

Applying this principle to the autocatalytic reaction $X+2Y \to 3Y$ which amplifies the first order reaction $X \to Y$ implies that the rate of decrease of X in one step is αxy^2, where α is a positive rate constant. Similarly, the rate of increase of Y is the same, i.e. αxy^2. Beginning with a positive concentration $f(t)$ of a precursor supplying the tank and zero initial concentration for x and y gives

$$\begin{aligned} \dot{x} &= \exp(-0.002t) - 0.08x - xy^2, \quad x(0) = 0 \\ \dot{y} &= -y + 0.08x + xy^2 \qquad\qquad\qquad y(0) = 0 \end{aligned} \tag{4.10}$$

where scaling the variables, coefficients and initial conditions was applied to get dimensionless quantities. Fig. 4.5a, b show diagrams of oscillations in the concentrations of the intermediates.

Note that after some initial rocking the concentrations change almost normally until about t =170 when the violent oscillations in the concentrations start. Oscillations end up about t = 600 and then concentrations of intermediates gradually decrease. The corresponding trajectory (Fig. 4.6c) self-intersects because the first rate function in (4.8) explicitly depends on time t. This implies that the orbit transits each point of self-intersection at different times. Moreover, the appearance of the orbital arc in Fig. 4.5c when it emerges from the splice of the oscillations suggests that the orbit is heading back to the origin as the reaction approaches accomplishment. The 3D plot of the time-state curve over time $0 < t < 250$ is shown in Fig. 4.5d.

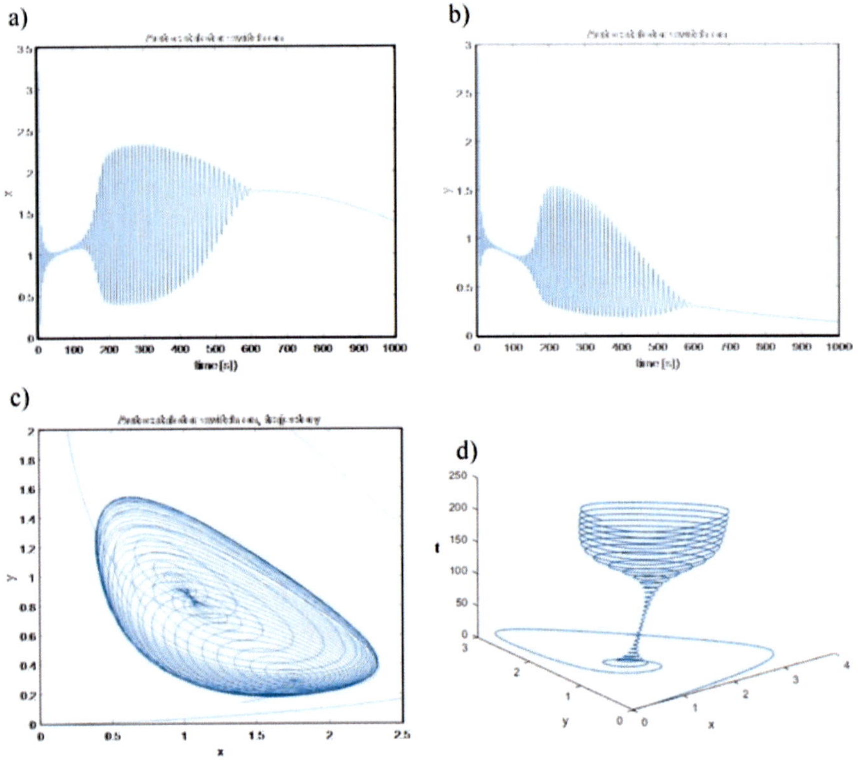

Fig. 4.5 Oscillations in autocatalytic tank: **a** intermediate x, **b** intermediate y, **c** trajectory in the plane (x, y), **d** 3D plot in the space (x, y, t)

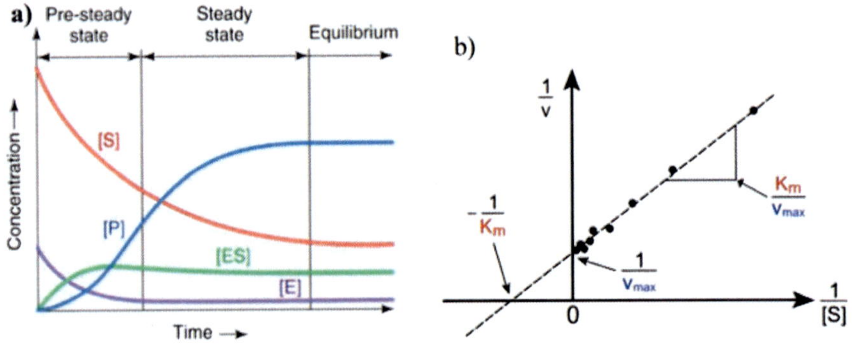

Fig. 4.6 Changes in time of an enzyme reaction: **a** particular components, **b** double-reciprocal plot

4.4 Enzyme Kinetics

Enzymes are the biological substances that appear as catalysts that speed up chemical reactions inside cells by which play crucial roles in every living organism. The catalytic activity of these protein molecules is widely recognized, accelerating reactions with great specificity and effectiveness in mild conditions. They help to carry out chemical reactions successfully. Enzymes catalyze (initiate) or accelerate chemical reactions that help convert raw materials into other products, such as fermentation for biopharmaceutical, fuel ethanol, and wine and beer production. Enzymes are incorporated in detergents and animal feed formulations. Regardless of the type of reaction or the finished product, the quality, activity and stability of the enzymes are essential for achieving the desired result. The importance of enzyme kinetics is invaluable as many diseases, such as cancer and HIV, raise questions that can largely be answered by good knowledge of enzyme kinetics. Also, with a large variety of industrial activities enzymes are crucial. Not only do they accelerate manufacturing, but they can improve quality, reduce waste and optimize product performance, thus ensuring greater profitability. The enzyme and the substrate fit together like a lock and key, and only substrates with the right shape will be processed by the enzyme. This makes the enzymes specific to their actions. The activity may be controlled by measuring changes in the concentration of substrates or products during the reaction. For instance, in medicine studying an enzyme's kinetics can reveal the catalytic mechanism of this enzyme, its role in metabolism, the manner in which its activity is controlled and the manner in which a medication or agonist could inhibit the enzyme. Depending on the type of enzymes and their sensitivity, many different analytical methods are used to fully determine their properties.

Most enzyme assays are based on spectroscopic techniques, with the two most used being absorption and fluorescence.

Over recent decades, the use of biocatalysts has rapidly expanded across multiple industrial processes. These enzymes are divided into six classes based on the types of reactions they catalyze—hydrolases, oxidoreductases, transferases, lyases, isomerases, and ligases/synthetases. An enzyme (E) is typically a *protein molecule* that promotes a reaction of another molecule, its *substrate* (S). This binds to the *active site* of the enzyme to produce an enzyme-substrate complex *ES,* and is transformed into an enzyme-product complex *EP* and from there to product P, via a *transition state ES**. The series of steps:

$$\mathrm{E} + \mathrm{S} \rightleftarrows \mathrm{ES} \rightleftarrows \mathrm{ES}^* \rightleftarrows \mathrm{EP} \rightleftarrows \mathrm{E} + \mathrm{P} \tag{4.11}$$

is known as the ***mechanism***. The coupled arrows indicate on possibility of the reciprocal course of a reaction. This example is based on the simplest case of a response with a substrate and a product. The enzyme-catalyzed reaction uses the exact same reagents and produces the exact same products as the non-catalyst reaction. For a given enzyme concentration and for relatively low substrate concentrations, the reaction rate increases linearly with substrate concentration.

On the other hand, at relatively high concentrations of the substrate, the reaction rate is asymptomatically close to the theoretical maximum. Figure 4.6 shows the time-varying of particular components in relation (4.11). Then, the active sites of the enzyme are almost all occupied by substrates causing saturation, and the reaction rate is determined by the inherent turnover rate of the enzyme. The substrate concentration midway between these two limiting cases is denoted by K_M. According to the principle of microscopic reversibility the nature of a catalyst in itself means that it cannot catalyze just one direction. Let us consider the case of an enzyme that catalysis the reaction in both directions:

$$E + S \underset{k_{-1}}{\overset{k_1}{\rightleftarrows}} ES \underset{k_{-2}}{\overset{k_2}{\rightleftarrows}} E + P \tag{4.12}$$

where k_1, k_2, k_{-1} and k_{-2} are rate constants. The steady-state, initial rate of the reaction is given by

$$v_0 = \frac{d[\mathrm{P}]}{dt} = \frac{(\mathrm{k_1 k_2 [S] - k_{-1} k_{-2} [P]) [E]_0}}{\mathrm{k_{-1} + k_2 + k_1 [S] + k_{-2} [P]}} \tag{4.13}$$

The initial rate v_0 is positive if the reaction proceeds in the forward direction (S $\rightarrow$ P) and negative otherwise. The range of enzyme, substrate and product concentrations are denoted as [S], [E] and [P], respectively. The differences between steady state and pre steady state kinetics is based on the timescale in which the data are collected and the makeup of the reaction components during different stages of the enzyme-catalyzed reaction.

Equilibrium requires that $v = 0$, which occurs when

$$\frac{[P]eq}{[S]\mathrm{eq}} = \frac{k_1 k_2}{k_{-1} k_{-2}} = K_{eq} \tag{4.14}$$

This shows that the *thermodynamics* forces a relation between the values of the 4 rate constants. If measurements are made early in the reaction, then the concentration of products is negligible, i.e. $P \approx 0$ and the back reaction can be ignored. In such conditions, the reaction's rate is

$$\frac{d[P]}{dt} = k_2[ES] \tag{4.15}$$

Note that the first step of the reaction equilibrates rapidly, and thus the change in bound substrate can be described in the following equation

$$\frac{\mathrm{d[ES]}}{\mathrm{dt}} = k_1[E][S] - k_{-1}[ES] = 0 \tag{4.16}$$

This is a main principle in analyzing the kinetics of chymotrypsin and is an omnipresent mechanism in biological enzyme catalysis. Combining all of the above quantities, the catalytic rate constant can be established as

$$k_{cat} = \frac{k_2 k_3}{k_2 + k_3} \tag{4.17}$$

where k_3 denotes the rate constant of acetate formation.

For instance, in ester hydrolysis, $k_3 >> k_2$, so the resultant catalytic rate constant simplifies to:

$$k_{cat} = k_2 \tag{4.18}$$

The above presented relations concerning the enzyme kinetics although derived from single substrate reactions, they can be adapted to the description *Multisubstrate Systems*. Enzymatic reactions with multiple substrates yielding multiple products are more common than single substrate reaction. Multi-substrate reactions proceed complex rate equations that describe how the substrates link up and in what order. For an enzyme *E* that takes two substrates *A* and *B* and turns them into two products *P* and *Q* (corresponding scheme is below), there are two types of mechanism: ternary complex and ping–pong

$$\mathrm{A + B \overset{E}{\rightleftharpoons} P + Q} \tag{4.19}$$

To resolve the enzymes kinetics of these more complicated systems it suffices to keep one of the substrates (for example *B*) fixed, and vary the other substrate (A) and obtain a series of hyperbolic plots of v_0 vs. *A* at different fixed concentrations. This would give a series of linear 1/v vs. 1/A double-reciprocal plots, known as Lineweaver-Burk plots, as well. The pattern of such plots depends on how the reactants and products interact with the enzyme (Fig. 4.6b).

The activity of enzymes can be regulated in ways that either promote or reduce it. There are many different kinds of substances that inhibit or promote enzyme activity, and various mechanisms exist for doing so. For example, an inhibitor molecule can be similar enough to a substrate that it can bind to the active site and simply block the substrate from binding. In such situation the enzyme is inhibited through competitive inhibition, because an inhibitor molecule competes with the substrate for active site binding. Alternatively, in noncompetitive inhibition, an inhibitor molecule binds to the enzyme in a location other than an allosteric site and still manages to block substrate binding to the active site. An inhibitor interacting with an enzyme decreases the enzyme's catalytic efficiency. After removing the inhibitor, the enzyme's catalytic efficiency returns to its normal level. An irreversible inhibitor covalently binds to the enzyme's active site, producing a permanent loss in catalytic efficiency even if the inhibitor's concentration decreases. In a case of competitive inhibition the substrate and the inhibitor rival for the same active site on the enzyme. Then the enzyme's catalytic efficiency for the substrate decreases because the substrate cannot bind to an enzyme–inhibitor complex (EI). When noncompetitive inhibition takes place then the substrate and the inhibitor bind to different active sites on the enzyme. This forms an enzyme–substrate–inhibitor, or ESI complex. In consequence the formation of an ESI complex decreases catalytic efficiency because only the enzyme–substrate complex reacts to form the product. Therefore, in uncompetitive inhibition the inhibitor binds to the enzyme– substrate complex, forming an inactive ESI complex

4.5 BZ Oscillating Reactions

A Belousov–Zhabotinsky reaction, or BZ reaction, is a solution of malonic acid, citric acid, cerium(IV) sulfate, and brominated potassium. In peculiar conditions, the concentration ratio of cerium(IV) and cerium(III) ions changes, making the color of the solution oscillate between a yellow solution and a colorless solution. This is caused by the reduction of cerium(IV) ions by malonic acid into cerium(III) ions, which are then oxidized into cerium(IV) ions by bromate(V) ions. Presently, the BZ reaction is one of a class of reactions that serve as a classical example of non-equilibrium thermodynamics, resulting in the establishment of a nonlinear chemical oscillator. The only common element in these oscillators is the inclusion of bromine and an acid. Such reactions are far from equilibrium and remain so for a significant interval of time and may vary chaotically. It is worth noticing that the BZ reaction occurs in the liquid phase and exhibits spectacular spatio-temporal color variations of the solution [26, 31, 44].

An important feature of the BZ reaction is so called "excitability" because under the influence of stimuli, patterns may develop in what would otherwise be a perfectly quiescent system. In the BZ reaction, the size of the interacting components is molecular and the time scale of the reaction is minutes, but there are specific systems as, for example, the soil amoeba, when the size of the elements is typical of single-celled organisms and the times involved are on the order of days to years. In a closed system the BZ reaction can generate up to several thousand oscillatory cycles, which permits studying chemical waves and patterns without constant replenishment of reactants. Frequently the chemical rate equations for an oscillating Belousov-Zhabotinski reaction are written as

$$\begin{array}{ll}
\mathrm{BrO_3^- + Br^- \to HBrO_2 + HOBr}, & \mathrm{Rate} = k_1\left[\mathrm{BrO_3^-}\right]\left[\mathrm{Br^-}\right] \\
\mathrm{HBrO_2 + Br^- \to 2HOBr}, & \mathrm{Rate} = k_2[\mathrm{HBrO_2}]\left[\mathrm{Br^-}\right] \\
\mathrm{BrO_3^- + HBrO_2 \to 2HBrO_2 + 2M_{OX}}, & \mathrm{Rate} = k_3\left[\mathrm{BrO_3^-}\right][\mathrm{HBrO_2}] \\
\mathrm{2HBrO_2 \to BrO_3^- + HOBr}, & \mathrm{Rate} = k_4[\mathrm{HBrO_2}]^2 \\
\mathrm{OS + M_{OX} \to \tfrac{1}{2}CBr^-}, & \mathrm{Rate} = k_5[\mathrm{OS}][\mathrm{M_{OX}}]
\end{array} \tag{4.20}$$

where OS represents all oxidizable organic species and C is a constant. Observe that in the third equation, species $HBrO_2$ stimulates its own production, thus the process is autocatalytic. The concentrations of intermediate species $x = [HBrO_2]$, $y = [Br^-]$, and $z = [M_{OX}]$ result from the following reactions

$$\dot{x} = k_1ay - k_2xy + k_3ax - 2k_4x^2,$$
$$\dot{y} = -k_1ay - k_2xy + \frac{1}{2}Ck_5bz,$$
$$\dot{z} = 2k_3ax - k_5bz, \tag{4.21}$$

where $a = \left[BrO_3^-\right]$ and $b = [OS]$ are assumed to be constant, and $[M_{OX}]$ represents the metal ion catalyst in its oxidized form. Introducing dimensionless variables

$$X = \frac{2k_4x}{k_5a}, Y = \frac{k_2y}{k_3a}, Z = \frac{k_5k_4bz}{(k_3a)^2}, \tau = k_5bt, \tag{4.22}$$

but still keeping the same designations yields

$$\frac{dX}{d\tau} = \frac{qY - XY + X(1-X)}{\varepsilon_1},$$
$$\frac{dY}{d\tau} = \frac{-qY - XY + CZ}{\varepsilon_2},$$
$$\frac{dZ}{d\tau} = X - Z, \tag{4.23}$$

where $\varepsilon_1 = k_5b/k_3a$, $\varepsilon_2 = 2k_5k_4b/k_2k_3a$, and $q = 2k_1k_4/k_2k_3$. Taking into account appropriate rate constant $k_p, p = 1, 2, \ldots, 5$ with $a = b = 0.2$ and resulting values $\varepsilon_1 = 0.0099$, $\varepsilon_2 = 2.4802e^{-5}$, $q = 3.1746e^{-5}$ computer simulations have been performed. Selected calculation results are shown in Fig. 4.7. For certain parameter values, system (4.23) has a limit cycle that represents an oscillating Belousov-Zhabotinski chemical reaction, as in Fig. 4.8. Note that the trajectory (x, y) moves quickly up and down along the right and left branches of the limit cycle and moves relatively slowly in the horizontal direction. This takes into account the fast color changes and the time intervals between those changes.

It is worth to emphasize that the course of chemical reactions differ importantly from many other types of dynamical system in that closed chemical reactions cannot oscillate about their chemical equilibrium state. Admittedly the concentrations of some reactants of the solution pass repeatedly through the same value, but the energy-releasing reaction that drives the oscillations courses continuously toward exhaustion, which means that the oscillations will eventually finish.

4.6 Limit Cycle Oscillations in the BZ Reaction

Oscillating chemical systems can easily be characterized by their limit cycles as well as by the temporal profiles of the concentration of the intermediates. In general oscillating phenomena are characterized by several consecutive steps, at least one

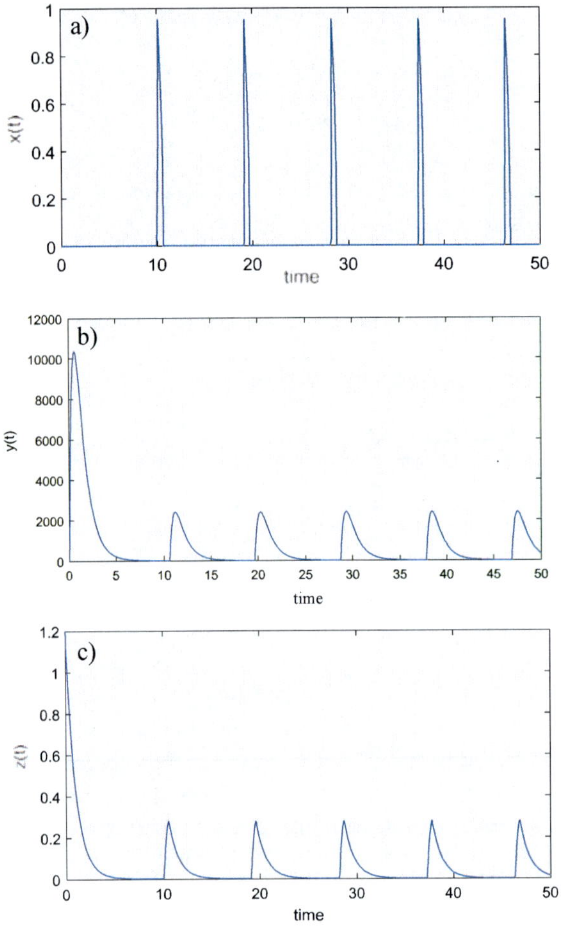

Fig. 4.7 Periodic behavior for the BZ reaction (4.13) when $x(0) = 0$, $x(0) = 0$, $x(0) = 1.2$, $\varepsilon_1 = 0.0099$, $\varepsilon_2 = 2.4802\mathrm{e}^{-5}|$, $q = 3.1746\mathrm{e} - ^{5}$, and C = 1. Note that the direct physical significance is lost and the graph shows relative concentrations of each of the concentrations of ions. State variables: **a** x(t), **b** y(t), **c** z(t)

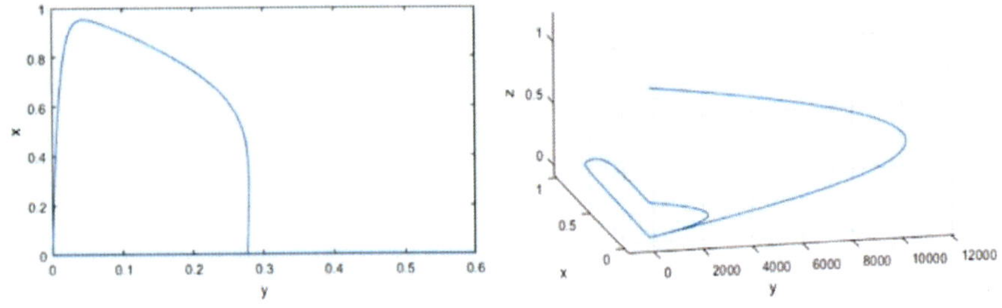

Fig. 4.8 Trajectories of the BZ oscillating reaction: **a** in the plane (x, y), **b** in the space (x, y, z)

feedback process and limit cycles. Limit cycles, or isolated periodic solutions are the most common form of steady state solution, observed when modeling physical systems in the plane. For instance, the differential equation, modeling the oscillation of a violin string, is given by

$$\ddot{x} + \epsilon\left(\frac{1}{3}\dot{x}^2 - 1\right)\dot{x} + x = 0, \tag{4.24}$$

Introducing new variable $\dot{x} = y$ it is possible to rewrite (4.24) into two first order equations as follows

$$\begin{aligned} \dot{x} &= y \\ \dot{y} &= -x - \epsilon\left(\frac{1}{3}y^2 - 1\right) \end{aligned} \tag{4.25}$$

Presenting the steady state solution of the above equation on the plane (x, y) gives the diagram which is shown in Fig. 4.9.

Belousov Zhabotinski (BZ) reactions, are characterized by the reaction between an organic acid substrate that can be easily brominated and oxidized, in the presence of a strong acid, a bromate ion and a transition metal catalyst. In what follows the mechanism of BZ reaction corresponds to any transfer of matter between the system and the surrounding. Moreover the system is also stirred to keep it homogeneous. Hence, it is assumed as a closed stirred system (CSTR). The Oregonator model [34] of the BZ reaction is a kinetic model of FKN mechanism proposed by Field and Noyes. The original three variables irreversible BZ model (4.23) which after appropriate elimination of variable results in a simple kinetics of relevant three variables [44, 45] as

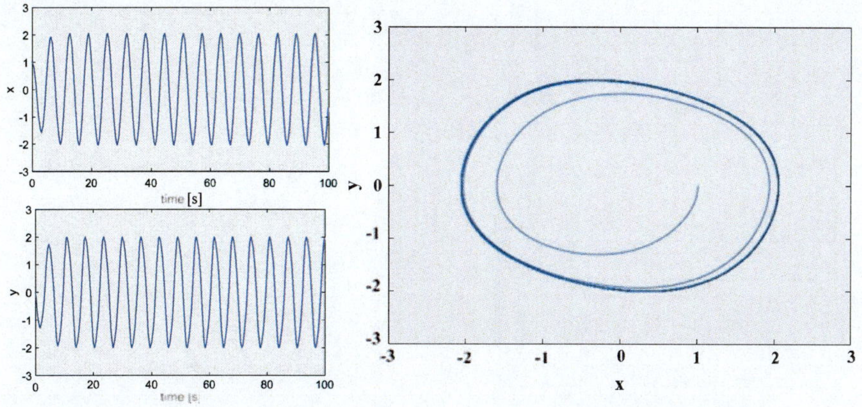

Fig. 4.9 Dynamics of (4.25) for $\epsilon = 0.6$: **a** variable $x(t)$, **b** variable $y(t)$, **c** limit cycle

$$\frac{dx}{d\tau} = \frac{x + y - cx^2 - xy}{a}$$
$$\frac{dy}{d\tau} = -y - xy + 2hz$$
$$\frac{dz}{d\tau} = \frac{x - z}{b}. \tag{4.26}$$

They are the *Oregonator* equations in dimensionless form suitable for computational analysis.

After solving these equations computationally, the values of the x, y, z with time are determined.

Writing x, y, z back in terms of X, Y, Z we get the concentrations of three main components of the BZ reaction at different times.

From the plots of $x(t)$, $y(t)$ and $z(t)$ given in Figs. 4.10a–c the concentrations exhibit oscillations in time. Next, the set of points given in Fig. 4.10d towards which the triplet (x, y, z) approach after a long period shows the phase portrait for this system. Note that ignoring the transients, the phase portrait is a closed loop, i.e. a limit cycle. Therefore, after a relatively short time, the variables (x, y, z) oscillate between the values on this loop only. Hence, by computational process it has been shown that for considered values of parameters and initial conditions, the above system of ODEs modelling the BZ reaction reaches a limit cycle. Note, that stable limit cycle is a structurally stable attractor. It means that any perturbation from it will disappear in time and in that manner it is different from oscillations in conservative systems, trajectories of which depend on initial conditions [40, 44, 45]. Therefore a limit cycle is stable to fluctuations, except fluctuations that are parallel to its trajectory, which will not disappear. This effect is called phase diffusion.

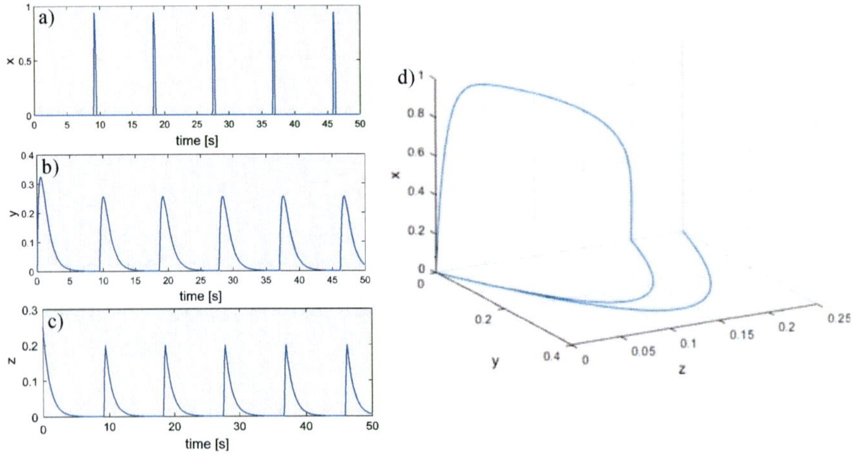

Fig. 4.10 Plots of simulation results: **a–c** dynamical variables, **d** phase trajectory

4.7 Numerical Simulations of Kinetic Bistability of Reactions

Numerical simulation is a powerful tool in understanding complex solution structure of autocatalytic reaction problem in systems like (4.5) involving two chemical species, a reactant A and an autocatalyst species B. Currently, this is met with considerable attention in everyday chemical practice, which is likely to persist over a somewhat longer period, because it allows the use of computer models to identify deficiencies of processes realized in practice. Thanks to accurate computer simulations, there are today possibilities of a detailed identification of the local chemical reaction patterns and of their relations with the vessel instantaneous states of particular reactants in the system. Varying several parameters of the problem leads to obtain bifurcation surfaces in the parametric space that separate parametric regions differing by the number and nature of states and/or their stability. Any change of the parameters of the system gives the bifurcation when the corresponding trajectory in the parametric space intersects a bifurcation surface. In the sequel the main attention is focused on conditions leading to generation of MMOs with organizing the results obtained by straightforward numerical integration.

To illustrate this phenomenon let us consider a continuously stirred tank reactor (CSTR) where the Iodate-Arsenous Acid (IAA) reaction occurs. The kinetics term of the IAA reaction under the conditions presented above may be written, with a trivial change of variable as $f(x) = ax + bx^3 + c$. If a kinetic coupling with another substance y takes a place, then such system introduces the following description

$$\frac{dx}{dt} = ax + bx^3 + c - y$$
$$\frac{\mathrm{dy}}{\mathrm{dt}} = d(x - gy) \tag{4.27}$$

where constants a, b, c, d and g determine properties of reacting substances and reaction conditions. Here $x = x(t)$ and $y = y(t)$ represent the dynamical variables in the system. To identify the behavior of the IAA reaction the situation where arsenic acid is in stoichiometric excess and when the pH is buffered forms the base for detailed simulations. Solutions of this equation are easy to investigate by numerical approach. We have shown above that there is a subcritical one Hopf branch at $c > 0$ in this model; see Fig. 4.11. On the other hand if the control variable c takes a time-dependent "adaptation term" C_{ad} to the external drive c, governed by equations of the form

$$C_{ad}(t + 0) = C_{ad}(t - 0) - \gamma \tag{4.28}$$

with $\gamma > 0$ and on the order of tenths then the unstable limit cycle is appearing which grows as c decreases and suddenly "explodes," colliding with and destroying the stable limit cycle. Thus, similarly to supercritical canards, subcritical ones can give rise to mixed-mode oscillations (Fig. 4.11). An open system exhibiting kinetic bistability is active at a particular pumping rate in either of two stationary states, and concentrations of the intermediate components may differ by orders of magnitude in each of the states. Kinetic bistability of reactions is of practical importance in chemical production processes which utilize pumped tank reactors, and may also be important in biological systems as an on/off mechanism. It is worth recalling that chemical reactions constitute the base in which biological oscillation originates, and the most numerous and important chemical oscillators are found in living systems [17, 46]. Many researchers are studying chemical systems by investigating the effects of a pharmaceutical on a chemical oscillator and examine the reaction mechanism. Different product concentrations in one stationary state versus another may be of appreciable practical importance in chemical manufacturing processes utilizing pumped tank reactors [47, 48].

More recently, the research interest has been focused on heterogeneous catalytic reactions occurring on solid surfaces. Initial studies in this field concerned such reactions as CO oxidation on Pt and the H_2–O_2 reaction on Pt and Ni. The scarce body of knowledge in this field can be enriched by further in-depth analysis in order to understand the general factors which may result in oscillations, and to simulate the experimental findings, particularly the temporal and spatial variations in rate and surface coverage. At present, the branch of surface kinetics dealing with oscillations is flourishing. Note that due to the Arrhenius-type temperature dependence of the rate constants for elementary steps at nonisothermal conditions, oscillations in the

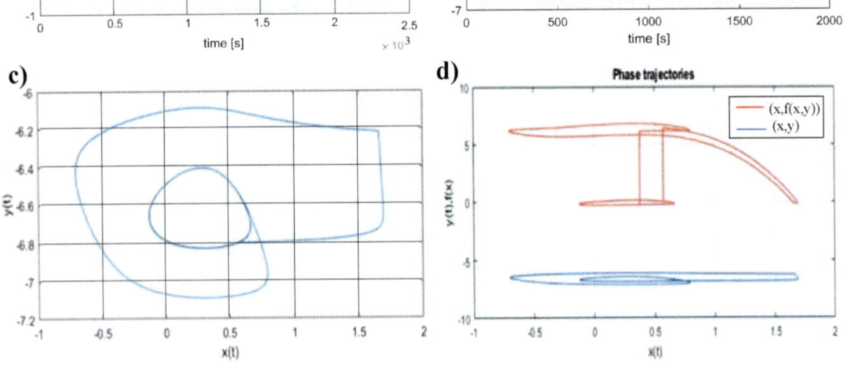

Fig. 4.11 Simulation results of an electrochemical reactor: **a** 1^{13} MMOs, **b** changes of the parameter C_{ad}, **c** limit cycle in the plane (x, y), **d** phase trajectories

reaction kinetics are possible even for the simplest reaction mechanisms. However, if the temperature is constant, oscillatory behavior is restricted to more sophisticated reaction schemes including, e.g., (i) a coverage dependence of the activation energies for some reaction steps, (ii) the activation energy should increase with increasing coverage, (iii) empty-site requirement for the reaction, (iv) buffer steps, (v) oxide or carbon formation, or (vi) surface reconstruction. In all of the above cases the question arises: if oscillatory kinetics, under isothermal conditions, can be produced through the coupling between surface kinetics and diffusion limitations in the gas phase near a catalyst, in the case when the surface kinetics alone cannot produce oscillations? In general, the oscillatory behavior is attributed to nonlinear coupling between different elementary processes. In what follows it is assumed that the type of surface catalyzed reaction is explored to exhibit bistability, and to be coupled with the diffusive mass transport of reactants to the catalyst. The mass transport takes place in the gas phase surrounding the catalyst. In the gas phase surrounding the catalyst takes place the mass transport. The reaction occurs only at the surface of the catalyst. When the surface kinetics exhibit bistability, the rate may suddenly decrease, due to a changing gas-phase composition near the catalyst, caused by diffusion limitations. Then the reactant concentrations near the catalyst will increase again and there exists subsequently a chance to restore the high-reactive state.

To get a bit of quantitative information about the progress of the process consider the $2A+B2 \rightarrow 2AB$ reaction on the surface of a spherical catalyst. The reaction mechanism involves reversible monomolecular A adsorption, irreversible dissociative $B2$ adsorption, and AB reaction between adsorbed species to form product AB molecules which desorbs rapidly. The corresponding mean-field kinetic equations for the adsorbate coverages take the following form

$$
\begin{aligned}
\frac{dQ_A}{dt} &= k_1 p_A(1 - Q_A) - k_2 Q_A - k_{32} Q_A Q_B, \\
\frac{dQ_B}{dt} &= k_4 p_{B2}(1 - Q_A - Q_B)^2 - k_3 Q_A Q_B. \qquad (4.29)
\end{aligned}
$$

where p_A and p_{B2} are the reactant pressures, k_1 and k_4 the rate constants for adsorption, k_2 is the rate constant for A desorption, and k_3 the rate constant for AB formation on the surface, which for simplicity is considered to be irreversible. For realistic values of the rate constants, k_1 and k_4 corresponding to $p \sim 1$ bar, a typical reaction rate in the absence of mass-transfer limitations is about 105 ML/s (ML stands for monolayer). To determine oscillations under such conditions, we need in fact only the steady-state solutions to (4.29). The bistability of these solutions is their most interesting property. Figure 4.12 shows results of computer simulations of heterogeneous catalytic reactions occurring on platinum surfaces at isothermal conditions. If a pressure P_A is given, Eqs. (4.29) have a unique steady-state solution both at sufficiently low and high p_{B2}.

On the other hand, at medium $B2$ pressures, $p^0_{B2} < p_{B2} < p^1_{B2}$ Eq. (4.29) have three solutions, of which the intermediate one is unstable and the other two are stable. The typical critical $B2$ pressures are $p^0_{B2} = 50.11$ bar and $p^1_{B2} = 50.29$ bar at $T =$

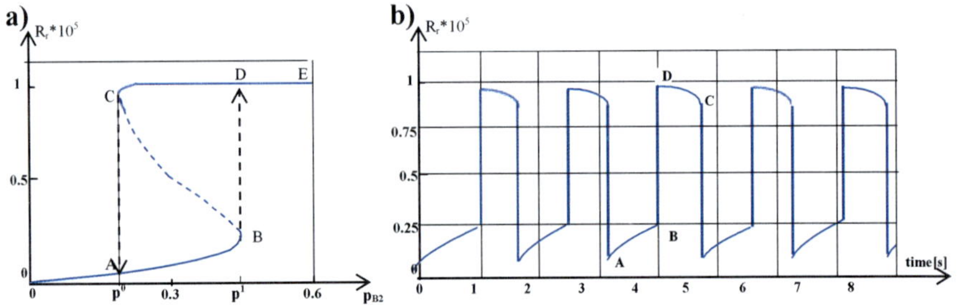

Fig. 4.12 Steady state oscillations in system (4.29): **a** hysteretic changes of the reaction rate, **b** time-varying oscillation of the reaction rate

500 K and $p_A = 1$ bar. The necessary conditions for generation of oscillations can be obtained by analyzing Fig. 4.12a in combination with the following equation for $B2$ molecules

$$J = D\left[n^* - n(R)\right]/R \tag{4.30}$$

where *J, D, n, R* denote the diffusion flux near the surface, diffusion coefficient, gas-phase concentration, pellet radius, respectively. The concentration at $R \to \infty$ is denoted by n^*. Note that the $B2$ pressure far from the surface should be larger than p^1. This condition makes it possible to reach all the bistable points and to create oscillations, provided that the kinetic parameters are of appropriate magnitude. For instance, let consider that the $B2$ pressure in the unperturbed reactant mixture corresponds to the point *D* in Fig. 4.12a, and that at $t = 0$ there are no concentration gradients. In such situation, the reaction rate is initially high (about 105 ML/s with ML standing for monolayer), i.e., the system is in the high-reactive state. With the passage of time, the local $B2$ pressure will decrease due to rapid consumption and limited transport of $B2$, while no gradients are created in p_A, since the *A* diffusion is assumed to be fast. Thus the system will start to move to the left along the line *DB*. In order to produce oscillations, it must reach the point *C*, where a kinetic phase transition will occur to *A* i.e., to the low-reactive state. Physically, this means that the diffusion limitations for $B2$ must be so strong that the steady-state diffusion flux is not able to maintain the high-reactive state on the line *DB*.

If the system reaches the point *C*, the reactive state becomes kinetically unstable, and we will have a rapid transition from the point *C* to point *A* and the system will then be in the low-reactive state. Here, the consumption of $B2$ is much smaller and, provided that D_{B2} is not too small, the local concentration $B2$ will recover. The system then starts to move along the line *EC* immediately after the kinetic phase transition $C \to A$, and may eventually reach the point *B*. The condition for this $A \to B \to D$ path is that the diffusion limitation of $B2$ is not too severe. Then a rapid transition from *D* to *A* takes place, i.e., the system will again be in the high-reactive state and will proceed once more along $D \to C$. Consequently, we have reached a state of repeated oscillations along $D \to C \to A \to B \to D$. Such round trips can be sustained indefinitely if the value of the $B2$ diffusion coefficient is inside the "oscillatory" window. Sustained oscillations of the reaction rate are shown in Fig. 4.12b.

4.8 Mathematical Modeling of Electrochemical Reactors

Electrochemical processes have received recently increasing attention in rapidly growing fields of science and technology, mainly in such domains as nanosciences (nanoelectrochemistry) and life sciences (organic and biological electrochemistry). This has a bearing mainly to such industries as the aerospace, automotive, military, microelectronics, computers, energy, biotechnology, medicine and related others. During the last twenty years, electrochemical processes have become the basis for techniques comprehensively being relevant for the production of nanostructured solid materials and surface layers, which are widely applied in many industries that use the top technologies. Presently, challenges in relations to electrochemical processes used in the practice are becoming more and more expressive in terms of better adapting them to the special technological requirements with meeting the directives for reducing negative impact on the surrounding environment. The electrocrystallization process is exceptionally suitable for manufacturing of specific materials, mainly the protective layers, and in the near future, it may be more effective than previously involved approaches in technologies of surface modification of known materials, as well as newly produced ones [46, 49–51].

Taking into account the specificity of electrochemically produced materials, it is possible the formation of such structures of materials and their properties that cannot be obtained by other technologies. The possibility of conscious control of the structures in the nanometric scale leads up to producing the useful materials with new properties and overcome previously insurmountable barriers to the development of technology. Nanomaterials possess at least one dimension sized from 1 to 100 nm [4, 6] and not only unique geometric, mechanical, electronic and chemical properties, but also properties different from macroscopic materials, such as quantum effect, surface effect, small size effect, etc. These properties have greatly prompted a broad range of applications of nanomaterials in medicine, electronics, biotechnologies, environmental science, energy production and biosensors. All chemical processes in which there is a net change in atomic charge occur as the result of oxidation–reduction reactions also known as redox reactions. They can be represented by a diagram shown in Fig. 4.13.

Electrocrystallization processes generated as a result of oxidation (losing electrons) and reduction (gains electrons) reactions occur at the electrodes: anode and

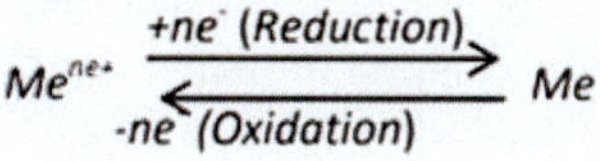

Fig. 4.13 Scheme for redox reaction

cathode, respectively. The processes of oxidation and reduction occur simultaneously and cannot happen independently of one another. Overall, the manufacturing of nanostructured materials is too complex for a correct theory, so it becomes the subject of computer simulations. The most frequently investigated electrocrystallization process is cathodic metal deposition on foreign and native substrates from electrolytes containing simple and/or complex metal ions [3, 12]. Typical examples are the electrocrystallization of Ag from electrolytes containing Ag^+ [13], the cathodic deposition of Ag on *n-Si* from electrolytes containing $[Ag(CN)_2]^-$ ions [3], and the electrodeposition of Cu [4], from electrolyte containing $[Cu_2SO_4]^+$ ions which has recently becoming of significant technological importance for the fabrication of Cu interconnects on integrated circuit chips.

A simplified scheme of appliances being widely applied as electrochemical reactors and automated wafer processing equipment is presented in Fig. 4.14. Electrodes charged by supplying generator are able to carry out chemical reactions for realizations of electrocrystallization processes. A variety of materials meeting the features in the art can be designed by choosing the type and composition of the electrolyte solution and the process parameters such as type and density of the current, temperature, electrolyte mixing, time duration of the process. In establishing the matrix of mechanisms that affect the electrochemical deposition process and its variation a great help can be obtained by applying the mathematical modeling approach. Development of a reaction model is an essential first step to obtaining a reactor model, which in turn is a suitable tool for process optimizations. It has to be noted that the electrochemical deposition process appears as a competition between nucleation and grain growth of the produced materials. In charging and discharging of a pulse ion flux, especially for short pulses, the double layer of the interface between electrolyte and cathode distorts the pulse current [10], and affects the over potential response acting on the electrolyte. The results of the experimental electrochemical impedance spectroscopy [8] show that both the reactor current frequency and the magnetic flux density are important factors for efficiency of electrochemical reactors. The Lorentz force acting on the electrolyte in a double layer at electrode increases as the current density increases and improves the flow and conductivity of the electrolyte, what increases the efficiency of electrochemical deposition in terms of accuracy and smoothness of the surface of the layer. When the supplying current is time varying, the charged particles are attracted to each other under the influence of the Lorentz force, which causes shrinkage of the charged particle flux, i.e. the pinch effect appears. This effect leads to increase the density and pressure of ions. A current flowing through the electrolyte squeezes smaller cylinder, until the magnetic force does not balance the pressure. The magnetic properties of electrolytes and the interactions of internally induced magnetic fields on electrolyte properties and ionic transport characteristics have significant influence on the form of the mathematical model of the processes occurring in the electrochemical reactor. It should be based on the questions concerning the changes for mass, energy and momentum.

The instantaneous state of an electrochemical reactor usually depends on several variables and is described by one or several state functions of one or several variables. For instance, considering the electrode variables we must take into account the kind

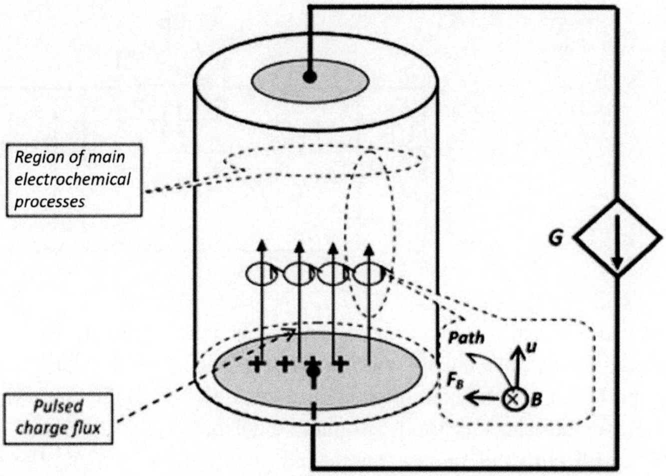

Fig. 4.14 Scheme of an electrochemical reactor: G—pulsed current generator, u—charge drift velocity, B—magnetic field density, FB—Lorentz force

of the material, surface area, geometric and surface conditions. Such factors as mode transfer (diffusion, convection, migration, …), surface concentrations and adsorptions have influences on the mass transfer variables. The temperature, pressure and time appear as external variables. Moreover, there are electrical variables such as the charge, current and potential. As solution variables we have the bulk concentrations of electroactive species, concentrations of other species and the kind of the solvent. It can be easily verified that the aspect of current distribution can have a major effect on the performance of electrochemical processes. The derivation of a mathematical model taken from diffusion-migration transport equations (the Maxwell equations and mass balance equations) is an error prone and time-consuming tasks normally requiring specific expertise. Following appropriate procedure it is possible to select the variables that govern performance of the reactor. A start point leads to considering simulations of electrochemical depositions of nanostructures, and will then turn toward current–voltage relations to gain insights into the charge—electrolyte concentration and coupling mechanisms ultimately responsible for different pattern of the reactor output quantities and to identify the relationship between them. It should be emphasized that the circuit modeling not only aids manufacturers of nanostructural species in immediate returns on investment like increased yield and decreased consumable costs, but circuit models are essential to understand the mechanisms that enable of material and cause process variation, which affects device performance and process efficiency. Therefore, by imposing appropriate restrictions which are satisfied under corresponding electroplating conditions it is possible elaborate models which lead to satisfied results when matching them with experimental data. Such models have a two-fold purpose: first, they can account all the facts discovered experimentally, and second, they can be able predict the system behavior under various

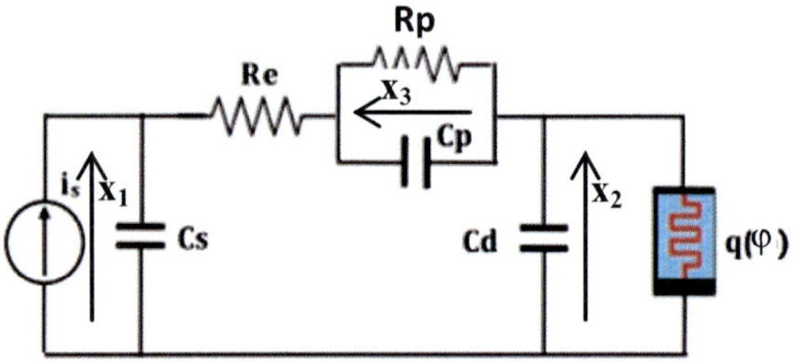

Fig. 4.15 Circuit model of electrochemical deposition processes

conditions of process realizations. Starting from this general imperative, the elaboration of a model for the interface phenomena can be performed by making a certain number of hypotheses which taken into account generally yield the governing equations. The representative circuit model predicts the capacitive and Faradaic currents of the process, and with the same average and peak current density, the ramp up waveform has higher instantaneous peak current to charge transfer, which results in an improvement in the microstructure of the nanocomposites. The circuit model depicted in Fig. 4.15 makes it possible to quantify the evolution of the electrochemical reactor processes with respect to time.

Applying circuit laws we can establish the mathematical description of the above model and obtain

$$\begin{aligned}
\dot{x}_1 &= k_1\left(\frac{x_2 + x_3 - x_1}{R_e} + i_s\right),\\
\dot{x}_2 &= k_2\left(-\frac{x_2 + x_3 - x_1}{R_e} - \left(a + 3bx_4^2\right)x_2\right)\\
\dot{x}_3 &= k_3\left(-\frac{x_2 + x_3 - x_1}{R_e} - \frac{x_3}{R_p}\right),\\
\dot{x}_4 &= x_2
\end{aligned} \tag{4.31}$$

where $k_1 = 1/C_s$, $k_2 = 1/C_d$, $k_3 = 1/C_p$ and $x_4 = \varphi$. The dot is taken as a symbol for differentiation with respect to time. The charge–magnetic flux characteristic of the memristive element representing the relation between the charge and its own magnetic field takes the form: $q(\varphi) = a\varphi + b\varphi^3$, where a and b denote constant parameters. The established circuit model can lead to methods for fast treating large quantities of electrocrystallizator data and extracting from them kinetic parameters importantly influenced the structure of the produced materials. Two types of periodic supplying currents play a significant role in the electrodeposition of nanostructured coatings: bipolar pulse current and sinusoidal rectified current. Both were taken into account in the sequel. To examine the effect of various system parameters on the electrocrystallization process, we can perform numerical calculations applying a computer program MATLAB with using effective numerical integration procedures. Results of performed computer simulations of electrocrystallization processes with the focus on mixed mode oscillations (MMOs) as the dynamical switches between small amplitude oscillations (SAOs) and large amplitude oscillations (LAOs) are presented in the next section.

4.9 MMOs in Electrochemical Reactors

The modeling of the nucleation and material growth mechanism during the electrocrystallization process is extremely important due to its use in effective production of thin layers of pure metal or in the form of alloys and nanostructured composites with a high degree of precision. Generally, the manufacturing of nanostructures is too complicated for proper theory because the differential equation (4.31) are nonlinear, their analytical solution is often unknown and one has to use the computer to solve the problem numerically. From the extensive set of solutions, in what follows the attention will be focused on the particular type of voltage and current waveforms that can be accordingly transformed on the effects of the electrochemical processes. The three basic modes of operation are illustrated in Fig. 4.15 which shows the solutions of (4.31) with $I_s \in \{0.15, 1.5, 15\}$ A, $a \in \{-1.25, -1.75, -2.75\}$, $b = 0.15$ and $T \in \{1, 2.5\}$ s. In the SAOs only case, the small amplitude oscillations around the origin (0, 0, 0) are due to Hopf bifurcation for $a = 0$. In the LAOs only case, a trajectory passing close to the origin bypasses the region of small amplitude oscillations. The MMOs case is in some sense a combination of the previous two cases. The mechanism in which SAOs and LAOs occur is quite complex and has been the topics of [14–16]. In the MMOs case, a series of SAOs around the origin (considered canard solutions) undergoes a rapid canard explosion yielding an LAO, which, through a special return mechanism brings back the system into the vicinity of the origin. The canard explosion described in detail in [14] is triggered when a trajectory leaves a fold point of a cubic nonlinearity ending a series of SAOs and entering the relaxation mode with one or more LAOs. This explosion occurs, for example, in Fig. 4.16, bottom right part, when two or four SAOs transform into an LAO. Depending on the parameters, the system may continue with one (or more) LAOs, or may go through a new series of SAOs after which trajectory leaves again the vicinity of the origin (a fold) and the phenomenon repeats.

Taking into account a small amount of perturbation in the characteristic $q(\varphi)$ it is possible to get MMOs exhibiting in each period a train of SAOs separated by a large amplitude pulse (Fig. 4.17a).

Varying pairs of control parameters, one can observe how the dynamical behaviour of the system changes and map out "dynamical phase diagrams" showing regions of the parameter space in which qualitatively different MMOs occurs, or, by varying a single parameter, one can trace out hysteresis loops of the sort shown in Fig. 4.17b. This type of time-varying electric currents and magnetic fields are most probably generated by an electric double layer within individual particles and a temporal excess of ion carriers within the sample. Perhaps the shape and magnitude of the induced temporal signals depend on the reaction zone propagation mode, reaction mechanism, and reactant properties. Therefore, by imposing appropriate restrictions,

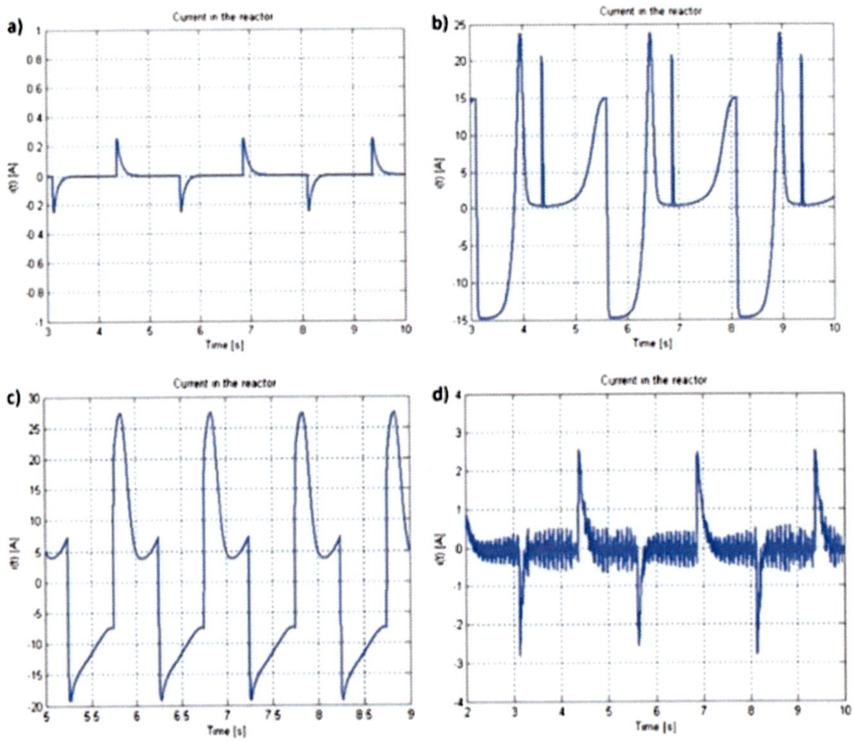

Fig. 4.16 Solutions of (4.31) for various sets of parameters I and b with unchanged other parameters: **a** SAOs at $I_s = 0.15$ A, $a = -1.75$ and $b = 0.15$, **b** LAOs at $I_s = 15$ A, $a = -2.75$ and $b = 0.15$, **c** LAOs at $I_s = 15$ A, $a = -1.25$ and $b = 0.15$, **d** MMOs at $I_s = 1.5$ A, $a = -1.75$, $b = 0.15$ and $T = 2.5$ s

which are satisfied under corresponding electroplating conditions, it is possible elaborate models which lead to satisfied results when matching them with experimental data. This is of great importance the case of manufacturing nanostructured materials whose final structure is very sensitive to rapid changes in the conditions of the manufacturing process. A particularly interesting context is that of current waveforms supplying the electrocrystallizator because their influences on mixed mode oscillations seem most important for the efficiency of realized processes (Fig. 4.17).

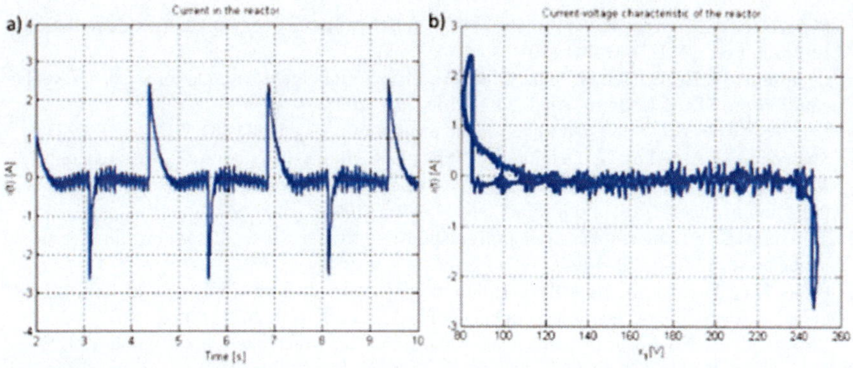

Fig. 4.17 MMOs at $I_s = 1.5$A (bipolar pulse), b = 0.5 and $T = 2.5$: **a** reactor current, **b** voltage–current characteristic of the reactor

References

1. Silberberg, M.: Chemistry—The Molecular Nature of Matter and Change. McGraw-Hill Science/Engineering/Math, New York (2008)
2. Epstein, I.R., Pojman, J.A.: An Introduction to Nonlinear Chemical Dynamics: Oscillations, Waves, Patterns, and Chaos. Oxford University Press, New York (1998)
3. Laganà, A., Gregory, A.P.: Chemical Reactions. Basic Theory and Computing. Springer, New Yok (2018)
4. Trzaska, M., Trzaska, Z.: Nanomaterials produced by electrocrystallization method. In: Aliofkhazraei, M., Makhlouf, A. (eds.) Handbook of Nanoelectrochemistry. Springer, Cham (2016)
5. Lona, F., Maria, L.: Step by Step Approach to the Modeling of Chemical Engineering Processes. Using Excel for Simulation. Springer, New Yok (2018)
6. Kreysa, G., Ota, K.-I., Savinell, R.F. (eds.): Encyclopedia of Applied Electrochemistry. Springer, New York (2014)
7. Cieślak, G., Trzaska, M.: Tribological properties of nanocomposite Ni/graphene coatings produced by electrochemical reduction method. Compos. Theory Pract. **2**, 79–83 (2016)
8. Trzaska, M., Trzaska, Z.: Electrochemical Impedance Spectroscopy in Materials Engineering. Publisher Office of the Warsaw University of Technology, Warsaw (2010)
9. Sangeetha, S., Kalaignan, G.P., Anthuvan, J.T.: Pulse electrodeposition of self-lubricating Ni–W/PTFE nanocomposite coatings on mild steel surface. Appl. Surf. Sci. **359**, 412–419 (2015)
10. Kamnerdkhag, P., Free, M.L., Shah, A., Rodchanarowan, A.: The effects of duty cycles on pulsed current electrodeposition of ZneNieAl$_2$O$_3$ composite on steel substrate: microstructures, hardness and corrosion resistance. Int. J. Hydrogen Energy **42**, 20783–20790 (2017)
11. Abdul, J., Yasin, G., Khan, W.Q., Anwar, M.Y., Korai, R.M., Nizam, M.N., Muhyodin, G.: Electrochemical deposition of nickel graphene composite coatings: effect of deposition temperature on its surface morphology and corrosion resistance. Royal Soc. Chem. **7**, 31100–31109 (2017)
12. Djokic, S.S.: Electrodeposition: Theory and Practice. Springer, New York (2010). https://doi.org/10.1007/978-1-4419-5589-0
13. Sagués, F., Epstein, I.R.: Nonlinear chemical dynamics. Dalton transactions (March 2003) 14. Focus issue: mixed-mode oscillations: experiment, computation, and analysis. Chaos **18**, 015101 (2008)
14. Marszalek, W., Trzaska, Z.W.: Mixed-mode oscillations in a modified Chua's circuit. Circ. Syst. Signal Proc. **29**, 1075–1087 (2010)

15. Marszalek, W., Trzaska, Z.W.: Properties of memristive circuits with mixed-mode oscillations. Electron. Lett. **51**(2), 140–141 (2015)
16. Beardmore, R.E., Laister, R.: The flow of a differential-algebraic equation near a singular equilibrium. SIAM J. Matrix Anal. **24**(1), 106–120 (2002)
17. Inaba, N., Kousaka, T.: Nested mixed-mode oscillations. Physica D **401**, 132–152 (2020)
18. Shabunin, A., Astakhov, V., Demidov, V., Provata, A., Baras, F., Nicolis, G., Anishchenko, V.: Modeling chemical reactions by forced limit-cycle oscillator: synchronization phenomena and transition to chaos. Chaos Solit. Fract. **15**, 395–405 (2003)
19. Tornheim, K.: Are metabolic oscillations responsible for normal oscillatory insulin secretion? Diabetes **46**, 1375–1380 (1997)
20. Chou, H.F., Berman, N., Ipp, E.: Oscillations of lactate released from islets of Langerhans: evidence for oscillatory glycolysis in beta-cells. Am. J. Physiol. **262**, E800–E805 (1992)
21. Strogatz, S.H.: Nonlinear Dynamics and Chaos, 2nd edn. Westview Press, Boulder (2015)
22. Gray, P., Scott, S.K.: Chemical Oscillations and Instabilities. Oxford Clarendon Press, Oxford (1990)
23. Physical Chemistry for the Biosciences. UC Davis Office of the Provost, California State University (2023)
24. Chance, B., Pye, E.K., Ghosh, A.K., Hess, B.: Biological and Biochemical Oscillators. Academic Press, New York, 1973; Selko'v, E.E.: Self-oscillations in glycolysis model. Eur. J. Biochem. **4**, 79–86 (1968)
25. Arnold, N.: Chemical Chaos. Hippo, London (1997)
26. Zhabotinskii, A.M.: Biological and Biochemical Osciihtws. Academic Press, New York (1972)
27. Zhabotinsky, A.M., Buchholtz, F., Kiyatkin, A.B., Epstein, I.R.: J. Phys. Chem. **97**, 7578 (1993)
28. Zhang, D., Gyorgyi, L., Peltier, W.R.: Deterministic chaos in the Belousov-Zhabotinsky reaction: experiments and simulations. Chaos **3**(4), 723–745 (1993)
29. Ali, F., Menzinger, M.: Stirring effects and phase-dependent inhomogeneity in chemical oscillations: the Belousov-Zhabotinsky reaction in a CSTR. J. Phys. Chem. A **101**, 2304–2309 (1997)
30. Zhabotinsky, A.M.: Belousov-Zhabotinsky reaction. Scholarpedia **2**, 9, 1435 (2007)
31. Belousov, B.P.: Periodically acting reaction and its mechanism. Сборник рефератов по радиационной медицине. **147**, 145 (1959)
32. Rachwalska, M.: Mixed mode, sequential and relaxation oscillations in the BelousovZhabotinsky system. Zeitschrift Naturforschung **62a**, 41–55 (2007)
33. Lynch, D.T., Rogers, T.D., Wanke, S.E.: Chaos in a continuous stirred tank reactor. Math. Model. **3**, 103–116 (1982)
34. Feng, J.M., Gao, Q.Y., Li, J., Liu, L., Mao, S.C.: Current oscillations during the electrochemical oxidation of sulfide in the presence of an external resistor. Sci. China Ser. B Chem. **51**(4), 333–340 (2008)
35. Green, B.J., Wang, W., Hudson, J.L.: Chaos and spatiotemporal pattern formation in electrochemical reactions. Forma **15**, 257–265 (2000)
36. Bartuccelli, M., Gentile, G., Wright, J.A.: Stable dynamics in forced systems with sufficiently high/low forcing frequency. Chaos **26**, 083108 (2016)
37. Sriram, K., Bernard, S.: Complex dynamics in the oregonator model with linear delayed feedback. Chaos **18**, 023126 (2008)
38. Segel, I.H.: Enzyme Kinetics: Behavior and Analysis of Rapid Equilibrium and Steady-State Enzyme Systems. Wiley, New York (1993)
39. Field, R.J.: Oregonator. Scholarpedia **2**, 5, 1386 (2007)
40. Monzon, L.M., Coey, J.M.D.: Magnetic fields in electrochemistry: the Lorentz force. A mini-review. Electrochem. Commun. **2014**, 1–13 (2014)
41. Estevez-Rams, E., Estevez-Moya, D., Aragon-Fernandez, B.: Phenomenology of coupled nonlinear oscillators. Chaos **28**, 023110 (2018)
42. Willy, C., Neugebauer, E.A.M., Gerngroß, H.: The concept of nonlinearity in complex systems. Eur. J. Trauma **200**(1), 11–22

43. Lapicque, F.: Electrochemical reactors. In: Chemical Engineeering and Chemical Process Technology, vol. III. Encyclopedia of Life Support Systems (2011)
44. Belousov, B.P.: In Oscillations and Travelling Waves in Chemical Systems. Wiley, New York (1985)
45. Epstein, I.R., Showalter, K.: Nonlinear chemical dynamics: oscillations, patterns, and chaos. J. Phys. Chem. **100**, 13132–13147 (1996)
46. Białostocka, A.M.: The electrochemical copper structure forming in the presence of the magnetic field. Electrotech. Rev. **89**(10), 254–256 (2013)
47. Trzaska, Z.: Mixed Mode Oscillations (MMOs). Springer, Cham Switzerland (2021)
48. Goryachev, A., Strizhak, P., Kapral, R.: Slow manifold structure and the emergence of mixed-mode oscillations. J. Chem. Phys. **107**, 2881–2889 (1997)
49. Goldbeter, A.: Biochemical Oscillations and Cellular Rhythms: The Molecular Bases of Periodic and Chaotic Behaviour. Cambridge University Press, Cambridge, UK (1996)
50. Sulaymon, A.H., Abbar, A.H.: Scale-up of electrochemical reactor. In: Electrolysis, Chapter 9. InTech, Rijeka (2011)
51. Drews, T.O., Krishnan, S., Alameda, J.C., Jr., Gannon, D., Braatz, R.D., Alkire, R.C.: Multiscale simulations of copper electrodeposition onto a resistive substrate. IBM J. Res. Dev. **49**(1) (2005)

Chapter 5
Oscillations in Biological Processes

5.1 Motivation, Brief History and Background

Biosystems are extremely myriad natural systems of great diversity. All biosystems are extremely complex, i.e. highly organized, polyphasic and store enormous amounts of information. The idea of understanding how the biosystem is working has fascinated almost all societies for many centuries. Nowadays, thanks to achievements of informatics and modern scanning techniques, it has become possible to generate superb images of such structural components and processes in biosystems as neurons and fibers, blood flow and energy metabolism, and changes in neuronal activity. Advanced information systems, such as living cells, provide the information needed to transform incoming energy into useful work—transformation under certain circumstances impossible in the case of physical systems that lack comparable levels of organization. Automation and miniaturisation are particularly important at the interfaces between chemistry, biotechnology, biology and medicine and it is likely that the rise of genomic and post genomic technologies will drive the need for chemical and biochemical reactions on surfaces. There have been many successful applications in diagnostics, omics studies, sensors, plant sciences, enzyme kinetics, enzyme specificity, cell biology, and organic (synthetic) chemistry [1–11]. Many biological systems consist of networks of coupled oscillators generating a large multiplicity of widespread behaviors that include periodic, quasiperiodic, and aperiodic rhythms. They putatively underlie empirical phenomena such as physiological and pathophysiological oscillations, cardiac arrhythmia, fluid turbulence, spiral waves and cluster patterns in excitable media, and deterministic chaos in many forms. Using principles from chemistry, physics and mathematics with applications of informatics, it can be seen that the highly complex behavior of biological systems is caused by a multitude of coupled feedback and feed-forward loops in the underlying auto control neuronal networks.

Realized up to date biological research has accumulated a powerful amount of detailed and inestimable knowledge about the structure and function of the brain.

Z. Trzaska, *Mathematical Modelling and Computing in Physics, Chemistry and Biology*, Studies in Systems, Decision and Control 495,
https://doi.org/10.1007/978-3-031-39985-5_5

The human brain is the most complex organ of the body and probably the most complicated system in the nature. It is a unrepeatable organ that thinks and feels, produces behavioral interactions with the environment, maintains bodily physiology relatively stable, and enables regeneration of the species being its most important role from evolution's impressive perspective, capable of exchanging genes or interbreeding. The elementary active units in the central nervous system are neurons, which are connected to each other in an entangled complex (Fig. 5.1a). A neuron network diagram (Fig. 5.1b) of cells (blue) and axons (red), with axon extent indicating its strength, portrays the information streamline density linking each cell pair. For clarity, only the strongest axons are shown in Fig. 5.1b. Neurons are formed up of a cell body connected with two types of protrusions: axons transferring stimuli and dendrites receiving information (Fig. 5.1c). An important role in passing and getting impulses between neurons play special connections called synapses. There are three main types of neurons: sensory neurons, motor neurons and interneurons. Sensory neurons are linked to receptors responsible for getting and reply to various internal and external environment stimuli. They are sensitive to changes in lighting, sound, and they support mechanical and chemical stimulation the senses of sight, hearing, touch, smell and taste. Motor neurons control muscle activity, and are responsible for all forms of behavior, including speech. The activities of interneurons are located between those that characterize sensory and motor neurons. Their role lays in connection two brain regions, i.e. not direct motor neurons or sensory neurons. Interneurons are the central nodes of neural networks, enabling communication between sensory or motor neurons and the central nervous system. They mediate in the transmission of the simplest stimuli, but they are also responsible for the most complex brain functions. Support glial cells have a great influence on the development of the nervous system as well adult brain functioning. They are a lot more numerous than neurons, but they do not conduct stimuli in a characteristic way for the former ones. The central nervous system, consisting of the brain and spinal cord, functions to receive, interpret, and respond to the nerve impulses of the peripheral nervous system. The peripheral nervous system serves as the connection between the central nervous system and the body, transporting sensory and motor information to and from the central nervous system, respectively. Neurons are organized into complex chains and networks, which are the routes of transmitting information in the nervous system. Each neuron can connect up to 10,000 other neurons. The entire nervous network is in a state of continuous electrical and chemical activities. Approximately mentioning, the nervous system is built mainly from nerve cells called neurons and glial cells as supporting cells, including Schwann cells, satellite cells, oligodendrocytes, microglia, astrocytes, and ependymal cell. The human brain is made of neural networks, contains approximately 100 billion nerve cells, 3.2 million kilometers of "wiring" formed by axons and dendrites, one million billion connections (synapses), and a volume of 1.5 L, weighs about 1.5 kg, and works with power only 10 W. Understanding the development and homeostasis of the nervous system, the many factors responsible for maintaining and regulating its proper functionality, and the implications of nervous system abnormality, remain most important within research concerning the nervous system and neurodegenerative diseases. Research

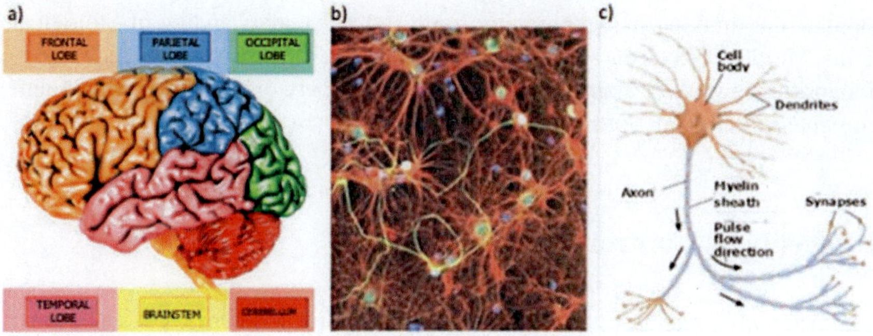

Fig. 5.1 Components of the human central nervous system: **a** brain, **b** bit of neural network, **c** single neuron

aimed at the biological processes of the nervous system and their influences on both human behavior and activity offers extensive possibilities for the comprehension as well as treatment, of various neurological diseases and ailments, such as Alzheimer's disease, multiple sclerosis, Parkinson's disease and epilepsy. The idea of understanding how the brain is working has fascinated almost all societies for many millennia and mainly during last centuries [12–15]. Mathematical modeling has a great potential for increasing our understanding of the physiological processes underlying the function of the human body. For example, modeling of the electrical properties of excitable neuron cells may provide insight into the complex electrical signaling revealed in a number of important activities, like transfer of information through neurons and coordination of the cardiovascular system.

Oscillations in neuronal networks have been of great interest to neuroscientists for many decades. They arguably represent the simplest form of coherent network behavior in the brain. One of the most interesting facts about brain oscillations is that they provide a possible mechanism by which Hebbian cell assemblies, thought to be the basic unit of information storage in the brain, can be created and held together. The so-called Hebbian cell assemblies are produced, in accordance with the well-known hypothesis of synaptic plasticity and memory [16–18], by self connecting neurons that strengthen synapses depending on their neuronal activity [16]. These sets of cells are believed to form the basis of long-term memory. It should be emphasized that models as simple as those discussed in this chapter obviously don't come close to reproducing the complexities of real human brain. The complexity of the brain is astonishing, and we currently cannot follow it faithfully in mathematical models and computational simulations. The presentation of the biological background in this chapter is therefore highly selective and focuses on those aspects that are needed to highlight the biological background of the theoretical work presented in this book. For an in-depth discussion of neurobiology, the reader is referred to the literature listed at the end of the book chapter. It is worth noting that many functions of the brain are simply not well understood experimentally, and computers are not (yet) powerful enough to reveal all significant problems. But, during the last 50 years computational thinking has importantly improved our understanding of biological systems. Moreover, in recent years major steps have been taken to understand biological algorithms by synthesizing biological regulatory networks de novo, which aim to compute specific functions [19–21].

5.2 Feedback Control Mechanisms

The human body's responses to adverse stimuli, ensuring maintenance of an optimal physiological environment control the homeostasis. Mechanisms of the homeostatic control have at least three interdependent components: a receptor, integrating center, and effector. The receptor feels environmental stimuli, and sends the information to the integrating center. Then the integrating center, generally a region of the brain called the hypothalamus, signals effectors (e.g. muscles or an organ) to respond to the stimuli. In a biological sense, a feedback mechanism involves a biological process, a signal, or a mechanism that tends to initiate (or accelerate) or to inhibit (or slow down) a process. The feedback principle can be illustrated by the block diagram shown in Fig. 5.2a and a chain of activities shown in Fig. 5.2b.

Two types of feedback mechanisms may function, depending on whether the input changes or the physiological parameters deviate from their limits. It may reveal itself positive feedback enhances or accelerates output of a system created by an activated stimulus (Fig. 5.3). Platelet aggregation and accumulation in response to injury is an example of positive feedback. While the negative reaction brings an organism back to its normal functional level. Adjustments of blood pressure, metabolism, and body temperature are all negative feedback. A system corresponding to element connection shown in Fig. 5.2 forms a structure known as feedback loop. Feedback loops are commonly divided into two main types: opened-loop mechanism and closed-loop mechanism.

A positive feedback mechanism or positive feedback homeostasis is a pathway that, in response to an output variation, causes the output to vary even more in the direction of the initial deviation. A positive feedback system amplifies deviations and causes output state changes. Because it distances the biological process from homeostasis, its reaction gradually increases. A single component that amplifies its

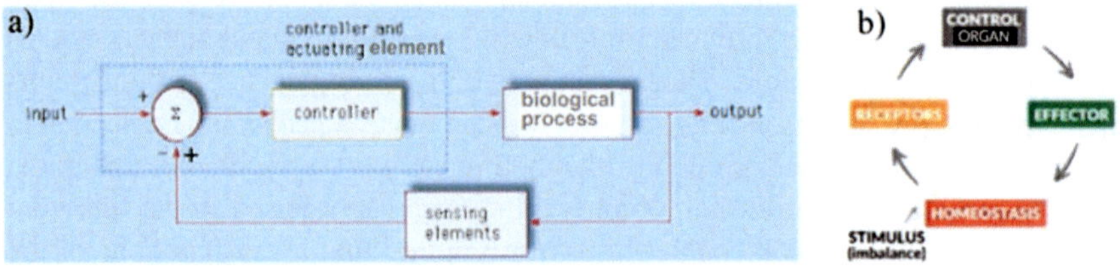

Fig. 5.2 Illustration of the feedback principle: **a** feedback loop, **b** chain of activities in a feedback loop

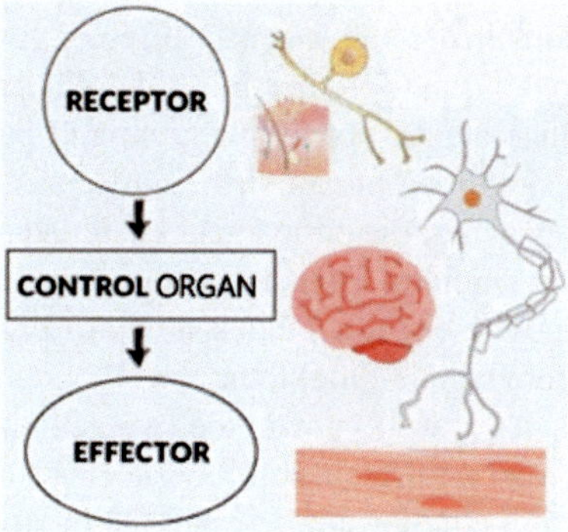

Fig. 5.3 The flow of signals from the receptor to the effector

own activity or numerous components with direct and indirect interactions might make up a positive feedback loop. Positive feedback loops in biological processes are common in processes that need to happen fast and efficiently, as the output tends to magnify the stimulus' influence. Positive mechanisms are rare in living systems such as the human body, but they can be found in the environment, such as in the instance of fruit ripening.

It is worth noticing that there are positive feedback loop examples resulting in uncontrolled conditions since a change in an input generates reactions that cause further modifications in the same manner. Even if the components of a loop (receptor, control center, and effector, Fig. 5.3) are not immediately recognizable, the term "positive feedback" is widely accepted when a variable has the ability to increase itself [22]. However, positive feedback is often damaging, and there are a few occasions where it can help people function normally when used in moderation.

It should be mentioned that both in nature and in technology there are usually systems with negative feedback loops occurring when a change in one direction produces a change in the other. For instance, a rise in a substance's concentrations produces feedback, which causes the substance's content to reduce. Negative feedback loops are mechanisms that seem to be naturally stable. When combined with the many stimuli that can affect a system variable, negative feedback loops usually result in the value oscillating about the set point. Negative feedback loop examples include temperature and blood glucose level regulation.

Between systems that illustrates the roles of feedback and feedforward control in nature are simple bacteria intrinsically closed-loop systems that exist in biological circuits, from gene level through cellular level. These biophysical systems display the same rich character as those encountered in process systems engineering: multivariable interactions, complex dynamics, and nonlinear behavior [23, 24].

A representative example of networked biological control is the circadian clock, which coordinates daily physiological behaviors of most organisms. The term circadian comes from the Latin meaning "approximately one day," and the circadian clock is vital to regulation of metabolic processes in everything from simple fungi to humans. These networks have attracted a great deal of attention at the level of gene regulation, where dozens of input connections may characterize the regulatory domain of a single gene in an eukaryote, as well as at the protein level, where hundreds to thousands of interactions have been mapped in protein interactome[1] diagrams that illustrate the potential coupling of pairs of proteins. To focus the attention consider a simplified model of the *Drosophila melanogaster*[2] circadian clock involving two key genes: the period gene (denoted *per*) and the timeless gene (denoted *tim*). Recall that the term circadian comes from the Latin and means "approximately one day," and the circadian clock is vital to regulation of metabolic processes in everything from simple fungi to humans.

As is well known, the two genes can be lumped together, as well as their corresponding proteins and the nuclear and cytoplasmic forms of the dimer. Assuming rapid equilibrium between the monomer and dimer, a second-order set of balances can be developed for the mRNA state M and the protein state P. The dynamics of the feedback controlled circuit is captured by the resulting pair of differential equations

$$\begin{aligned} \frac{dM}{dt} &= \frac{v_m}{1+(P(1-z)/2P_{cr})^2} - k_m M, \\ \frac{dP}{dt} &= v_p M \frac{k_{p1} z P + k_{p2} P}{J_p + P} - k_{p3} P \end{aligned} \tag{5.1}$$

where

$$z = \frac{2}{1 + \sqrt{1 + 8k_{eq} P}} \tag{5.2}$$

denotes the transcription rate on the protein concentration P. System parameters: v_m, v_p, k_m, k_{p1}, k_{p2}, k_{p3}, J_p, P_{cr} are constant computer simulation with using such program package as Simulink/MATLAB and the defined system parameters, with initial values of M and P equal to $[2.0;\ 2.0]'$, it was possible the gene regulatory circuit determining. Results of computing for $v_m \in \{1;\ 1.1;\ 1.5;\ 2;\ 4\}$ are shown in Fig. 5.4. It is clearly evident that the period of the clock lengthens as v_m is increased. At the extreme value of $v_m = 4.0$ the system settles to a stable equilibrium and the stability of the oscillations is quite remarkable for such large perturbations in v_m. If a living organism will be put into action of an external signal, for instance of sunlight, then the period of the oscillations of M and P match exactly the period of the external signal. In this manner, the organism's clock is reset to a period of precisely 24 h. Note that the same simulation model with altering k_{eq}, between 100 and 200, can be applied to observing the period of the driven system.

---

[1] In molecular biology, an **interactome** is the whole set of molecular interactions in a particular cell. The term specifically refers to physical interactions among molecules (such as those among proteins, also known as protein–protein interactions, PPIs) or between small molecules and proteins but can also describe sets of indirect interactions among genes (genetic interactions).

[2] Drosophila melanogaster (DM) is a species of fly in the family Drosophilidae. The species is often referred to as the fruit fly or lesser fruit fly, or less commonly the "vinegar fly" or "pomace fly". Starting with Charles W. Woodworth's 1901 proposal of the use of this species as a model organism, DM continues to be widely used for biological research in genetics, physiology, microbial pathogenesis, and life history evolution. As of 2017, six Nobel Prizes have been awarded to drosophilists for their work using the insect.

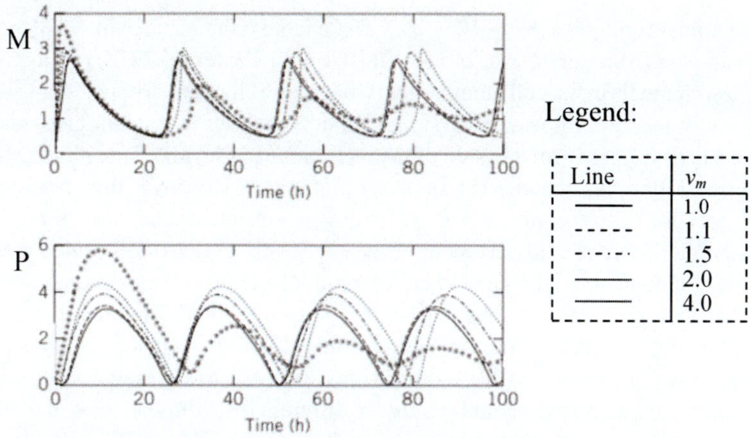

Fig. 5.4 Simulation of circadian clock model for varying values of v_m

5.3 The Hodgkin–Huxley Neuron

5.3.1 Basic Mathematical Model

In biological systems interesting phenomena take place close to the neuron cell membrane. Neurons are highly polarized cells, meaning that they form distinct subcellular zones that perform different functions. From morphological point of view, a typical neuron can be divided in three major parts: (1) the cell body containing the nucleus and the major cytoplasmic organelles; (2) a tree of dendrites, which originate from the perikaryon and ramify over a certain region of gray matter and which differ in size and shape, depending on the neuronal type; and (3) a single axon, which, in most cases, takes a form of a truncated cone extended more farther from the cell body than the dendritic tree (Fig. 5.5). Neurons and other cells are filled and surrounded by water in which ions such as sodium (Na^+), potassium (K^+), chloride (Cl^-), and calcium (Ca^{2+}) are dissolved. The superscripts indicate electrical charge, i.e., the

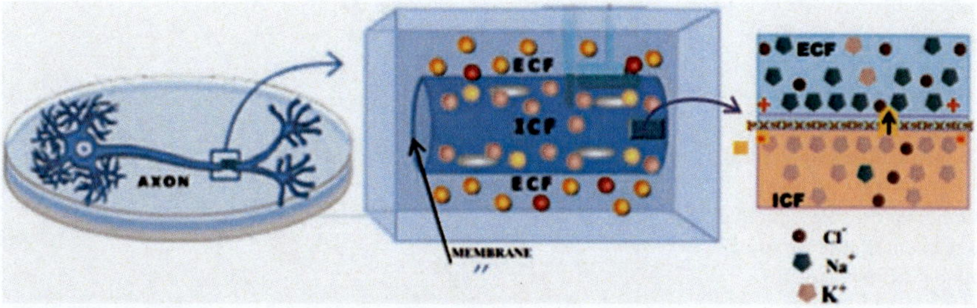

Fig. 5.5 Scheme of components of nervous systems

charge of a positron: $q \approx 1.60 \times 10^{-19}$ °C. The charge of an electron is -q. Most ion species can pass from the interior of the cell (ICF) to its exterior (ECF) or vice versa through ion channels in the cell membranes. Often these channels are specific: sodium channels only let sodium pass, potassium channels only let potassium pass, and so on. Since ions can pass from one side of the membrane to the other, one might expect their concentrations to become the same on both sides. However, this is not what happens because of two complications. First, some ions are not capable of crossing the membrane. There are, in particular, large negatively charged molecules present only in the cell interior, not in the extracellular fluid. Second, cell membranes contain pumps that actively transport ions from one side to the other. The most famous of these is the sodium–potassium pump, which removes sodium from the cell and brings in potassium -2 potassium ions for 3 sodium ions. Suffice it to say that in equilibrium, ion concentrations in the extracellular fluid are different from those in the cell interior, and there is an electrical potential jump across the cell membrane. The difference in electrical potential between the interior and the exterior of the membrane is called the membrane potential, denoted by v. For a nerve cell in equilibrium, a typical value of v might be -70 mV, i.e., the potential on the interior side of the membrane is 70 mV below that in the extracellular fluid (Fig. 5.6). From 1938 to the publication in 1952 of five landmark, among seven papers published between 1949 and 1955 A. L. Hodgkin and A. F. Huxley with their collaborators exposed the results of many of their works, both theoretical and experimental, in which they attempted to explain the basic physical mechanism by which, in humans and animals, electrical impulses are generated by nerve cells, and transmitted along axons. They clarified the mechanism underlying the action potentials in the spaceclamped squid axon, and summarized this action in the form of a system of ordinary differential equations (ODEs). For their discoveries concerning the ionic mechanisms involved in excitation and inhibition in the peripheral and central portions of the nerve cell membrane they, together with J. C. Eccles, won the 1963 Nobel Prize in Physiology and Medicine. The Hodgkin-Huxley

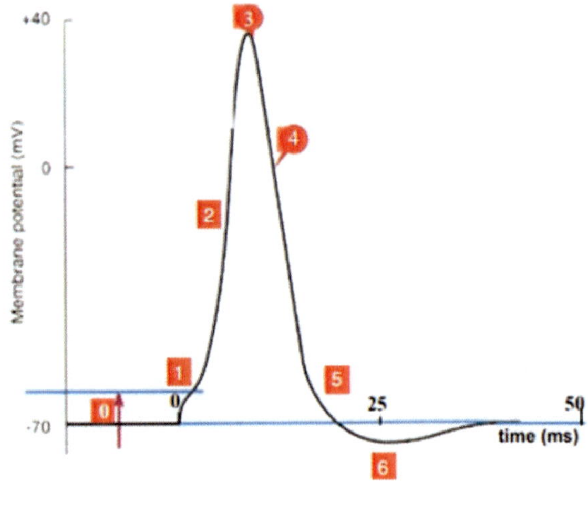

Fig. 5.6 Time-varying action potential: ⓿—threshold of excitation; ❶ Na^+ channels open, Na^+ begins to enter cell; ❷K^+ channels open, K^+ begins to leave cell; ❸Na^+ channels become refractory, no more Na^+ enters cell; ❹K^+ continues to leave cell, causes membrane potential to return to resting level; ❺K^+ channels close, Na^+ channels reset to enter cell; ❻extra K^+ outside diffusive away

theory allowed the foundation for modern computational neuroscience providing the adequate model system. Viewed as a neuron model with concentrated parameters it describes the way a neuron deals with an input voltage to produce or not to produce an action potential. However, such models do not address the more complex features of neurons that may influence the formation of networks and dissemination of an action potential. Therefore the Hodgkin-Huxley model simulates only the biological functioning of a neuron. Different parameters of the model represent specific biological components of a neuron. The morphology of the axon and its course through the nervous system are correlated with the type of information processed by the particular neuron and by its connectivity patterns with other neurons. At the interface of axon terminals with target cells are the synapses, which represent specialized zones of contact consisting of a presynaptic (axonal) element, a narrow synaptic cleft, and a postsynaptic element on a dendrite or perikaryon containing the nucleus and a variety of cytoplasmic organelles. The Hodgkin-Huxley model [4] is based on the assumption that the membrane contains proteins that selectively conduct sodium and potassium ions in a time- and voltage-dependent manner. The elaborated model focuses particularly on the electrical attributes of the neural membranes during signal generation and propagation in the nerves. When a current enters the cell it changes the membrane potential. An action potential, the rapid movement of ions, starts when the membrane potential reaches a prescribed threshold due to external sources of input current. When the membrane potential reaches said threshold, Na^+ channels open allowing Na^+ to enter the cell (flood the intracellular fluid) which causes the membrane potential to spike. K^+ also have channels to allow K^+ ions to leave the intracellular fluid but these channels require a higher membrane potential voltage than the Na^+ channels. Therefore, first Na^+ enters the cell and creates a sharp increase in the membrane potential. This increase then causes the K^+ channels to open, allowing K^+ to leave the intracellular fluid. At this point, Na^+ is entering faster than K^+ is leaving because of the electrostatic and diffusion forces exerted on it, so there is a still a net gain in membrane potential. When the peak occurs, Na^+ channels become refractory (blocked) so Na^+ can no longer enter the cell, but K^+ can still leave, thereby making the membrane potential decrease (Fig. 5.7). Once the resting potential of the membrane is reached K^+ channels close and Na^+ channels reset so that they can eventually be opened again. At the end of the action potential, the mix of Na^+ and K^+ in the intra- and extracellular fluid is not the stable electrolytic solution. Therefore the forces of diffusion and electrostatic pressure are not balanced, producing a refractory period where the cell membrane goes below the resting point before increasing again as the ions move into their stable states. Ion movements and the subsequent action potentials form a background of the neurons working. The Hodgkin-Huxley model treats the nerve axon as an electrical circuit in which the proteins are resistors and the membrane is a capacitor. Ion currents flow through the membrane and along the nerve axon leading to a propagating pulse. Assuming that a nerve cell membrane acts like a capacitor, separating two charge layers of opposite signs, it is possible to derive the equation describing how the action potential v evolves with time. It equals to a jump in the electrical potential across the cell membrane with the separated charges $\pm q$. The equation governing the instantaneous state of the capacitor has the form

$$C\frac{dv}{dt} = i_{tot} \tag{5.3}$$

where C is the membrane capacitance and i_{tot} denotes the total current. Three different ion currents contribute to the voltage signal of the neuron, that is, sodium current, a potassium current, and a leak current that consists mainly of Cl^- ions. The flow of these ions through the cell membrane is controlled by their respective voltage-dependent ion channels. The total current is given by

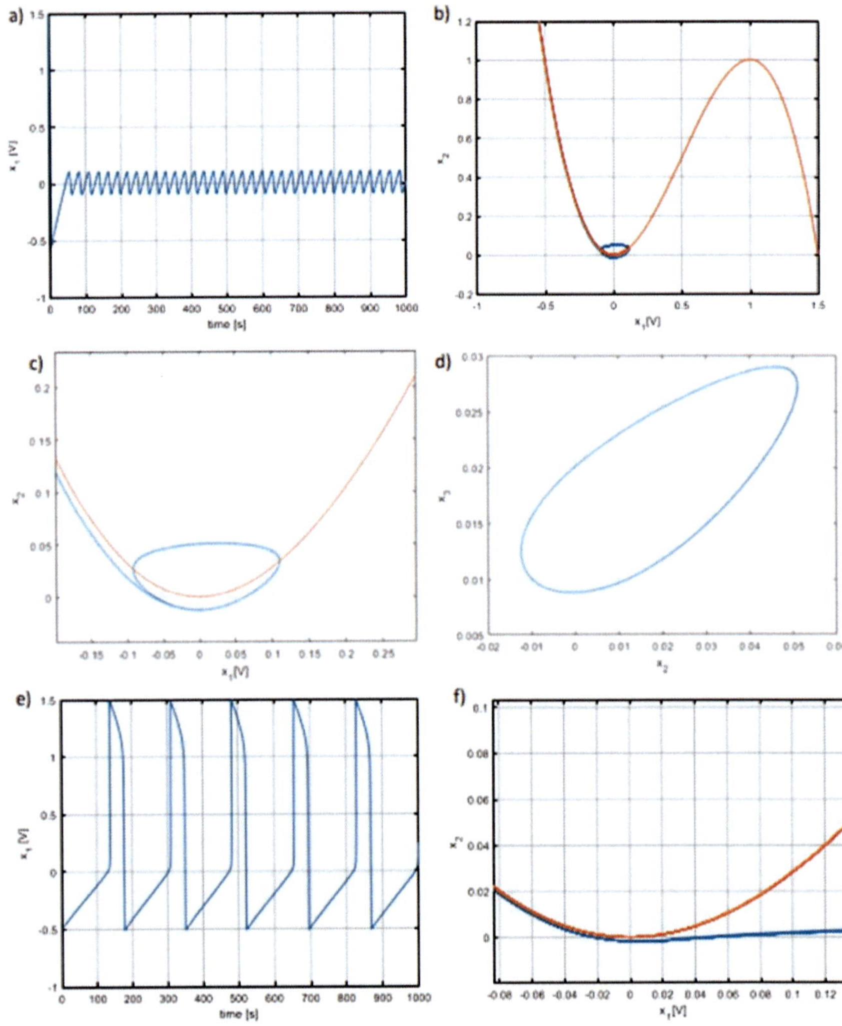

Fig. 5.7 Illustration of solutions of (5.3) for different neuron parameters: **a** spikes, **b** phase space $x_2(x_1)$, **c** transition to STO, **d** phase space $x_3(x_2)$, **e** relaxation oscillation, **f** phase space $x_2(x_1)$

$$i_{tot} = -(i_K + i_{Na} + i_{lek} + i_{stim}) \tag{5.4}$$

where i_K, i_{Na}, i_{lek} and i_{stim} are, respectively, the potassium ion current, the sodium ion current, the leakage current and the stimulus current. The i_K component is given by $g_K h^4(v, t)(v - E_K) dh(v, t)/dt = -(\alpha_h + \beta_h)h(v, t) + \alpha_h$. The i_{Na} component is given by $g_{Na}m^3(v, t)(v - E_{Na})$ with, $dm(v, t)/dt = -(\alpha_m + \beta_m)m(v, t) + \alpha_m$, where α_p and β_p, $p = h, m$ are voltage dependent parameters. The i_{leak} component given by $g_{leak}(v - E_{leak})$ with $g_{leak} =$ const is small value and will be omitted in the following. The quantities E_{Na}, E_K and E_{leak} are the Nernst potentials of different ions [1]. The terms g_{Na} and g_K are the conductances of the ion channel proteins for the respective ions. They are assumed to be functions of voltage and time. Taking into account the functions m and h, called gating variables, ranging between 0 and 1 and related to the likelihood that the ion channel is open which fulfill the above simple linear differential equations and combining them with Eqs. (5.3) and (5.4) yields

$$\begin{aligned} \frac{dv}{dt} &= f_1(v, m, h, t) \\ \frac{dm}{dt} &= f_2(v, m, h, t) \\ \frac{dh}{dt} &= f_3(v, m, h, t) \end{aligned} \tag{5.5}$$

where $f_1(\cdot)$, $f_2(\cdot)$ and $f_3(\cdot)$ are nonlinear functions of the action potential and gating variables. Their explicit forms depend on many factors, and most often are fitted to the experimental data. The model could reproduce and explain a remarkable range of data from squid axon, including the shape and propagation of the action potential, its sharp threshold, refractory period, anode-break excitation, accommodation and subthreshold oscillations. Such model focuses on the role of the ionic currents in shaping the MMO patterns (generation of STOs and the origin of spikes) or the dynamic actions that give evolution to MMOs.

5.3.2 *Periodic Neuron Firing*

Most of the neurons are excitable, i.e., they show quiescent behavior; however when they are stimulated by input impulse they can also fire spikes. The complexity of spikes or their trains can be controlled by external stimuli, e.g. by injected electrical currents.

Very often the neurons continue to fire a train of spikes when there is an input stimulus by injecting a pulse of current and this is called tonic spiking. There exist different types of spiking patterns depending on the nature of the intrinsic dynamics of the neuronal network. One of the interesting complex firing patterns emerge from the activity of neurons is the mixed-mode oscillations (MMOs) [15–17], which are a combination of small amplitude (sub-threshold) oscillations interspersed with large amplitude relaxation type oscillations what is the main focus of this section. The generation of MMOs requires the coordinated action of various mechanisms: (i) a mechanism for the generation of subthreshold oscillations (STOs); (ii) a spiking mechanism, including the onset of spikes and the description of the spiking dynamics; and (iii) a return mechanism from the spiking regime back to the subthreshold regime. It should be emphasized that phenomena realized in neurons can be depicted as the interaction of rudimentary subprocesses. It can be also stated that very often different subprocesses evolve at distinct time scales. It is known that the separation of variables changing with different time scales may introduce interesting and complicated system dynamics, including exceptional responses known as canards. In the case where there is just one slow variable in a system, then canard responses appear as explosive growths in oscillation amplitude of a response to a very small change in system parameters. However, when in a system there are two slow variables the existence of canard responses can lead to complicated oscillatory patterns in the form of MMOs [25–27].

Applying the model (5.5) with $f_1(\cdot) = 3v^2 - 2v^3 - h; f_2(\cdot) = 0.008v - 0.002h - 0.003m$ and $f_3(\cdot) = 0.02\text{v} - 0.0135\text{m} + 0.00005$ gives MMO patterns presented in Fig. 5.7 for $x_1 = v(t)$, $x_2 = h(t)$ and $x_3 = m(t)$, where v is the membrane potential (mV), C is the membrane capacitance (μF/cm^2), I_{app} is the applied bias (DC) current (μA/cm^2), $I_{Na} = G_{Na}m^3h\ (V - E_{Na})$, $I_K = G_K n^4\ (V - E_K)$, $I_L = G_L(V - E_L)$, $I_{Nap} = G_p p_\infty(V)\ (\text{V} - E_{Na})$, $I_h = G_h\ (\text{cf} \cdot rf + cs \cdot rs)\ (V - E_h)$. The parameters G_X and E_X ($X = Na, K, L, p, h$) are the maximal conductances (mS/cm^2) and reversal potentials (mV), respectively. The unit of time is *ms*. The variables *rf* and *rs* are the *h*-current fast and slow gating variables and the parameters *cf* and *cs* represent the fraction of the total *h*-current corresponding to its fast and slow components respectively. The values of the parameters have been used [28, 29]: $E_{Na} = 55$, $E_K = -90$, $E_L = -65$, $E_h = -20$, $G_{Na} = 52$, $G_K = 11$, $G_L = 0.5$, $G_p = 0.5$, $G_h = 1.5$ and $C = 1$. In the diagrams presented in Fig. 5.7 all the MMO mechanisms are described by the same model. The generation of STOs (Fig. 5.8a) and the onset of spikes (Fig. 5.8e) are governed by the locally parabolic portion of the v-nullcline, near its minimum. The spiking dynamics and the return mechanisms are primarily governed by the right and left branches of the v-nullcline respectively (Fig. 5.8b). The abrupt transition from STOs to spikes is due to the 2D (Fig. 5.8a, b) and 3D canard phenomena (Fig. 5.8c) [30]. The locally parabolic nonlinearity at the minimum of the v-nullcline and the time scale separation between v and the remaining dependent variables are key for these mechanisms [31, 32]. There have been various attempts to simplify the complexity of the equation, for instance by FitzHugh and Nagumo [33, 34] (called the FitzHugh–Nagumo (FHN) model), Hindmarsh–Rose [35] and Rajagopal [30]. Such models are widely used in simulating the neural networks in order to rationalize experimental data.

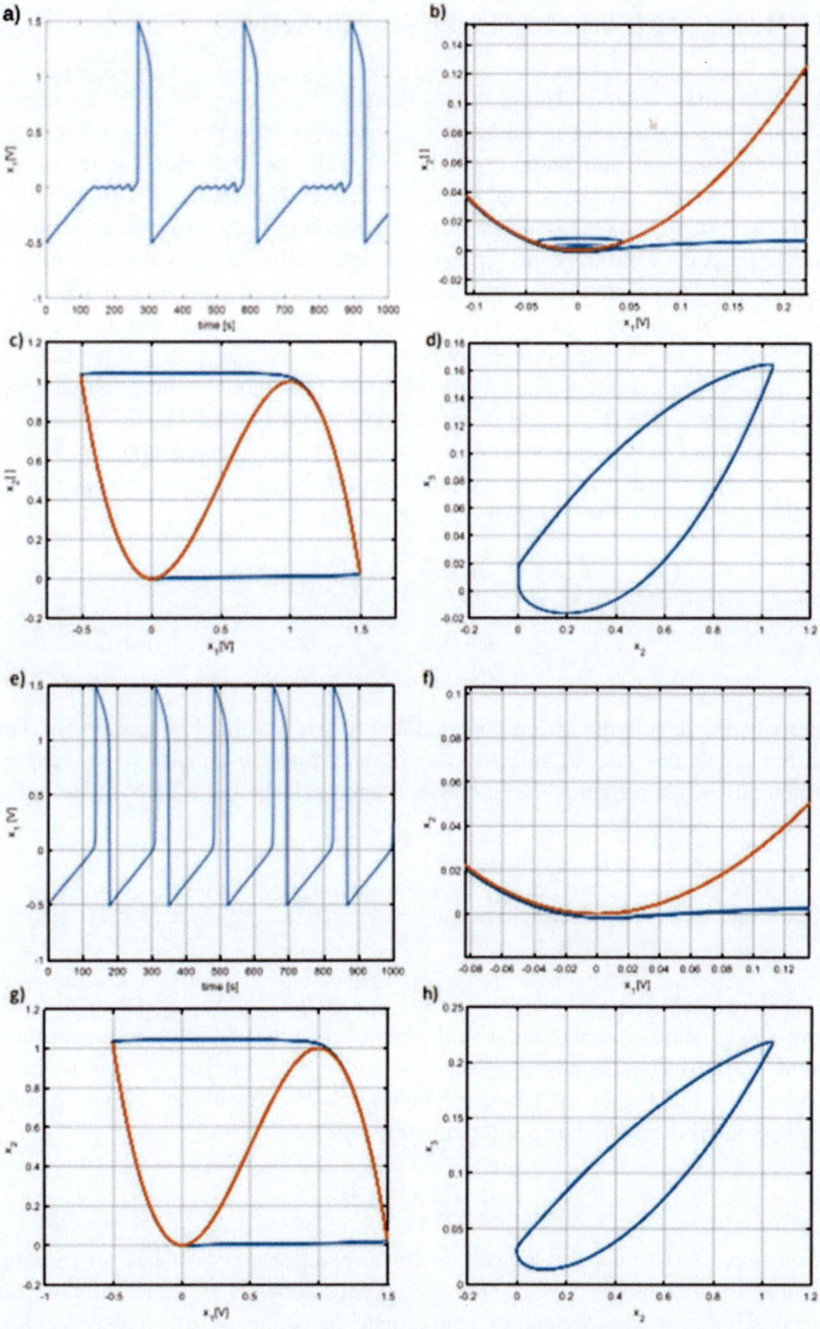

Fig. 5.8 MMOs with (5.5) for different neuron parameters: **a** 1^5 MMOs, **b** nullcline $x_2(x_1)$ and phase portrait for SAOs, **c** transition to STO, **d** trajectory in phase space $x_3(x_2)$, **e** relaxation oscillation, **f** trajectory in phase space $x_2(x_1)$, **g** transition to STO, **h** trajectory in phase space $x_3(x_2)$

5.4 Reduced Model of a Single Neuron Activity

Since the seminal work of Hodgkin and Huxley, there has been various attempts to simplify the complexity in the dynamical systems view point of a neuron, for instance by FitzHugh and Nagumo (FHN) [33, 34]. The FHN model is frequently used as a reduced two-dimensional model for the description of neural oscillations and of other phenomena in fast–slow systems. Models of FHN type capture the basic dynamic properties observed in various conductance-based approaches to neuron activities. They represent the minimal models being able to generate intrinsic oscillatory neuronal behavior and involve the dynamics of the voltage v and a recovery variable w. In such a model, the selfcoupling is provided by an extra synaptic variable whose evolution alternates between fast and slow modes. These models may generate either STOs (Fig. 5.8a) or spikes (Fig. 5.8e) but not MMOs. Assuming similar dynamics of slow sodium and potassium ions and replacing $m(t)$, $h(t)$ in (5.5) by one effective current $w(t)$ leads to the FitzHugh-Nagumo model represented by the following two nonlinear ordinary differential equations

$$\begin{aligned}\frac{dv}{dt} &= f(v, w)\\ \frac{dw}{dt} &= g(v, w)\end{aligned}$$

where v and w denote the action potential and recovery variable, respectively. The field flow functions $f(v, w)$ and $g(v, w)$ are determined by relatively low degree polynomials of their arguments. In classical approach the basic FHN model takes the form

$$\begin{aligned}\frac{dv}{dt} &= v - \frac{1}{3}v^3 - w + i_{stim}\\ \frac{dw}{dt} &= \frac{av - bw}{\tau_w}\end{aligned} \tag{5.6}$$

where a, b, τ_w and i_{stim} denote the scaling parameters, time constant and the stimulus current, respectively. Usually $a > 0$, $b < 0$ and $\tau_w \gg 1$. FitzHugh-Nagumo made the observation that when the Hodgkin/Huxley equations exhibit periodically firing action potentials, the following approximately holds

$$n(t) + h(t) \approx 0.8 \tag{5.7}$$

Equation (5.7) should be viewed as an observation only—it has no rigorous mathematical or biological basis. However, it does allow us to further simplify the Hodgkin/Huxley model. We choose to eliminate the gating variable h by taking $h(t) = 0.8 - n(t)$. All variables and parameters should be considered dimensionless here. FitzHugh's choices, translated into our notation are: $a = 1.25$ and $\tau_w = 15.625$.

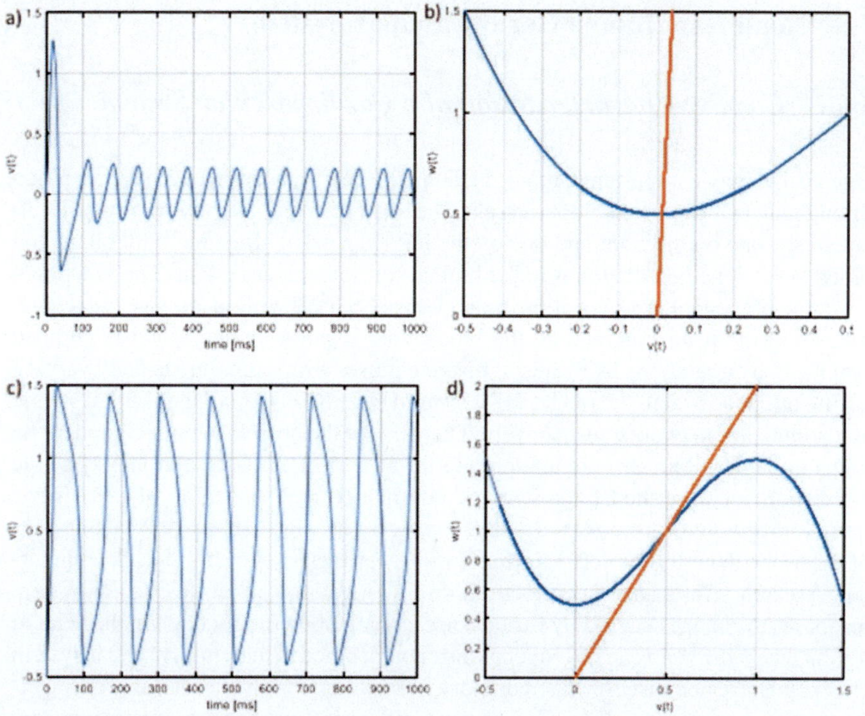

Fig. 5.9 Simulation results with (5.6) and (5.7): **a** spikes, **b** nullcline, **c** relaxation oscillation, **d** nullcline

The left panel of Fig. 5.9 is the analogue of Fig. 5.7a, but for the FitzHugh-Nagumo model, with $a = 1.25$ and $\tau_w = 25$. The right panel shows w as a function of v.

In spite of its simplicity, the FitzHugh-Nagumo model can reproduce aspects of the dynamics of the classical Hodgkin-Huxley model surprisingly well. The generation of STOs (Fig. 5.9a) and the onset of spikes (Fig. 5.9c) are governed by the locally parabolic portion of the v-nullcline, near its minimum. The spiking dynamics and the return mechanisms are primarily governed by the right and left branches of the v-nullcline respectively (Fig. 5.9b, d). The abrupt transition from STOs to spikes is due to the 2D (Fig. 5.9a, c) and 3D canard phenomena.

5.5 Nonlinear Human Cardiovascular System

5.5.1 Concise Characterization of a Cardiovascular System

One of the most important organs in human body is heart which pumps blood throughout the body using blood vessels. The heart works as two pumps acting simultaneously, one on the right and one on the left (Fig. 5.10). Together with blood and blood vessels the heart belongs to the human cardiovascular system. The functional and structural characteristics of the blood vessels change with successive branching. Thanks to the heart contraction, the blood flow is pulsatile and blood is pumped into the two circulations by means of discrete pulses with a pressure usually varying during one heartbeat in the ranges 70–130 mmHg and 20–30 mmHg for the systemic and pulmonary networks, respectively. The work of the heart is to pump blood to the lungs through pulmonary circulation and to the rest of the body through systemic circulation accomplished by systematic contraction and relaxation of the cardiac muscle in the myocardium. Normally, for the right ventricle, work is about 16% of the work done by the left ventricle [36]. Over the past several decades, important research efforts have been focused on investigations of cardiovascular system responses to changes caused by various diseases affecting malfunction in the heart to a point where the heart can no longer pump enough blood to the cardiovascular system [14–18]. Untreated diseases like coronary heart disease, valvular disease, or hypertension have important influences on heart malfunction through different factors that lead eventually to potential clinical impacts. To determine the factors causing the insufficient functioning of the cardiovascular system and its deficiencies, one can reach the significant help in the form of mathematical modeling allowing isolating the contribution of several key parameters from the whole cardiovascular process, and trying to predict the dominant factors, as well as changes in the patient state. In this way, clinical examinations can be minimized, especially at the expensive stage of the pilot investigations, and then the costs of developing a practical procedure of treatment combination of cardiotherapy are significantly reduced.

To determine the correct diagnosis and administer high-quality treatment, it is necessary to identify accurately the physiological processes that establish the flow of blood in the cardiovascular system. Accurate knowledge of the interplay between the blood dynamics and physiological rhythms, such as heart rate, is of crucial importance in treatment applying [37]. In what follows the focus is put on a computational model of the cardiovascular system that represents a resulting human response on such health deficiencies as a myocardial infarction, congestive cardiac failure, chest pain, severe headache requiring hospitalization, as well as other cardiovascular system illnesses, and the use of this model in the analysis of experimental observations from a specific group of people who suffer from transient adaptation after undergoing clinical therapy (Fig. 5.10).

Due to the space limits we cannot and do not require for giving an extensive exposition of such a complex domain as cardiology. The subjects presented in this section are therefore highly selective and focus on those aspects needed to appreciate the biological background of the dynamical processes generating MMOs. For

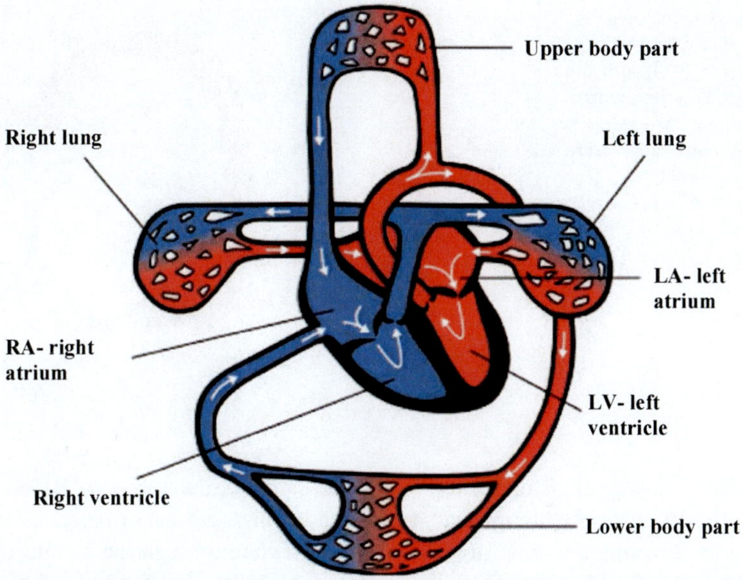

Fig. 5.10 Scheme of a human cardiovascular system

an in-depth discussion of remaining aspects of cardiology we refer the reader to the corresponding literature positions mentioned at this chapter end. One of the fundamental roles of the heart is to forcing blood venous return by maintenance right atrial pressure low and adopts left ventricle output to the body demand for oxygen supply. An improper physiological response to changes in different segments of the cardiovascular system is revealed by variations in blood pressure, which often cause a significant deterioration of the health of the person touched by these changes. It is worth point out that the electrical stimulation that generates the heartbeat resides within the cardiac muscle (Fig. 5.11).

The heart beats in the absence of any nervous connections. Many different factors affect and control the cardiac output including the atrial and ventricular reflexes, the autonomic nervous system, hormones, blood ion concentrations, and emotions (Fig. 5.11). These factors affect cardiac output by changing heart rate and stroke volume. After initiation, the electrical signals spread throughout the heart, reaching every cardiac cell rapidly as possible. They go from the top to the bottom of the heart. The actions of all individual cells are strictly coordinated. Through selective openings and closings of plasma membrane channels for sodium, potassium, and calcium ions, the electrical activity of cardiac cells is excited. Through a conductive system, electrical impulses are sent to the chambers that through contractions pump blood into the circulatory system. The most important from point of view of the human body metabolism is the left ventricle with the aorta and large arteries forming a bridge between the heart and the arterioles, serving firstly as a conduit, and second as a shock absorber. Knowledge of instantaneous state of such systems focuses an important attention in clinical practice and research. For this reason, we will

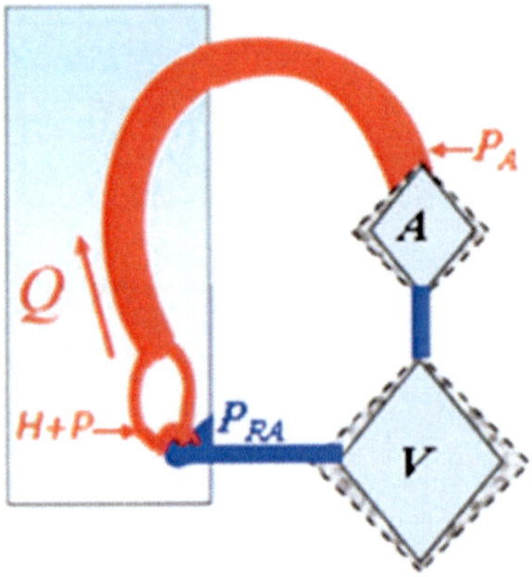

Fig. 5.11 Scheme of systemic circulation in diastole: H + P—hearth and pulmonary, Q—flow, equals cardiac output, A—arterial blood, PA—arterial pressure, V—venous blood, P_{RA}—right atrial pressure

concentrate considerations on the left part of the cardiovascular system. Blood flow through the circulation system occurs from areas of high pressure to areas of lower pressure by lowering the downstream left atrial pressure through the actions of the left heart rather than by increasing the upstream pressure. The feed blood back to the heart occurs from the upstream veins and venules. The pressure in this region is determined by the volume of these vessels and by the elastance of their walls. The arterial pulse pressure is created by the single impact volume ejected by the heart and the resistance to its flow from the arterial chamber and by the volume remaining in the aorta at the diastole end. With the same arterial pressure, widely different stroke volumes can be observed, depending on the arterial conductance. The transmission of pressure processes along the arterial circuit and the degree of residual pulsate at the input of peripheries depends on inserted parameters of the arterial circuit. As a consequence, a total peripheral conductance (TPC) changes in complex and nonlinear form depending not only on mean pressure and flow values but also on other dynamical factors such as the shape of the systemic input wave. The stressed component of the whole blood volume determines a blood flow denoted by Q. The flow volume is all the volume that moves through the system on each beat. The blood flow, venous return, and right atrial pressure are determined by the interaction of the function of the heart, and the function of venous return. The systemic circulation is formed by the arteries, that carry the oxygenated blood pumped by the left ventricle to the living tissues, and the veins that permit the nonoxygenated blood to setback into the right ventricle. The replacement of oxygen between the blood and the body tissues takes place in the microvasculature, which really separates the systemic arterial structure from the venous systems. The human cardiovascular system (CVS) performs several important functions and the most important from them is carrying the oxygenated blood and other nutrients to the body's organs, particularly to the brain and central nervous system, and to remove carbon dioxide and other harmful products from them. In general, analyzing processes arising in the cardiovascular system is too complex in terms of theory, and therefore, it has become a field of adequate modeling [38, 39].

5.5.2 *Thermodynamic Model of the Cardiovascular System Dynamics*

Cardiovascular systems are composed of many different physiological components (subsystems) and they are effects of their mutual interaction that combine to produce heart rate variability (HRV) as the temporal variation between sequences of consecutive heart beats. Problems of the nature of blood dynamics, the coupling of the cardiac output motion, and the contribution of peripheral conductance to the blood pressure, respectively, could be analyzed in detail applying physical principles, offering the possibility of generating realistic data on flow, velocity, pressure, and blood volume. The cardiovascular system can be characterized in terms of its state variables, such as the blood pressure, volume and by the cardiovascular vessel parameters such as compliances and conductances or resistances of the corresponding system's elements. When considering appropriate models of the cardiovascular system, it is very useful in clinical practice to build a model based on physical principles, offering the possibility of generating realistic data on flow velocity, pressure, and blood volume. The cardiovascular system can be modeled using lumped parameters which are hydrodynamic resistance that opposes flow, the compliance which indicates how much volume changes with a given pressure differential, and the inertance which determines the pressure difference required to change the flow rate. Moreover, the laws of physics being applied broadly to the flow of all types of liquids and gases explain also the interaction of pressure, volume, flow, and resistance of the blood in the cardiovascular system. The blood circulation is accomplished in the arteries, that carry the oxygenated blood pumped by the left ventricle to the living tissues, and in the veins that permit the non-oxygenated blood to setback into the right ventricle. The exchange of oxygen between the blood and the body tissues takes place in the microvasculature, which really separates the systemic arterial structure from the venous systems. In healthy persons, the sinoatrial node, called also the pacemaker of the heart, rhythmically activates excitation pulses 70–80 times per minute, without any neural stimulation. A suitably accurate model of the cardiovascular system can be handy in identifying the pressure pulse waveforms produced at the aorta and then in the radial arteries. It can inform greatly about how a system operates in different conditions. Denoting instantaneous blood flow rate by $q = q(t)$ and the pressure difference by $p = p(t)$, the instantaneous vascular conductance may be expressed by

$$g = g(q,\ p,\ t) \tag{5.8}$$

with

$$p = q/g \tag{5.9}$$

what means that the cardiac load, i.e., blood volume per time, coming out of the heart, determines arterial pressure. Here and what it follows the time argument t has been dropped mostly for notational simplicity. The pumping action of the heart is represented by an impulsive source of the blood flow rate, $q(t)$, that deposits a stroke volume, q_n, into the arterial system during the nth cardiac beat (or cycle in 5.10)

$$q(t) = \sum q_n \delta(t - t_n) \tag{5.10}$$

where $\delta(t)$ is a unit Dirac impulse at time $t = 0$ and t_n is the delay time of the nth cardiac cycle. Because the blood is an inert fluid then under the pressure difference between the two ends of a vessel filled with blood the flow cannot change impulsively and the mass of the blood presents the tendency to inertness for moving. The inertness of blood can be modeled by an inertance L, which in respect of arteries relates the pressure drop p with the rate of the blood flow q as follows

$$p = L\frac{dq}{dt} \tag{5.11}$$

Considering the complete systemic arterial tree, we found that the inertial term results from the proper summation of all local inertial terms, and we call it total arterial inertance. It should be emphasized that due to viscous friction in individual parts of the cardiovascular system, kinetic energy of the entire blood flow is transformed into thermal energy due to both dissipative viscosity and turbulent losses. The heat produced by every component of the circulatory system during one heart beat is equal to energy generated by the left ventricle during the systole and is eventually transferred to adjacent tissues. The energy produced by left ventricle is simply.

$$E_T = \oint_\Gamma pq dt = \oint_\Gamma p d\theta \tag{5.12}$$

where θ means cardiac blood flow, T is the heart beat period and Γ denotes the closed loop on the (p, θ) plane. Note that the integral is only during systole as this is the only time when ventricular muscles impart new mechanical energy into the blood stream. The dissipation of energy E_v at any component of the circulatory system generally depends on the mean flow rate q, or cardiac output θ, the blood density ρ, the kinematic viscosity μ, the characteristic size of the geometry α, which is governed by the patient body surface area, wall elastic modulus ε, characteristic length l and a dimensionless quantity σ representing the shape of the geometry. Opposition of a vessel to blood movement is represented by resistance R, which is determined by the radius r and length l of the vessel, and blood viscosity μ, namely

$$R = (Pinp - P_{out})/\theta \approx ml/\pi r^4 \tag{5.13}$$

where P_{inp} and P_{out} denote the mean pressures at the input and output of vessel, respectively. The resistance to flow from the ventricle is therefore mainly found in the resistance vessels: the main aorta and the arteries. When all individual resistances in the ventricle circulation load are properly added, the resistance of the entire segment is obtained and we call this (total) arterial resistance R. It is worth mentioning that heart valves are made of thin leaflets. They do not cause any resistance to the blood during systole and sustain large pressure gradients during diastole. Moreover, unlike vascular vessel wall, they are subjected to very large displacements. These features make the mathematical and numerical modeling of the valve mechanics and of their interaction with the blood flow can be integrated directly to the ventricle activity. As already observed in clinical practice, in cardiovascular system there is mutual exchange of energy between blood and extensible vessel walls. These latter accumulate elastic potential energy under the forces exerted by the blood pressure, which is then transferred to the blood as kinetic energy. From the mechanical point of view, this gives rise to a blood–structure interaction problem. Therefore, apart from the total arterial resistance R, the main part of the arterial load of the heart also includes total arterial compliance C. It is proposed taking full advantage of the information concerning the entire waveforms of pressure and flow, and on this basis estimate the parameter C in accord to the relation

$$q = C\frac{dp}{dt} \tag{5.14}$$

Thus, the rate of change of blood volume in the vessel expressing the cardiac output q is related to a change velocity in time of pressure p inside the vessel with the compliance C. Several studies have shown that compliance C exhibits a strong nonlinear dependence on pressure, which falls sharply from low- to mid-pressure values and tending to an asymptotic value when approaching to high pressure values (like hyperbolic relation) [40]. All the above presented elements, with their nonlinear extensions, can be used in various forms to create models of the heart and its arterial system which together represent relatively simple models of the cardiovascular systems. Problems of the nature of blood dynamics, the coupling of the cardiac output motion, and the contribution of peripheral conductance to the blood pressure, respectively, could be analyzed in detail applying physical principles, offering the possibility of generating realistic data on flow velocity, pressure, and blood volume. Many aspects of functional interactions that determine the overall behavior of blood flow through the arterial and venous tree can be accurately identified and characterized by its state variables such as blood pressure and volume, as well as cardiovascular parameters such as compliances, inertances, and conductances or resistances of the relevant system components.

Recall, that the so called state variable description provides the system's dynamics as a set of coupled first-order ordinary differential equations in terms of internal variables known as state variables $x_i(t)$, $i = 1, 2, \ldots, n$. Knowledge of those variables at an initial time t_0 and the system inputs (forcing functions) for time $t \geq t_0$, are sufficient to predict the future system state and outputs for all time $t > t_0$ [37, 40–42]. In general, the choice of state variables is not unique, but usually it is convenient to fixe as state variables those variables which describe the energy stored in all of the independent energy storage elements in the system. However, in the case of cardiovascular system this rule must be slightly modified because apart from the main blood vessels such as arteries (plus smaller arterioles), there are also capillaries and veins (plus smaller veins). The arterial and venous systems are branched in a dendrite-like manner, without any loops. Through their walls they transport the blood without much substance exchange. On the other hand, the "capillary bed", is a dense structure of very thin vessels with perforated walls that allow substance exchange with the intercellular environment. This structure is not tree like, but has many nodes and parallel branches. The capillary structure presents as a porous medium which can transport the blood from any point to other neighboring points, based on the pressure gradient. This porous medium is supplied with blood through a dendrite-like system of arteries (and smaller arterioles), and then the blood is evacuated from it by the venous dendrite-like structure. The topology of the two dendrite-like networks is generated in accord to a minimum cost criterion, coupled with necessary pressure and flow constraints. The systemic capillary structure receives the oxygen-rich blood which is pumped by the heart, and then the oxygen diffuses into the intracellular environment which is very low in oxygen. Moreover, this structure is also the one with the least friction losses. All this energy is eventually transformed irreversibly to heat. Irreversibility found in the blood capillary structure and the arterial vessels are related to energy lost because of dissipation in laminar or turbulent flow. Thus, the heart and the blood circuitry need some energy to work. All this energy is eventually degraded irreversibly to heat. The heart is actually a pump, which first converts chemical energy into mechanical work (through the heart's muscle, the myocardium) and then uses this work to pump blood through the ventricle to the whole circuitry of the cardiovascular system. Due to the large number of parallel branches in the capillary structure, its action is reflected by conductance $g = g(\mathrm{t})$ called peripheral conductance, which is determined by the expression

$$g = G_{\min} + \left[1 - \exp\left(-\frac{q^2}{Q_0^2}\right)\right]\frac{pq - Kq^2}{P_C^2} + \exp\left(-\frac{q^2}{Q_0^2}\right)\left(\frac{q^2}{S_m}\right) - \upsilon\frac{dg}{dt} \tag{5.15}$$

where G_{min}, Q_0, K, P_C, S_m are real positive constants, and $\vartheta = \vartheta_0 + \vartheta_1 \mathrm{e}^{-\alpha|q|}$, with ϑ_0, ϑ_1, and α being constants, such that $\vartheta_0 \ll \vartheta_1$. When the cardiac output q is small, one can consider $\vartheta \approx \vartheta_1$, while for large cardiac output, $\vartheta \approx \vartheta_0$. Note that in accord to expression (5.15) the peripheral conductance depends not only on q and p but also on the velocity of its changes in time and as such it represents one of the state variables of the model. Thus, expression (5.15) is suitable for fast and large-scale simulation of electrophysiological dynamics of cardiac circuit and is also computationally more tractable for optimization. Taking into consideration the above findings, we obtain an equation of state variables representing a dynamic cardiovascular system using four first order ordinary differential equations

$$\begin{bmatrix} \dot{x}_1 \\ \dot{x}_2 \\ \dot{x}_3 \\ \dot{x}_4 \end{bmatrix} = \begin{bmatrix} C_p^{-1}(-x_3 + q_s) \\ C_d^{-1}(x_3 + x_2 x_4) \\ L^{-1}(x_1 - Rx_3 - x_2) \\ \upsilon^{-1}(G_{\min} + zh + u - x_4) \end{bmatrix} \tag{5.16}$$

where x_1, x_2, x_3, and x_4 are state variables as pressure at input of the aorta, pressure at input of peripheral vessels, blood flow rate in the circuit and conductance of peripheral vessels, respectively; the dot is taken as a symbol for differentiation with respect to time. The components

$$\begin{aligned} z &= 1 - \exp\left(-\frac{x_2^2 x_4^2}{a^2}\right); \\ h &= \frac{x_2^2 x_4 - k x_2^2 x_4^2}{b^2}; \\ u &= \exp\left(-\frac{x_2^2 x_4^2}{c^2}\right) x_2^2 x_4 / d \end{aligned} \tag{5.17}$$

are strongly nonlinear functions of state variables and threshold parameters of the cardiovascular peripheral elements: a, b, c, d and k. The time argument t has been omitted for notational simplicity. In the above model C_p and C_d represent compliances of the highly elastic arteries proximal to the heart and, the less elastic arteries distal from the heart, respectively. The compliance C_p was included to model of ventricle and aorta tissues and to act as a low pass filter whose role is to cutoff the high frequencies of the pulsate rigor of circulation in large vessels, so that the very low pulsation of microcirculation can be reconstructed. A systemic circulation compartment is represented with the fixed compliance C_d. The inertance L captures the effects of arterial blood pressure pulse propagation on the distal and proximal arterial blood pressure waveforms. The loss of energy in the main vessels of the circulatory system in result of dissipation of a significant part of the energy generated by the ventricle during its compression is mapped by the resistance R. The cardiac output is represented by the source $q_s(t) = q_s(t+T)$ with T denoting the cycle period. The presented model captures the low frequency dynamics of the central and peripheral arterial blood pressure waveforms quite well. Actually, it captures the morphology of the aortic blood pressure wavelet quite well, apart from the high-frequency dicrotic notch [28, 30]. An interesting point related to the presented model is that the problems of the nature of blood dynamics, the coupling of the cardiac output motion, and the contribution of peripheral conductance to the blood pressure, respectively, could be with ease analyzed in detail. Many features of the dynamics of such systems can best be understood in terms of the concepts and laws of thermodynamics. Such formulation of cardiovascular dynamics, which applies to both equilibrium and transient behaviors, can also explain the mathematical and physical foundations of the application of the principles of thermodynamics of nonequilibrium and irreversibility. Focusing for now strictly on the cardiovascular system, we note that the friction between blood and walls of vessels and heart presents some sources of irreversibility. The heart needs some energy to work. All this energy is eventually degraded irreversibly to heat. The "thermodynamic model" represents a mathematical description of the cardiovascular system in which internal states are not exactly known, although statistical information is available about them, and the focus is put on the behavior of the interbranch system, especially on the exchange of energy with the surrounding environment. The terminologies of cardiovascular models and thermodynamics are reflected clearly in the present framework because a distinction between "conservation of energy" and "the first law of thermodynamics" is made. The former term will be used for the samplepath statement of the conservation law, while the term "first law of thermodynamics" will be reserved for the ensemble averaged version. The framework for these considerations encompasses transient behavior, with equilibrium and steady states studied as special cases. Therefore, the first law is stated in terms of the derivatives of various quantities with respect to time.

Recall that a fundamental assumption, proved by experience, of statistical mechanics and thermodynamics is that, if the temperature of a system is strictly positive and its inputs are left undisturbed, the system will eventually settle to a unique state of thermodynamic equilibrium. One reason why the results in this section are not entirely general is that that we have adopted direct access to compliance and inertance to identify relevant state variables. Having such possibilities eliminates the limitation of the basis for explaining the cardiovascular system operation which is closest to the principles of classical thermodynamics. Lifting this limitation is not difficult, because in nonlinear systems the lack of a large signal for thermal noise in elements in which there is a non-linear transformation of potential and kinetic energy into thermal energy is widely accepted.

5.5.3 MMOs *as an Indicator of Illness in the Cardiovascular System*

Particularly significant value of the established model (5.16) is that it is capable of accounting for a notable amount of the dynamical behaviors exhibited by the cardiovascular system and especially the identification of the defective functioning of its different segments. Furthermore, one of many significant advantages of the model is that it gives possibilities for testing theories concerning a particular system. For instance, simulating the blood flow in an aorta, can improve identification of the effects of coupling between the left ventricle and arterial geometry on the flow and, in turn, medical treatment evaluation. This numerical approach is one aspect of a new paradigm in clinical practice known as "predictive medicine" in which the progress in computational fluid dynamics as well as the increase in computer potency has transferred the numerical experiments to the tools being daily in disposition of medical researchers. Thanks to accurate computer simulations there are today possibilities of a detailed identification of the local hemodynamic patterns and of their relations with the vessel instantaneous states of particular segments of the cardiovascular system.

Due to the complex form of the established model (5.16), the determination of its analytical solution is not an easy task and therefore, to examine the influences of diverse system parameters on the hemodynamic process, we can perform computer simulations applying program MATLAB with using effective procedures of numerical integration [4].

Simulations of the blood circulation enable studies of the impact of important parameter values of the cardiovascular model corresponding to the basic physico-chemical parameters on the dynamics of blood flow in the circulatory system. The supreme and ambitious purpose of mathematical models applied in medicine is to improve quality of patient life and prevent the health detriment, and then to achieve a right blood pressure at stable normal standard. Therefore, we start by considering the blood flow rate—pressure relations and then turn toward the load reactions and coupling mechanisms responsible for different patterns of the heart output quantities and to identify the relationship between them.

Responses of cardiovascular systems with some pathological changes in the arterial segments are shown in Fig. 5.12. Equations (5.16) and (5.17) were solved with the variable step ODE23 procedure from MATLAB with RelEr = AbsEr = 10^{-8} and $0 \leq t \leq 50$ s for different initial conditions. Figure 5.12 illustrates the steady-state pulsatile arterial blood pressure waveforms that result from simulating the model (5.16), (5.17) with $Q = 100$ mL/s, $T = T_n = 1$ s, $R = 1$ mmHg/(mL/s), $L = 0.001$ mmHg/(mL/s^2), $C_p = 0.25$ mL/mmHg, and $C_d = 0.025$ mL/mmHg. These parameters represent typical values for a 70 kg male human [7, 37], and, when used with Eqs. (5.16) and (5.17), result in reasonable approximations of the pressure waveforms during the cardiac cycle. Periodic orbits reflect the deterministic dynamics of the underlying system.

It is worth to notice that the generated waveforms of the blood pressures, inertial flow, and peripheral conductance change their values with the same period as the heartbeat and take the form of relaxation oscillations. The slopes of the pressure curves (Fig. 5.12a) in the aorta and at the exit of the arteries (Fig. 5.12b) are the same, but during their descents the slope of the pressure curve for the output of the arteries is steeper than at the aorta entrance. At the same time, the pressure drop in the systemic arterial system between the ascending of the aorta and the equivalent artery does not exceed 105–110 mmHg at the upper level and 40.5–41.5 mmHg at the lower level. The 2D and 3D phase portrait are formed by closed univalent curves which correspond very well to the relaxation oscillations. It is also visible that the instantaneous vascular conductance (Fig. 5.12d) changes its values in a slightly different manner and exhibits the form like the graph of rectified sine function with small magnitude superposed on relatively large constant value. The dependence of aortic pressure on blood flow rate (Fig. 5.12e) takes the typical human left ventricular pressure–flow rate loop.

Particularly dignified of attention is that that the cardiovascular model (5.16) with (5.17) is capable of accounting for a notable amount of the dynamical behaviors exhibited by the cardiovascular system and particularly the identification of the defective functioning of its different segments. Responses of cardiovascular systems with some pathological changes in the arterial segments are shown in Fig. 5.13. In this case, the steady-state response waveforms exhibit the form of mixed-mode oscillations (MMOs) caused by changes in compliance values ($C_p = 0.025$ mL/mmHg, $C_d = 0.025$ mL/mmHg), inertance ($L = 0.0001$ mmHg/(mL/s2)), resistance [$R =$

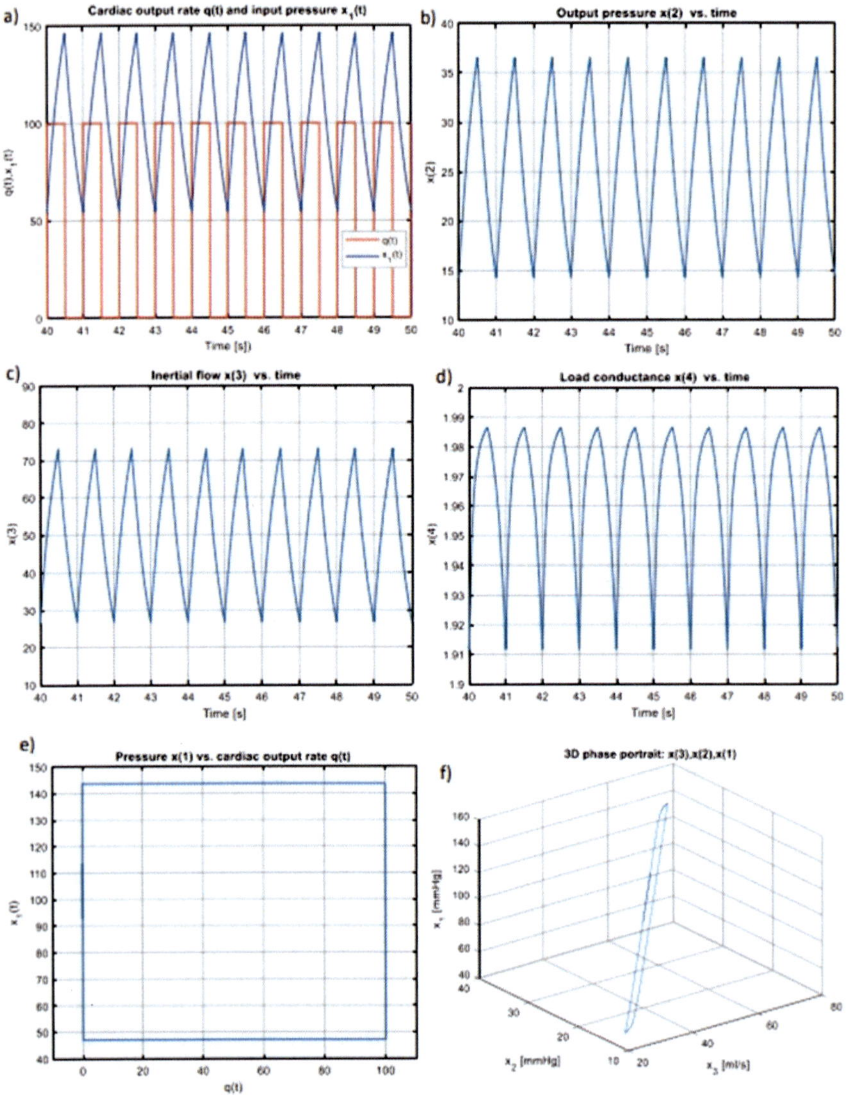

Fig. 5.12 Steady-state time evolution of cardiovascular variables generated by the nominal circulatory model of a heart: **a** cardiac output rate and pressures at input, **b** pressure at output, **c** inertial flow, **d** peripheral vascular conductance, **e** phase portrait $x_1(t)[q(t)]$, **f** 3D phase portrait

0.0001 mmHg/(mL/s)], magnitude of blood rate pulses ($Q = 70$ mL/s) and only slightly modified peripheral conductance. These waveforms express the role of the arterial flexibilities in shaping the MMO patterns (generation of STOs—subthreshold oscillations—and the onset of spikes) and the blood dynamic mechanisms that give rise to MMOs. Along with the change of the parameter of the element affected by the pathological change, MMOs may appear alternating between SAO and LAO [4, 5]. The MMO waveforms in Fig. 5.12a can be denoted as 3^3 MMOs. It is worth to mention that the generation of MMO patterns requires the coordinated action of such processes as: (i) generation of SAOs, (ii) a spiking, including the onset of spikes (LAOs) and a support of the spiking dynamics, and (iii) a return phenomenon from the spiking regime back to the sub-threshold regime. Moreover, generation of MMO patterns requires an additional mechanism that provokes the dynamic transition between these two regimes by an oscillatory forcing input (cardiac output) and by an additional nonlinear dependent variable (peripheral conductance) leading to the 3D canard phenomenon. The major problem related to mixed mode oscillations (MMOs), which has not been studied so far in relation to cardiovascular systems, is focused on determining the correlation between different forms of response identified by the model with the states of pathological changes in the vascular elements of a given patient. In some cases of nonlinear dynamics, the effect can be fairly complex, with a sequence of mixed-mode oscillations of different standing and numbers of SAOs and LAOs. In some cases of weakened action of cardiovascular elements with an increased non-linear effect, more complex disease states of the patient with a mixed mode oscillation sequence with different positions and numbers of SAO and LAO may occur. Therefore, the form of the MMOs may exhibit the degree of nonlinearity produced by a disease of the cardiovascular system and diagnosis of the patient's instantaneous state could be checked out via a function of heartbeats in accord to LAOs and SAOs (Fig. 5.13).

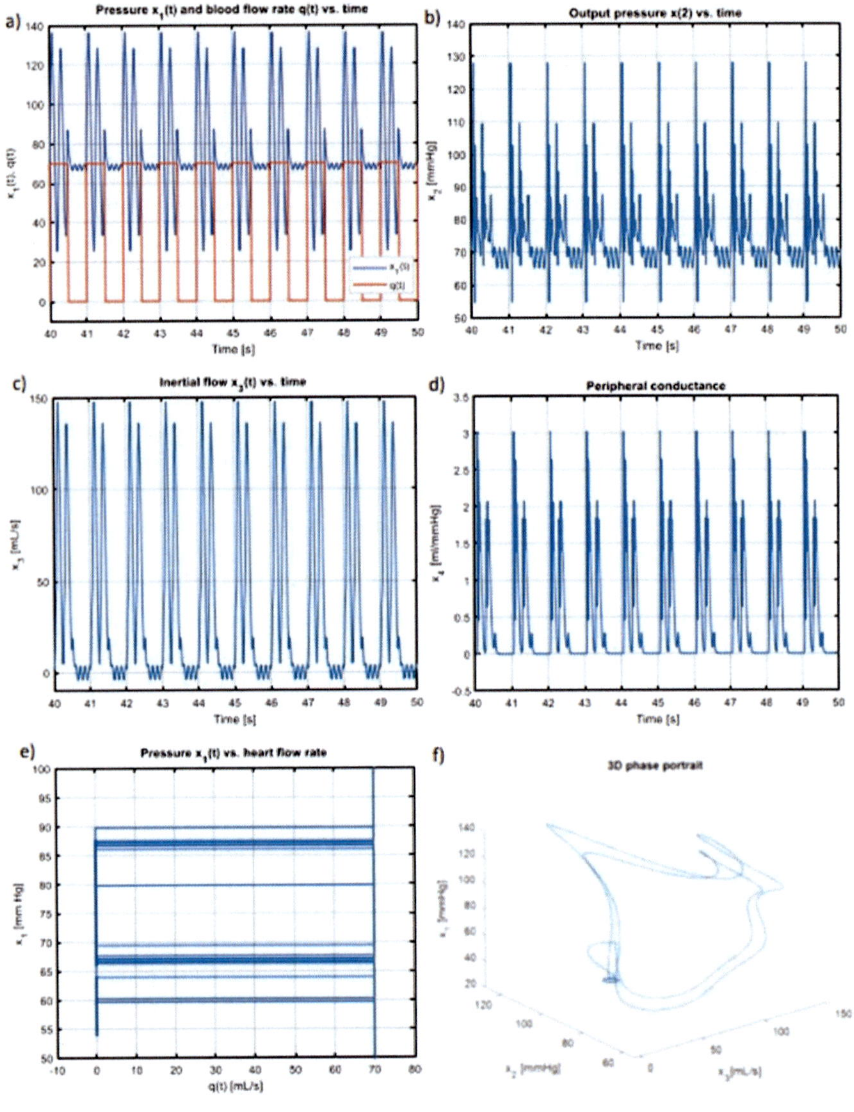

Fig. 5.13 Steady-state time evolution of cardiovascular variables generated by a pathological circulatory model of a heart: **a** pulsatile left ventricle blood output rate and MMOs pressure at aorta input, **b** pressure MMOs at arterial output, **c** MMO oscillations of inertial flow rate, **d** MMOs of peripheral conductance, **e** output pressure versus blood flow rate at ventricle output, **f** 3D phase portrait of MMO

References

1. Izhikevich, E.M.: Dynamical Systems in Neuroscience: The Geometry of Excitability and Bursting. MIT Press, Cambridge (2007)
2. Lu, Z., Mukkamala, R.: Continuous cardiac output monitoring in humans by invasive and non invasive peripheral blood pressure waveform analysis. J. Appl. Physiol. **101**(2), 598–608 (2006)
3. Mukkamala, R., Reisner, A., Hojman, H., Mark, R., Cohen, R.: Continuous cardiac output monitoring by peripheral blood pressure waveform analysis. IEEE Trans. Biomed. Eng. **53**(3), 459–467 (2006)
4. Heldt, T., Chang, J.L., Chen, J.J.S., Verghese, G.C., Mark, R.G.: Cycle-averaged dynamics of a periodically-driven, closed loop circulation model. Control Eng. Pract. **13**(9), 1163–1171 (2005)
5. Parlikar, T.A., Heldt, T., Verghese, G.C.: Cycle-averaged models of cardiovascular dynamics. IEEE Trans. Circuits Syst. I Fundam. Theory Appl. **53**(11), 2459–2468 (2006)
6. Shibao, C., Grijalva, C.G., Raj, S.R., Biaggioni, I., Griffin, M.R.: Orthostatic hypotensionrelated hospitalizations in the United States. Am. J. Med. **120**, 975–980 (2007)
7. Fedorowski, A., Stavenow, L., Hedblad, B., Berglund, G., Nilsson, P.M., Melander, O.: Orthostatic hypotension predicts all-cause mortality and coronary events in middle-aged individuals. Eur. Heart J. **31**, 85–91 (2010)
8. Kozlovskaya, I.B., Grigoriev, A.I.: Russian system of countermeasures on board of the International Space Station (ISS): the first results. Acta Astronaut. **55**, 233–237 (2004)
9. Walchenbach, R., Geiger, E., Thomeer, R.T., Vanneste, J.A.: The value of temporary external lumbar CSF drainage in predicting the outcome of shunting on normal pressure hydrocephalus. J. Neurol. Neurosurg. Psychiatry **72**, 503–536 (2002)
10. Arnold, A.C., Raj, S.R.: Orthostatic hypotension: a practical approach to investigation and management. Can. J. Cardiol. **33**, 1725–1728 (2017)
11. Low, Ph.A., Tomalia, V.A.: Orthostatic hypotension: mechanisms, causes, management. J. Clin. Neurol. **11**(3), 220–226 (2015). https://doi.org/10.3988/jcn.2015.11.3.220
12. Adamec, I., Junakovic, A., Krbot, M., Habek, M.: Association of autonomic nervous system abnormalities on head-up tilt table test with joint hypermobility. Eur. Neurol. **79**(5–6), 319–324 (2018)
13. Genecand, L., Dupuis-Lozeron, E., Adler, D., Lador, F.: Determination of cardiac output: a game of thrones. Respiration **96**(6), 1 (2018)
14. MacEwen, C., Sutherland, Sh, Pugh, D.J., Tarassenko, L.: Validation of model flow estimates of cardiac output in hemodialysis patients. Ther. Apher. Dial. Off. Peer-Rev. J. Int. Soc. Apher. Jpn. Soc. Apher. Jpn. Soc. Dial. Therapy **22**(4), 1 (2018)
15. Niewiadomski, W., Gasiorowska, A., Krauss, B., Mróz, A., Cybulski, G.: Suppression of heart rate variability after supramaximal exertion. Clin. Physiol. Funct. Imaging **27**(5), 309–319 (2007)
16. Strasz, A., Niewiadomski, W., Skupińska, M., Gąsiorowska, A., Laskowska, D., Leonarcik, R., Cybulski, G.: Systolic time intervals detection and analysis of polyphysiographic signals. In: Jablonski, R., Brezina, T. (eds.) Mechatronics Recent Technological and Scientific Advances. Springer, Berlin (2011)
17. Rajzer, M., Kawecka-Jaszcz, K.: Arterial compliance in arterial hypertension. From pathophysiology to clinical relevance. Arter. Hypertens. **6**(1), 61–73 (2002)
18. Dimitrakopoulos, E.G.: Nonsmooth analysis of the impact between successive skew bridge-segments. Nonlinear Dyn. **74**, 911–928 (2013)
19. Lesniak, M.S., Clatterbuck, R.E., Rigamonti, D., Williams, M.A.: Low pressure hydrocephalus and ventriculomegaly: hysteresis, non-linear dynamics, and the benefits of CSF diversion. Br. J. Neurosurg. **16**, 555–561 (2002)
20. Colombaro, I., Giusti, A., Mainardi, F.: On the propagation of transient waves in a viscoelastic Bessel medium. Z. Angew. Math. Phys. **68**(3), 62 (2017)

21. Podhaisky, H., Marszalek, W.: Bifurcations and synchronization of singularly perturbed oscillators: an application case study. Nonlinear Dyn. **69**, 949–959 (2012)
22. Torday, J.S.: Homeostasis as the mechanism of evolution. Biology **4**(3), 573–590 (2021)
23. Suen, J.Y., Navlakha, S.: A feedback control principle common to several biological and engineered systems. J. R. Soc. Interface **19**, 20210711 (2022)
24. Cinquin, O., Demongeot, J.: Roles of positive and negative feedback in biological systems. C. R. Biologies **325**, 1085–1095 (2002)
25. Trzaska, Z.: Nonsmooth analysis of the pulse pressured infusion fluid flow. Nonlinear Dyn. **78**, 525–540 (2014)
26. Trzaska, Z.: Dynamical processes in sequential-bipolar pulse sources supplying nonlinear loads. Electr. Rev. **90**(3), 147–152
27. Trzaska, Z.: Properties and applications of memristors—memristor circuits with innovation in electronics. In: Czy, Z.Z., Macia, G.K. (eds.) Contemporary Problems of Electrical Engineering and Development and Evaluation of Technological Processes, pp. 76–93. Publisher TYGIEL, Lublin (2017)
28. Rotstein, H.G., Oppermann, T., White, J.A., Kopell, N.: A reduced model for medial entorhinal cortex stellate cells: subthreshold oscillations, spiking and synchronization. J. Comput. Neurosci. **21**, 271–292 (2006)
29. Acker, C.D., Kopell, N., White, J.A.: Synchronization of strongly coupled excitatory neurons: relating network behavior to biophysics. J. Comput. Neurosci. **15**, 71–90 (2003)
30. Krupa, M., Szmolyan, P.: Relaxation oscillation and canard explosion. J. Differ. Eqs. **174**, 312–368 (2001)
31. Marszalek, W.: Bifurcations and Newtonian properties of Chua's circuits with memristors. DeVry Univ. J. Sch. Res. **2**(2), 13–21 (2015)
32. Marszalek, W.: Fold points and singularity induced bifurcation in inviscid transonic flow. Phys. Lett. A **376**, 2032–2037 (2012)
33. Hodgkin, A., Huxley, A.: A quantitative description of membrane current and its application to conduction and excitation in nerve. J. Physiol. **117**, 500–544 (1952)
34. Zhang, X.J., You, G.Q., Chen, T.P., Feng, J.F.: Readout of spike waves in a microcolumn. Neural Comput. **21**, 3079–3105 (2009)
35. Nayak, S.K., Bit, A., Dey, A., Mohapatra, B., Pal, K.: A review on the nonlinear dynamical system analysis of electrocardiogram signal. Hindawi J. Healthc. Eng. **69**, 1–19 (2018)
36. Rowell, L.B.: The cardiovascular System. In: Tipton, C.M. (ed.) Exercise Physiology. American Physiological Society, Rockville (2003)
37. Heldt, T., Verghese, G.C., Mark, R.G.: Mathematical modeling of physiological systems. In: Batzel, J.J., Bachar, M., Kappel, F. (eds) Included in Mathematical Modeling and Validation in Physiology: Applications to the Cardiovascular and Respiratory Systems. Springer, Berlin (2013)
38. Subramaniam, B., Khabbaz, K.R., Heldt, T., Lerner, A.B., Mittleman, M.A., Davis, R.B., Goldberger, A.L., Costa, M.D.: Blood pressure variability: can nonlinear dynamics enhance risk assessment during cardiovascular surgery. J. Cardiothorac. Vasc. Anesth. **28**(2), 392–397 (2014)
39. Wolf, A., Swift, J.B., Swinney, H.L., Vastano, J.A.: Determining Lyapunov exponent from a time series. Physica **16D**, 285–317 (1985)
40. Tharp, B.Z., Erdmann, D.B., Matyas, M.L., McNeel, R.L., Moreno, N.P.: The Science of the Heart and Circulation. Baylor College of Medicine, Houston (2009)
41. Solinski, M., Gieraltowski, J.: Different Calculations of Mathematical Changes of the Heart Rhythm. Center of Mathematics Applications, Warsaw University of Technology, Warsaw (2015)
42. Giusti, A., Mainardi, F.: A dynamic viscoelastic analogy for fluid-filled elastic tubes. Meccanica **51**(10), 2321–2330 (2016)

Chapter 6
Energy Flow Analysis of Nonlinear Dynamical Systems

6.1 Introduction and Short Historical References

A fundamental basis to investigate dynamical systems is created by the energy flow approach, for which the following essential features and relationships are worth to underline. First, the principle used in this approach rests on the universal law of energy conservation and transformation that underpins the action of dynamic systems. Therefore, it provides a common approach to analyze various types of systems, including mechanical, thermal and electrical/magnetic ones, as well, such as solid, fluid, acoustic and control systems, as more complex systems involving their couplings or interactions. Second, the variable studied in the energy flow analysis combines the effects from both strengths and rapidities, and it takes their product, named power, i.e. the change rate of energy, as a single parameter to characterize/to describe the dynamic behavior and responses of a system. It allows including and reflecting the full information on the equilibrium and dynamics of the system, and therefore overcomes the limitations emerging in study of force and dynamic responses separately. The approaches adopted in power flow analysis focus on a global statistical energy estimations, distributions, transmissions, designs and controls for dynamic systems or sub-systems rather than the detailed spatial pattern of the structural responses. It overcomes difficulties encountered while using finite element methods or experimental modal analyses of vibration responses at medium to high frequency regions, which requires extreme small size of elements to reach a necessary computational accuracy.

Most waveforms occurring in practice are continuous and single valued i.e. having a single value at any particular instant. However when sudden changes occur (such as in switching operations or in square waveforms), theoretically vertical lines could occur in the system answer waveform giving multi-values at these instants. As long as these multi-values occur over finite bounds, the waveform is single-valued and continuous in pieces, or said to be *piecewise continuous*.

Z. Trzaska, *Mathematical Modelling and Computing in Physics, Chemistry and Biology*, Studies in Systems, Decision and Control 495,
https://doi.org/10.1007/978-3-031-39985-5_6

Figure 6.1a shows such a waveform with vertical changes at $t_1 = T/2$ and $t_2 = T$. Analysis of a system without sudden changes can be carried out using the Fourier series for both continuous and piecewise continuous waveforms. However, it is worth to emphasize that in the case of piecewise continuous waveforms, the value calculated from the Fourier series for the waveform at the discontinuities would correspond to the mean value of the vertical region. The main theorem concerning the convergence of the Fourier series at a discontinuity implies that this series converges to $f(t)$ except at the point $t = t_o$, which is a point of discontinuity of $f(t)$. Indeed, Gibbs [1, 2] showed that if $f(t)$ is piecewise smooth on $[0, T]$ and t_k is a point of discontinuity, then the Fourier partial sums will exhibit the same behaviour, with the bump's height almost equal to

$$\Delta f(t_k) = 0.18(f(t_{k+}) - f(t_{k-})) \tag{6.1}$$

Recall that the notations $f(t_{k+})$ and $f(t_{k-})$ represent the right-hand limit and left-hand limit, respectively, of $f(t)$ at the point t_k.

Therefore, this is not a practical problem as actual waveforms do not exhibit exactly vertical changes but those occurring over very small intervals of time. Nevertheless due to the presence of non-linear devices in the system, waveforms resident

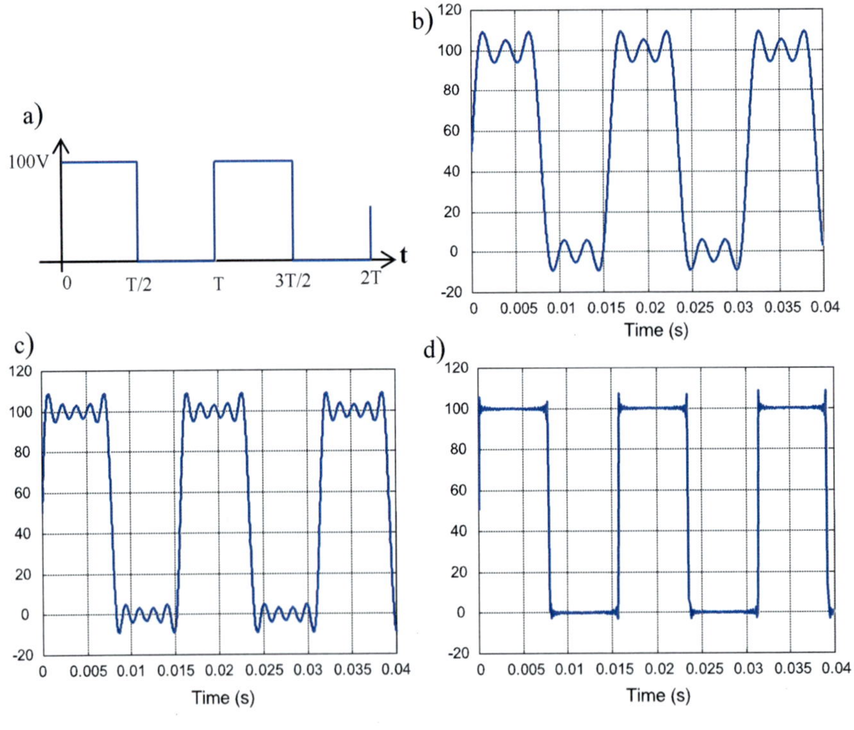

Fig. 6.1 Waveform with discontinuities: **a** vertical at T/2 and T, **b–d** Fourier series representations of a rectangular waveform ⓪ (**a**) using: **b** ③, **c** ⑦ and **d** ⑤⓪ sinusoidal components

in particular system components get distorted forms from the sinusoidal (Fig. 6.1b). For instance, the Fourier series representation of the square wave $f(t)$ takes the form

$$f(t) = \frac{4A}{\pi}\left[\cos(\omega t) - \frac{1}{3}\cos(3\omega t) + \frac{1}{5}\cos(5\omega t) - \frac{1}{7}\cos(7\omega t) + \cdots\right]$$
$$= \frac{4A}{\pi}\sum_{n=1}^{\infty}\frac{1}{n}(-1)^{\frac{n-1}{2}}\cos(n\omega t) \tag{6.2}$$

where n denotes successive odd numbers and $\omega = \frac{2\pi}{T}$ is the fundamental angular frequency. Most of the approaches adopted at present for the analysis and operation of the nonlinear dynamical systems—from the mathematical formulation of the equations to model the system behavior through the implementation of devices for compensating the distortion to the necessary measuring equipment—are undergoing a thorough revision. It should be emphasized that the analysis and design of such systems present significant challenges. The evolution of these topics appears as a tremendous turmoil of different approaches in the worldwide and it is arduous to find a general agreement among individual authors, even when they refer to identical concepts and quantities [3, 4]. Electrical networks and physical systems in general, are known to satisfy a power balance equation which states that the rate of change of the energy generated by sources in time equals the power at the loads minus the power dissipated in the network. From its beginnings in the late nineteenth century, electrical engineering has blossomed from focusing on electrical circuits for power, telegraphy and telephony to focusing on a much broader range of disciplines in which energy flows play a significant role. Although power creation and transmission as well as information have been the underlying themes of physics and especially of electrical engineering, then they are relevant today and require special care. However, the increased availability of powerful computing has made direct simulation widely accessible [1–12] and has enlarged the set of tractable modelling and analysis approaches. Simulations or computations (such as determination of component stresses) at this modelling level can frequently be implemented simply and conveniently in a spreadsheet program such as Excel or using a general computational environment such as Mathematica or MATLAB.

In the last few years the renewable energy, wind and solar energy as well as hydrogen energy, on the other hand, have attracted the attention of state and local authorities in many countries. To propagate the renewable energy sources and to place them in the top of community efforts it is necessary to develop and improve effective tools to evaluate generation, consumption and transfer of the energy in systems operating within non-sinusoidal conditions. However, in non-sinusoidal conditions apparent power definition of dynamical systems loads is inconsistent with the energy conservation law. Up until now, in the classical methods with complex numbers application, this mathematical model is only valid in the case of purely dissipative loads. Interestingly, although the formulation of every power theory relies on the classic apparent power S, which in the case of an electric circuit component takes the following form

$$S = U_{rms} I_{rms}, \tag{6.3}$$

where U_{rms} and I_{rms} denote root mean square values of the voltage and current at the component terminals.

The principle of conservation of energy does not apply to this quantity. Thus, it cannot be corroborated for unified systems where sources and loads operate simultaneously under non-sinusoidal conditions, and as a result, it can lead to erroneous conclusion. This power concept must be revisited because the classical tools based simply on complex numbers are not enough to achieve physically correct results.

Although many power theories have been proposed recently regarding non-sinusoidal operation, an adequate solution is yet to be found and no power theory has achieved general acceptance yet. Therefore, it has to be underlined that the typical linear/nonlinear behaviour of dynamical systems operating in non-sinusoidal conditions requires, for its complete analysis, a new mathematical structure that could guarantee the right character of different components.

6.2 New Standards for the Energy Avenue in Non-sinusoidal States

6.2.1 Preliminary

When the circuit waveforms are irregular, then the command signals are poorly controlled and instantaneous events occur that distort the usually "pure" or ideal fundamental waveforms. In such conditions the instantaneous state of the system is degraded and some superfluous losses are involved with respect to the normal utilization of the energy. For the above and other reasons studies of energy phenomena in linear and nonlinear systems have emerged as an important subject in recent years due to renewed interest in improving the quality of the performance and efficiency of systems of different kinds. To add an idea brick to this development, it is also valuable to throw energy harvesting processes whereby energy comes from external environments, captured and stored in small wireless autonomous devices, such as those used in wearable electronics and wireless sensor networks [13]. Thus the renewable energy, or even lower cost energy, is to become prevalent energy storage and is a critical component in reducing peak power demands and the intermittent nature of solar and wind power. Presently, the energy storage is a critical component of manufacturing, of the service industry, of the future renewable energy industry, and of all the portable electronics with which we have become obsessed.

As mentioned in the previous subsection, all up-to-date methods used for energy determination of a system element operating in periodic non-sinusoidal conditions have many insufficiencies which vary from one case to another and importantly depend on the assumed interpretation of particular components of the apparent power defined in the complex domain (6.3). It should be emphasized that no power theory has achieved general acceptance as yet. Despite of the fact that the concept of a "distorting" power has no physical significance it is, even today, widely used amongst practicing engineers. What's more, the mathematical formulation of reactive power may cause incorrect interpretations of the energy flow. Therefore the lack of a unique definition of a power under non-sinusoidal conditions makes that commercial measurement systems utilize different definitions, producing different results, and as a consequence, generates significant negative economic effects. Indeed, measurement systems may present different results, not only because of different principle of their operation, but because of the adoption of different quantities definitions as well.

The need for universally accepted power definitions has come to the fore now, like never before, in order that designs and energy accounts can be defined and used unambiguously and through which distortion compensation can be specified. The correct approach can be derived from the fact that, in non-sinusoidal conditions and with linear and/or nonlinear loads, the traditional apparent power definition is erratic, except for resistive loads. One of the main reasons for this imperfection is the specificity of the properties of the Fourier series, which does not accurately reproduce the course of the signal with its "value spikes". The second reason is the widespread use of defined variables such as voltage $v(t)$ and current $i(t)$ to simplify notations and calculations of energetic processes despite the fact that they are accepted to be "mathematically derived" rather than "physically basic".

It should be now considered that in order to present easy as much as possible, the relevant concepts, terminology and notations that are useful in the offered proposal, they will be examined as exemplary standard on an example of processes occurring in electrical circuits.

All the presented in this way relevant characteristics of energy flow analysis confirms its universal approach to investigate any dynamical systems in science and engineering fields. This will be demonstrated later in this chapter.

6.2.2 Hysteresis Loops on Energy Phase Plane

To defeat the above mentioned difficult problems we consider in what follows a much more general setting, i.e., hysteresis loops on energy phase plane. A common feature of these loops is that all problems concerning the non-sinusoidal periodic states in electrical circuits can be reported to studies of the somewhat complementary aspect of shaping the energy of the dynamical system on the energy phase plane. The established method is very straightforward and appears as a powerful broadly applicable technique that enables to characterise non-sinusoidal periodic oscillations from a perspective different than that obtained by the method resulting from the Fourier series approach. The main feature of the present approach is the total elimination of the harmonic analysis which, in turn, significantly simplifies the appliances used for measuring the energy rate. Taking into account the four classic circuit elements (resistor, inductor, capacitor and memristor) there are six different mathematical relations connecting pairs of the four fundamental circuit attributes: electric current i, voltage v, charge q, and magnetic flux φ. By accepting φ and q as basic (existing in nature), the other, $v = \frac{d\varphi}{dt}$ and $i = \frac{dq}{dt}$ should be viewed as "derived attributes" that are not independent. Physically, magnetic flux φ, and electric charge q, are fundamental features to describe a circuit component. In other words, φ and q are internal features associated with the device material and its physical operating mechanism. They are basic and intrinsic. Contrarily, voltage v, and current i, could be derived from and via Faraday's law and by definition. Therefore, the voltage and the current, which are although conveniently used in practice, are only mathematically defined and exhibit external measures of a component, only.

To support the statement that φ and q are physically intrinsic and basic we can find at least three evidences, namely. Voltage and current should not be chosen to generate new elements as they are not physically basic because voltage v is always a "difference" when measuring it, and current i is always the velocity of electric charge. Note, however, that in such case, $I \equiv 0$ does not imply $q \equiv 0$, and thus such device can store the energy.

Any primary two-terminal component of electronic circuits should link two attributes, at least one of which should be physically basic.

In these links the charge q being the time integral of the current i is determined from the definitions of two of the attributes, and another, the flux φ as the time integral of the electromotive force, or voltage, is determined from Faraday's law of induction.

Taking into account the above observation that φ and q are basic physical attributes, we will focus our study on the φ–q plane and its transformation to the v–q and i–φ planes. It is worth to underline that although v and i are considered to be "mathematically derived" rather than "physically basic," they are conveniently used in everyday life as a result of the historical development of electrical sciences, so the relationships represented on v–i plane are yet commonly and extensively practiced (e.g. Ohm law).

When switching operations occur or driving signals take square waveforms, theoretically sudden changes in instantaneous values of other waveforms could occur giving those multi-values at corresponding instants. A systematic technique that facilitates the energy computations of system components when the switch occurs can lead to determination of effective values of the voltages and currents.

The idea of effective values, often known as the root-mean-square values, arises from the need to measure the effectiveness of a voltage or current source in delivering energy to a resistive load. They are found as

$$V_{rms} = \sqrt{\frac{1}{T}\int_0^T v^2(t)dt}, \quad I_{rms} = \sqrt{\frac{1}{T}\int_0^T i^2(t)dt} \tag{6.4}$$

In general, for any periodic function $f(t) = f(t + T)$, the *rms* value is given by

$$F_{rms} = \sqrt{\frac{1}{T}\int_{0}^{T} f^2(t)dt} \tag{6.5}$$

The importance of (6.4) lies mainly in the fact that just as in all areas of physics, biology and chemistry, power is the rate at which energy is produced or consumed. Taking into account a circuit as a causal system we can assume that for $t < 0$ the element was devoid of energy, i.e. $w(t) = 0$ for $t \leq 0$ and obtain

$$w(t) = \int_{0}^{t} v(\tau)i(\tau)d\tau \tag{6.6}$$

with $w(0) = 0$. This means that the energy of the circuit element is determined in terms of two mathematically derived quantities $v(t)$ and $i(t)$. It should be emphasized that voltage $v(t)$ and current $i(t)$ are interrelated with each other by means of the equations of the circuit state, i.e., Kirchhoff voltage and current laws, and appropriate relationships that define the individual elements, i.e., generalized Ohm law. Of special importance is the definition of the energy under non-sinusoidal conditions. Hence, there are frequent problems with the use of (6.6) for calculations the energy and especially for the power of circuit elements in the case of discontinuous signals. Although the electrical scientific community has been seeking for more than a hundred years of precise power theory for interpreting the energy flow within electric circuits under non-sinusoidal conditions, so far there is no satisfactory result. It has to be noted that many power theories have been up-to-date proposed regarding non-sinusoidal operation but any adequate solution is yet to be found. Moreover, the evolution of these topics is a tremendous turmoil of different approaches and it is very difficult to find a general agreement among individual authors, even when they refer to identical concepts and quantities [14, 15]. The advocates of each of these theories claim unique as well as universal advantages for their definitions, but none of these are truly able to cover the whole spectrum of requirements. In what follows a new approach for the determination of the energy of circuit elements operating in periodic non-sinusoidal conditions is presented. Exact calculations for such problems are indeed possible and accurate. To meet this requirement new standards for one-period energy in non-sinusoidal states are involved and well justified through suitable applications of appropriate results following properties of Riemann–Stieltjes integral [16, 17].

The context for all the following on will be a finite interval $[a, b] \in \Re$ and a pair of real-valued functions $f(t)$ and $g(t)$ that are defined and bounded on it. In most of applications, $g(t)$ will be a function of bounded variation, but the definition does not require it.

Definition 1 A Riemann–Stieltjes integral is of the form

$$I_{RS} = \int_a^b f(t)dg(t) \tag{6.7}$$

Functions $f(t)$ and $g(t)$ are called the integrand and integrator, respectively. If the integral (6.7) exists, then $f(t) \in \Re(g(t))$ on $[a, b]$. Function $f(t)$ is Riemann–Stieltjes integrable with respect to $g(t)$ if and only if $g(t)$ is Riemann–Stieltjes integrable with respect to $f(t)$. This insures that if $f(t) \in \Re(g(t))$ on $[a, b]$ then $f(t) \in \Re(g(t))$ on $[a, t]$ also, for all $t \in [a, b]$. Thus the function $F(t) = \int_a^t f(\tau)dg(\tau)$ is well defined on $[a, b]$.

Theorem 1 *If $\int_a^b f(t)dg(t)$ exists in the sense of either definition, then so does $\int_a^b g(t)df(t)$ and their values are related by*

$$\int_a^b f(t)dg(t) + \int_a^b g(t)df(t) = f(b)g(b) - f(a)g(a) \tag{6.8}$$

Proof It follows immediately from properties of corresponding telescoping and Riemann–Stieltjes sums and more details in this direction are exposed in [16].

The next theorem is devoted to the problem of making all propositions about applications of the Riemann–Stieltjes integrals because when they exist then appear to be nonvacuous.

Theorem 2 *Suppose that f(t) and g(t) are continuous on [a, b] and g(t) ≥ 0 on [a, b] (or g(t) ≤ 0 on [a, b]). Then there exists a number ξ in [a, b] such that*

$$\int_a^b f(t)dg(t) = f(\xi)\int_a^b dg(t) \tag{6.9}$$

Moreover, if $\dot{g}(t)$ is not increasing (or not decreasing) on [a, b], then there exists a number γ in [a, b] such that

$$\int_a^b f(t)dg(t) = \dot{g}(a)\int_a^\gamma f(t)dt + \dot{g}(b)\int_\gamma^b f(t)dt \tag{6.10}$$

where $(\cdot) = d(\ )/dt$.

Next, if to above assumptions we add $\dot{g}(t) \geq 0$ on [a, b], then there is a such number γ in [a, b] that

$$\int_a^b f(t)dg(t) = \begin{cases} \dot{g}(a)\int_\gamma^b f(t)dt \text{ if } \dot{g}(t) \text{ is not decreasing} \\ \dot{g}(a)\int_a^\gamma f(t)dt \text{ if } \dot{g}(t) \text{ is not increasing} \end{cases} \tag{6.11}$$

Proof Details of the proof are beyond the scope of this chapter for the sake of compactness of presentation. Let us be allowed only to indicate that it can be found in [17].

It has to be noted that the above result is both general interest and appears as a useful tool for proving some of the properties often seek in applications of dynamical systems. Taking into account these possibilities, we can go back to solving problems generated in electrical circuits that operate in non-sinusoidal periodic states.

At first, consider as a guide for further analyses the use of Riemann–Stieltjes integral to determine the *rms* values of periodic voltage $v(t)$ with discontinuities. Taking into considerations the particular values $a = 0$ and $b = T$ we can evaluate the *rms* values applying the definition relationship (6.4). After expressing the voltage $v(t)$ in terms of the *physically intrinsic quantity* $\varphi(t)$ (magnetic flux) by $v(t) = \frac{d\varphi(t)}{dt}$ yields

$$V_{rms} = \sqrt{\frac{1}{T}\int\limits_{\varphi(0)}^{\varphi(T)} v(t)d\varphi(t)} = \sqrt{\frac{S_l}{T}} \tag{6.12}$$

where S_l denotes the delimited surface on (v, φ) plane of the circuit two terminals component.

The importance of (6.12) lies mainly in the fact that it can easily be mapped on the plane (v, φ) in the form of a hysteresis loop that delimits the surface S_l. By referring this surface the period T and then calculating the square root, we set the desired *rms* value. Here, we will study periodic voltages $v(t)$ with period $T = 2\pi/\omega$, amplitude A, and zero mean values, i.e. $\frac{1}{T}\int\limits_0^T v(\tau)d\tau = 0$. Such voltages are standard test signals for electronic devices and lead to the classic characterization of the loads through hysteresis and Lissajous-type $i(v)$ characteristics. In particular, we consider four test signals widely used in engineering settings, namely, the sinusoidal, bipolar square (Heaviside), triangular and saw tooth waveforms, all defined over a period $0 \leq t \leq T$ as follows:

(i) Sinusoidal voltage:

$$v(t) = A\sin(\omega t), \tag{6.13a}$$

(ii) Bipolar square (Heaviside) voltage:

$$v(t) = A\frac{abs(\cos(\omega\, t))}{\cos(\omega\, t)}, \tag{6.13b}$$

(iii) Triangular voltage:

$$v(t) = \frac{2A}{\pi}\arcsin\left(\sin\left(\frac{2\pi}{T}t\right)\right) \tag{6.13c}$$

(iv) Saw tooth voltage (based on the *floor function* of time t):

$$v(t) = 2A\left(\frac{t}{T} - floor\left(\frac{1}{2} + \frac{t}{T}\right)\right). \tag{6.13d}$$

The four voltage waveforms with $T = 2$ s and $A = 1$ V are shown in Fig. 6.2a. Using the expression (6.12), we can easily compute the *rms* value of the above voltages (6.13a)–(6.13d) applying the procedure *plot* or *trapz* from the MATLAB package. The hysteresis loops mapped on the plane (v, φ) for the voltages shown in Fig. 6.2a are presented in Fig. 6.2b. Note that the effect of the voltage shape on its *rms* value is prominent and easily recognizable. In such cases, the surfaces delimited by the corresponding loops are: $S_1 = 2.0000$, $S_2 = 1.0000$, $S_3 = 0.6667$ and $S_4 = 0.6667$, respectively, measured in V^2s. Thus on the base of expression (6.12) we get the following *rms* values: $V_{1rms} = 1.0000$, $V_{2rms} = 0.7071$, $V_{3rms} = 0.5774$ and $V_{4rms} = 0.5774$ (all in volts).

Observe that the bipolar square voltage is characterized by the highest *rms* value but the triangular and saw tooth voltage waves exhibit the same *rms* values although they exhibit quite different shapes. It should be added that the above result is obtained without any recourse to the Fourier series being widely up to date in using. Moreover, the recommended method leads to entire relinquishment of classic frequency analysis.

The above considerations are only an entry into *rms* values calculation in cases of more complex signal shapes, and above all, those for which the analytical description is difficult or impossible to determine. We will consider this problem on the example of the circuit shown in Fig. 6.3 which contains a memristor.

In this case, and many others also, the memristor $M(\varphi)$ represents nonlinear function (e.g. Chua's diode) and together with other elements (voltage source, resistor, capacitors, inductor and current controlled current source) realizes mixed mode oscillations (MMOs) [17]. Nevertheless, in what follows we are not concentrated on memristor elements realization itself. Its nonlinear and dynamical properties are used

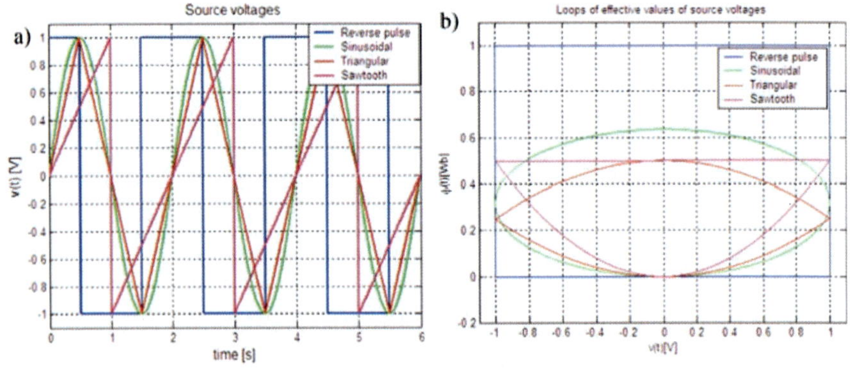

Fig. 6.2 Results for *rms* values: **a** testing waveforms, **b** hysteresis loops on energy phase plane

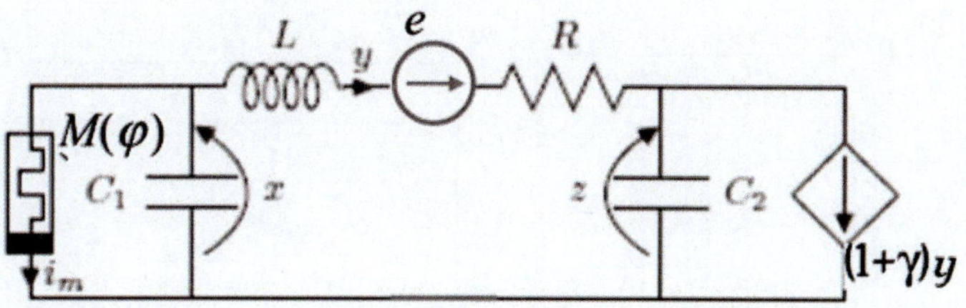

Fig. 6.3 Memristive circuit

for the realization of a simple mixed mode oscillator, where the memristor function is important part of the circuit. This type of interesting periodic oscillations is revealed by dynamic resistive nonlinear circuits (usually having cubic current–voltage characteristics) and in various chemical, biological and astrophysical systems [18–22]. It is worth mentioning that no results concerning energy's characterizations of memristive circuits with MMOs responses have been published yet.

The circuit shown in Fig. 6.3 is described by the following system of equations

$$\begin{aligned} C_1\dot{x} &= -y - (a + 3bw)x \\ \dot{y} &= \alpha(x - Ry - z + e) \\ \dot{z} &= -\beta y \\ \dot{w} &= x \end{aligned} \tag{6.14}$$

with $w = \varphi$ (flux), $0 < C_1 \equiv \varepsilon \ll 1$, $\alpha = 1/L$, $\beta = \gamma/C_2 > 0$, and $q = a\varphi + b\varphi^3$ represents the memristor characteristic. Several typical steady-state MMOs responses of the above circuits for various values of a and b (parameters of the memristive element) can be established. The appearing MMOs are sensitive not only to the chosen initial conditions (ICs) but also to the parameters in (6.14). For instance, taking into account the following parameters and ICs: $\varepsilon = 0.01$, $R = 2.0$, $\alpha = 1.55$, $\beta = 0.15$, $e = -0.04$, $a = -0.34$, $b = 0.05$ and $x(0) = y(0) = z(0) = w(0) = 0.0$ results in MMOs 2^{11}. The obtained large- and small-amplitude periodic oscillations (LAOs and SAOs), denoted in brief by L^s, of the memristor voltage are shown in Fig. 6.4a in the time interval 275 s $\leq$ t $\leq$ 300 s.

In order to determine the *rms* value of this voltage the expression (6.12) was used and after representing the integral of the internal expression in the square root on the plane (v, w), we obtain the *rms* hysteresis loop, which is shown in Fig. 6.4b. Calculating the surface delimited by this loop with applying procedure *trapz* from MATLAB yields $S_l = 392.0512$ V s and in accord with (6.12) gives $V_{rms} = 7.0823$ V. The period $T = 7.8160$ s of changes in the instantaneous voltage has been determined using the procedure described in [17].

Applying the above procedure appropriate MMOs responses can be obtained for other parameter values and ICs. Changing ICs to $x(0) = 2.22$, $y(0) = -1.105$, $z(0) = -0.00628$, $w(0) = 0.375$ and accepting $a = -0.5$ with unchanged other parameters gives the course of MMOs $2^4 2^6$ shown in Fig. 6.5a. The corresponding hysteresis

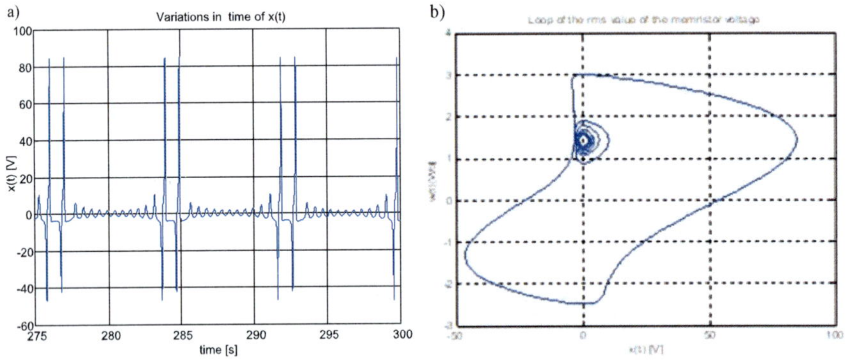

Fig. 6.4 Mixed mode oscillation 2^{11} in the memristive circuit: **a** time varying voltage, **b** loop of the *rms* value of the voltage

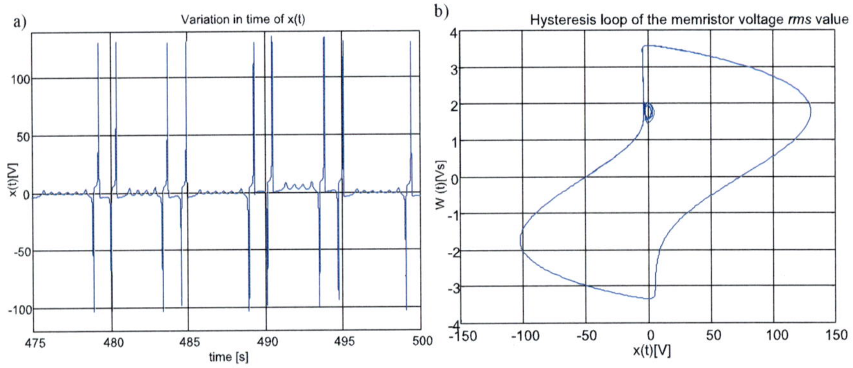

Fig. 6.5 Mixed mode oscillation $2^4 2^6$ in the modified memristive circuit: **a** time varying voltage, **b** loop of the *rms* value of the voltage

loop of the memristor voltage *rms* value is presented in Fig. 6.5b. The computed loop's area equals $S_l = 791.0177$ $\mathrm{V}^2\mathrm{s}$ and hence $V_{rms} = 8.8939$ V.

Applying the above procedure we can calculate the *rms* values of all other state variables in the circuit but for the sake of presentation compactness they are not reported here.

6.2.3 Quantitative Measures of the Energy Hysteresis Loop

The never-ending drive towards correct descriptions of energetic processes occurring in dynamical circuits and systems operating under periodic non-sinusoidal conditions poses severe challenges for the efficient supply designer.

In view of the discussion in the precedent sections controlling and designing the energy generated or consumed by a circuit element is crucial for its use both individually or as a part of larger circuits working under periodic non-sinusoidal conditions. It is worth to mention that the energy flow in complex dynamical circuits and systems is denoted as one of the outstanding unsolved problems of modern technologies. The analysis of the interaction between voltage and current waveforms is indispensable to determining the electrical energy flow by interpreting appropriate power components [23–27]. Therefore, it is important to have an appropriate method to control the energy effects of the circuit individual elements in order to achieve the intended data for practical implementations.

If problems related to the energy are under considerations, then as is well known in non-sinusoidal conditions, some quantities can conduct to wrong interpretations, and others can have no meaning at all. For this reason the need for universally adopted energy identifications has come to the fore now, like never before, in order that designs and energy accounts can be defined and used unambiguously and through which distortion compensation can be specified. Hence, research for a single theory in this area is very hot up to date and should meet the requirements of modern science and technology.

In order to overcome the problems highlighted, here is proposed a quantitative measure of the energy hysteresis on the (v, q) or, equivalently, (i, φ) planes as the work carried out by the driving non-sinusoidal input on the device during a desired multiplicity of one period. We start on a well known and up-to-date implemented extensively notion of the instantaneous power $p(t)$ generated by a two terminal source or consumed by a two terminal load which is given by the product of the voltage $v(t)$ and current $i(t)$, affiliated correspondingly with clamps of the element, namely

$$p(t) = v(t)i(t) \tag{6.15}$$

Power $p(t)$ is an additive quantity and, therefore, if it is supplied to an object via many sources, then its total power is equal to the sum of the power supplied by all particular sources. In periodic states, it is the average electric power P which is produced during the period T of a voltage and current waveforms. It is named as active power expressed by

$$P = \frac{1}{T}\int_0^T p(t)dt = \frac{1}{T}\int_0^T v(t)i(t)dt \tag{6.16}$$

which has a clear physical meaning even for non-sinusoidal conditions. On the other hand, the mathematical formulation of reactive and distortion powers may cause incorrect interpretation, aggravated when the analysis is extended to three phase systems operating under non-sinusoidal periodic conditions [28–30].

For any periodic voltage and current waveforms the following relation holds:

$$P \leq V_{rms} I_{rms}, \tag{6.17}$$

where V_{rms} and I_{rms} denotes the root mean square values of the distorted voltage and current, respectively.

In view of the major discrepancies in interpretation and frequent errors occurring in the application of definitions for different powers, the most appropriate approach to base the assessment of energy transfer in a circuit to be on the definition of one-period energy appears. The basic strategy lays on the construction of a geometric interpretation of the hysteresis loop on the energy phase plane.

Consider the steady state energy $W(\Delta t)$ transferred from or delivered to a circuit element during the time interval $\Delta t = nT$, where $n \gg 0$ denotes a positive integer. It is expressed by

$$W(\Delta t) = nW_T \tag{6.18}$$

where W_T denotes the energy flow during one period of the input and output waveforms. Therefore, in the periodic non-sinusoidal state it is sufficient to evaluate W_T and then multiplying it by n yields the energy generated (or consumed) by a circuit element during the given time interval $\Delta t = nT$.

Thus, from here and on, the goal is to establish W_T. In order to deal with it our analysis will follow similar line of argument to the one presented in the precedent section. Therefore, as mentioned previously, the reliable relationship between the energy of an element and its attributes should be unique and not synthesized. Hence, in accordance with this statement, also the energy of the circuit element should be expressed by two physical attributes, at least one of which should be basic.

To meet these requirements new standards for one-period energy in non-sinusoidal states can be involved and well justified through suitable applications of appropriate results following the Riemann–Stieltjes integral properties. In this new approach the main emphasis is put on the somewhat central aspect of shaping the energy of the circuit elements, which directly involves the hysteresis loop on the energy phase plane, as opposed to the widely up-to-date used concept of the active and different non-active powers.

Using the expression (6.4), the energy of a circuit two-terminal element over a period T is:

$$W_T = \int_0^T v(t)i(t)dt \tag{6.19}$$

where the condition that $W(0) = 0$ is taken into account. The derivation of the alternative expression for W_T leads to

$$W_T = \int_0^T v(t)i(t)dt = \int_0^T v(t)\frac{d}{dt}\left(\int_0^t i(\tau)d\tau\right)dt$$
$$= \int_{q(0)}^{q(T)} v(t)dq(t) = \int_{\varphi(0)}^{\varphi(T)} i(t)d\varphi(t) \tag{6.20}$$

where

$$q(t) = \int_0^t i(\tau)d\tau \text{ and } \varphi(t) = \int_0^t v(\tau)d\tau \tag{6.21}$$

denote the instantaneous electric charge and magnetic flux, respectively.

It follows from expressions (6.20) and (6.21) that the area enclosed by a loop on the energy phase plane with coordinates $(q(t), v(t))$ or, equivalently, $(\varphi(t), i(t))$ determines the one-period energy W_T delivered from, or absorbed into, respectively, a one-port network being under periodic non-sinusoidal conditions. Thus for any two-terminal circuit component operating in a periodic non-sinusoidal regime it is possible to produce directly the one-period energy W_T without recourse to any forms neither of its power nor the Fourier series approach. The area within the hysteresis loop corresponds to the hysteresis losses, i.e. to the energy dissipated in the load during one cycle of changes of circuit variables. It is worth mentioning that in the case of conservative circuit elements their one-period energy loops reduce to a line segment.

Fundamental properties of the Riemann–Stieltjes integral in the context of the energy characteristics of electrical dynamical systems give possibilities for introducing new quantities describing circuit elements from a quite different point of view then all up-to-date used ones. First, for a two-terminal circuit element operating under periodic non-sinusoidal conditions we can establish a new quantity called "*energetically effective value*" V_{effT}. To obtain this quantity we take into account the findings of Theorem 2 and the relation (6.9), and after applying them to the two-terminal circuit element, in accordance with (6.20), we determine

$$W_T = \int_0^T v(t)dq(t) = V_{efT}\int_0^T i(t)dt = V_{efT}TI_{av} \tag{6.22}$$

where

$$I_{av} = \frac{1}{T}\int_0^T i(t)dt \tag{6.23}$$

denotes the average value of the element current $i(t)$. Thus we can define energetically effective values of periodic non-harmonic waveforms at terminals of a two-terminal element as follows.

Definition 1 A bounded periodic non-sinusoidal voltage $v(t)$ exhibiting a finite number of jump discontinuities on the whole period interval $[0; T]$ and exciting in an element of dynamic circuit a current with non-zero average value I_{av} is characterized by the value

$$V_{efT} = \frac{W_T}{T I_{av}} \tag{6.24}$$

which can be called *energetically effective value* of the periodic non-sinusoidal voltage $v(t) = v(t + T)$. Similarly, for a bounded source current $i_s(t) = i_s(t + T)$ exciting a voltage $v(t)$ at terminals of a circuit element with nonzero average value V_{av} the energetically effective value is given by

$$I_{efT} = \frac{W_T}{T V_{av}} \tag{6.25}$$

where the same notations as above ones are used.

To concentrate the attention consider a circuit operating in sinusoidal conditions with instantaneous voltage and current at given two-terminal element determined by

$$v(t) = \sqrt{2}|V|\cos(\omega\, t),\, i(t) = \sqrt{2}|I|\cos(\omega\, t - \alpha) \tag{6.26}$$

where $|V|$ and $|I|$, ω and α denote the *rms* values of the voltage and current, angular frequency and angle lag, respectively.

We assume the energy phase plane with coordinates $(i(t), \varphi(t))$, where the flux $\varphi(t)$ is expressed as follows

$$\varphi(t) = \int_0^t v(\tau)d\tau = \frac{\sqrt{2}}{\omega}|V|\sin(\omega t) \tag{6.27}$$

Thus the one-period energy of the two terminal elements can be expressed on the energy phase plane by

$$\left(\frac{i(t)}{a}\right)^2 + \left(\frac{\varphi(t)}{b}\right)^2 = 1 \tag{6.28}$$

where

$$a = \frac{\sqrt{2}|I|\cos(\alpha)}{\sqrt{1 + \sin(\alpha)}}, \quad b = \frac{\sqrt{2}|V|\cos(\alpha)}{\omega\sqrt{1 - \sin(\alpha)}} \tag{6.29}$$

Equation (6.28) represents an ellipse on the energy phase plane with semi-axes defined by expressions (6.29). Since the area of an ellipse is defined by the relation $S = \pi ab$, then taking into account (6.29) yields

$$S = \pi ab = \frac{2\pi}{\omega} \frac{|V| \cdot |I| \cos^2\alpha}{\sqrt{1 - \sin^2 \alpha}} = T|V| \cdot |I| \cos(\alpha) = W_T \tag{6.30}$$

Hence, the one-period energy of a two-terminal element operating in sinusoidal conditions can be expressed as the surface of a corresponding ellipse plotted on the energy phase plane.

For instance, series connected elements $R = 1\Omega$ and $L = 5/\pi$ mH supplied by voltage source $v = 220\sqrt{2}\cos(2\pi f t)$ V exhibit the one-period energy ellipses shown in Fig. 6.6a. The ellipses have been established for different frequencies f taken successively from the set {50, 100, 150, 200, 250, 300, 600} Hz. This pattern changes with changing parameter L. It should be noted that the set of vertices of the individual ellipses forms a curve that is similar to the basic magnetized curve of a ferromagnetic material when it is magnetized first time.

Since sinusoidal signals are considered as widely used basic signals in most electrical circuits and systems, then the ellipse representing the energy of two-terminal elements operating under sinusoidal conditions can be accepted as a standard pattern for which the energies of all elements operating in non-sinusoidal states should be compared. On this basis it is possible to determine the energetic properties of the circuit elements by means of such factors as the coefficient χ of energy efficiency of the one-period energy loop and the coefficient k as a measure of its deformation with respect to the standard one-period energy ellipse.

The coefficient χ of the energy efficiency of a circuit element is defined as the ratio of the square of the perimeter $\mathcal{O}$ to the surface S of the loop, namely

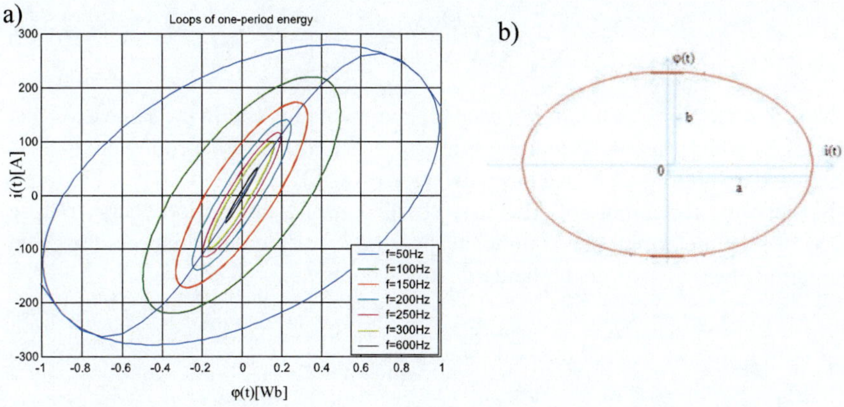

Fig. 6.6 One-period energy phase plane: **a** illustration of frequency influences on energy ellipse form, **b** standard form of an ellipse

$$\chi = \frac{\mathcal{O}^2}{\mathcal{S}} \tag{6.31}$$

It characterizes the dominance of the perimeter of a given one-period energy loop with respect to a square root of its surface. This allows to possibilities of comparing different loops of the same perimeter but with as large surfaces as possible. The absolute minimum is in the case of an ellipse with the same semi-axes (i.e. circle) for which $\chi_{\min\,\mathrm{abs}} = 4\pi$.

However, considering the nominal standard ellipse (Fig. 6.6b) as a one-period energy loop in the sinusoidal state the more efficient approach is to refer a given loop to the elongated ellipse, elongation η of which is determined by

$$\eta = \frac{a-b}{a+b} \tag{6.32}$$

Depending on the η, the perimeter and the area of the ellipse are defined as

$$\begin{aligned} \mathcal{O}_e &= \pi(a+b)\left[1 + \frac{\eta^2}{4} + \frac{\eta^4}{64} + \frac{\eta^6}{256} + \frac{\eta^8}{1638} + K\right] \\ \mathcal{S}_e &= \pi ab \end{aligned} \tag{6.33}$$

where K denotes a small remainder of the power series.

Taking into account the shape coefficient of the ellipse $\lambda = \frac{a}{b}$ we get the estimation

$$\mathcal{O}_e > \pi(a+b) = \pi b(1+\lambda) \tag{6.34}$$

Hence, the minimum efficiency coefficient is for a one-period energy ellipse, namely

$$\chi_{e\min} = \min\left(\frac{\mathcal{O}_e^2}{\mathcal{S}_e}\right) = \pi\frac{(1+\lambda)^2}{\lambda} \tag{6.35}$$

Note that the absolute minimum is obtained for an ellipse with the same axes (i.e. a circle) for which this coefficient is 4π. So, for every ellipse representing one-period energy and exhibiting a > b, we have always $\chi_e > \chi_{e\min} > 4\pi$.

By applying the minimum efficiency coefficient of a one-period energy ellipse (6.35), we can determine the significant measure k_λ of the deformation of a given one-period energy loop versus standard pattern energy expressed by

$$k_\lambda = \frac{\mathcal{O}^2/\mathcal{S}}{\chi_{emin}} = \frac{\mathcal{O}^2/\mathcal{S}}{\pi(1+\lambda)^2/\lambda} \tag{6.36}$$

where $\mathcal{O}$ and $\mathcal{S}$ denote the perimeter and area of one-period energy loop of the given two-terminal circuit element operating under periodic non-sinusoidal conditions.

For instance, if a circuit element is characterized by one-period energy loop taking the form of a rectangle with sides $c = 2a$ and $d = 2b$ (the rectangle is described exactly on the ellipse) then

$$k_{\lambda rect} = \frac{4(2a + 2b)^2/4ab}{\pi(1 + \lambda)^2/\lambda} = \frac{4}{\pi} \cong 1.2732 \tag{6.37}$$

This result refers to a complex *RLC* circuit in which the source voltage takes alternatively the rectangular and sinusoidal waveforms of the same period and the same amplitudes. The value of this deformation measure being greater than the unity indicates that the rectangular supplying voltage provides the circuit with greater energy in comparison to sinusoidal power supply. If the rectangular voltage is replaced by a triangular voltage of the same period and the same amplitude as the sinusoidal voltage, then the deformation measure will take the following value

$$k_{\lambda tri} = \frac{\mathcal{O}_{tri}^2/\mathcal{S}_{tri}}{\pi(1 + \lambda)^2/\lambda} = \frac{\pi(1 + \lambda_{tri})^2/\lambda_{tri}}{\pi(1 + \lambda)^2/\lambda} = \frac{11.66}{70.71} \cong 0.1649 \tag{6.38}$$

In this case, the deformation measure is significantly less than unity, indicating a significant deformation of both the source-related signals and the smaller one-wave energy produced by the triangular voltage source.

Therefore, the one-period energy loop allows determining not only the element's energy, but also the degree of distortion of signals with respect to sinusoidal waves.

To exploit further the insight provided by the framework, we apply a quantitative measure of the energy hysteresis of the memristor in terms of the work done by the driving signal e, and we use the computer solution of Eq. (6.14) to show that the hysteresis of the one-period energy exhibits a specific double pinched shape that combines all the dynamic characteristics of the model. For the above given parameters of the memristive circuit (Fig. 6.3) the resulting loop of the one-period energy on the (i, w) plane is shown in Fig. 6.7. This gives $W_T = 15.4127$ J as energy consumed by memristor during one period of MMOs. As our result shows, the hysteresis pinches correspond to the i–v curve of the memristor exhibiting only one pinched point at the plane origin, where $i = 0$ always excites $v = 0$ [31, 32].

However, notice that in the case of MMOs the one-period energy hysteresis loops are much more rich because they exhibit both LAOs and SAOs and contain two pinch points, both at $i_n = 0$ for φ_p and $-\ \varphi_p$, respectively. Moreover, small partial loops corresponding to SAOs are fixed at the point $(0, \varphi_p)$.

It is now clear that, by corresponding selection of the circuit elements parameters, it is possible controlling the losses in the memristor, what is important for the device to achieve the highest performance in the target application. Notice also that the shape of the hysteresis loop of one-period energy presents all dynamical properties (LAOs and SAOs) of the given memristor under MMOs conditions. Such properties can be crucial for the use of memristor devices both individually or as part of larger circuits.

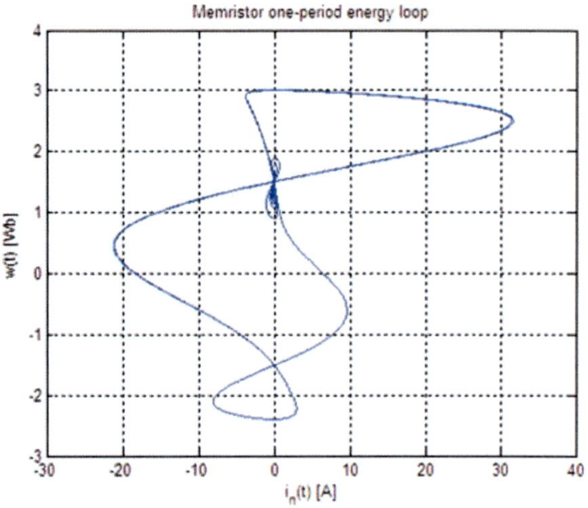

Fig. 6.7 One-period energy loop for the memristor

It is believed that a further insight into the one-period energy phenomena is possible by studying the concepts of various meminductive and memcapacitive circuits and systems [9]. Those concepts, being important in modelling of nanostructural systems, will add extra complexity to systems (6.14) and allow for better and more realistic models of lossy nanodevices [33].

6.2.4 Estimates for One-Period Energy Loops

In the study of electrical circuits and systems arise often problems of determining the range and estimation of a one-period energy of sources or loads for assumed changes in periodic values of non-sinusoidal excitation signals. This problem can be solved by considering the lower and upper limits of the instantaneous values of the corresponding signals associated with the given element. The appropriate estimate can be established by applying the statement of the theorem of Biernacki, Pidek and Ryll-Nardzewski (in brief theorem of B.P.R.-N.) [34].

Theorem of B.P.R.-N. If functions $f(x)$ and $g(x)$ with $x \in \Re$ are Riemann–Stieltjes integrable in the interval (a, b) and moreover $m_1 \leq f(x) \leq M_1$ and $m_2 \leq g(x) \leq M_2$, where m_1, M_1, m_2, M_2 denote constants, then the inequality

$$\left|\frac{1}{b-a}\int_a^b f(x)g(x)dx - \frac{1}{(b-a)^2}\int_a^b f(x)dx \int_a^b g(x)dx\right| \leq \frac{1}{4}(M_1 - m_1)(M_2 - m_2) \tag{6.39}$$

is fulfilled and the constant 1/4 is sharp in the sense that it cannot be replaced by a smaller one.

The relatively simple and effective proof of this theorem is given in [35]. Applying the above theorem to the one-period energy expression (6.20) yields

$$|W_T - \frac{1}{T}\int_0^T v(t)dt \int_0^T q(t)dt \leq \frac{1}{4}(v_{\max} - v_{\min})(q_{\max} - q_{\min}) \tag{6.40}$$

where $v_{\max}$, $v_{\min}$, $q_{\max}$, $q_{\min}$ denote maximum and minimum values of the voltage and charge, respectively.

Taking into account the definition of a mean value of a periodic function we get

$$|W_T - V_0 Q_0| \leq \frac{1}{4}(v_{\max} - v_{\min})(q_{\max} - q_{\min}) \tag{6.41}$$

where V_0 and Q_0 denote the mean values of the voltage and charge, respectively. In the case of a system element exhibiting $V_0 = 0$ or $Q_0 = 0$ we get

$$W_T \leq \frac{1}{4}(v_{\max} - v_{\min})(q_{\max} - q_{\min}) \tag{6.42}$$

Applying the relation (6.42) in the case of an element operating in a sinusoidal circuit with amplitudes of the voltage V_m and the current I_m for a given element, leads to the inequality

$$W_T < \frac{1}{4} 2V_m 2Q_m = TV_m I_m = 2T|S| \tag{6.43}$$

where $|S| = V_{rms} I_{rms}$ denotes the absolute value of the complex power.

This result is consistent with the physical interpretation of the considered problem, since the actual one-period energy of such an element is determined by the expression

$$W_T = TP = TV_{rms} I_{rms} \cos(\alpha) = T|S|\cos(\alpha) \tag{6.44}$$

where P and $\cos(\alpha)$ denote the active power and power coefficient of the given element, respectively.

It is important to note that the relation (6.40) in a general case and the relation (6.42) in a particular case allow to estimate the maximal value of the one-period energy of a given two-terminal element on the basis of knowledge of extreme values of its voltage and charge. There is no need to calculate their instantaneous values over the entire period of changes. This eliminates the often frustrating process of solving a differential equation that describes a given circuit operating in a non-sinusoidal state.

6.2.5 Analysis for Energy Aspects in Dynamical Systems with Switches

In this subsection the effectiveness and computational efficiency of the established analysis for energy aspects in dynamical systems with switches is demonstrated. First, a circuit model of the energy recovery during an electric car breaking [36–38] is considered. To see how the theory of one-period energy can be suitable to effectively realizations of this problem a controlled charger for a supercapacitor installed on the car board in cooperation with a Faraday disk generator integrated with the car brake is taken into account. Such an innovation leads to a significant improvement in the energy efficiency of the electric car and extends its travel range.

The circuit shown in Fig. 6.8 carries out the braking energy of the car in which the energy produced by the disc Faraday generator appropriately enough mounted in the brake system is transmitted to the supercapacitor by a switched network with an induction coil. The supercapacitor is under repetitive charging with a very small time constant.

The switches SW_1 and SW_2 operate in such a way that the charger provides a maximum charging current for a short period and then reduce to zero until next charging cycle. The charger operates at a fixed frequency of switching (Fig. 6.9) under a working cycle determined by V_{in} as voltage generated by braking (Fig. 6.10).

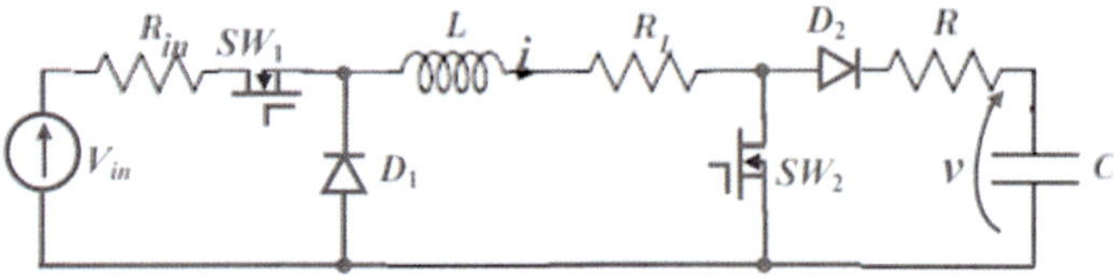

Fig. 6.8 Charging the supercapacitor C

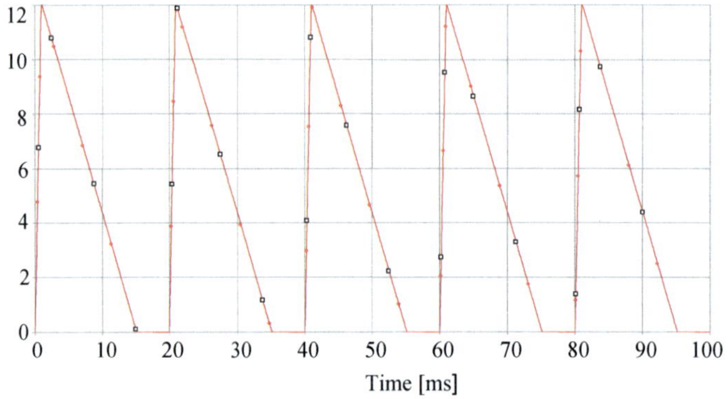

Fig. 6.9 Voltage V_{in} at the braking generator

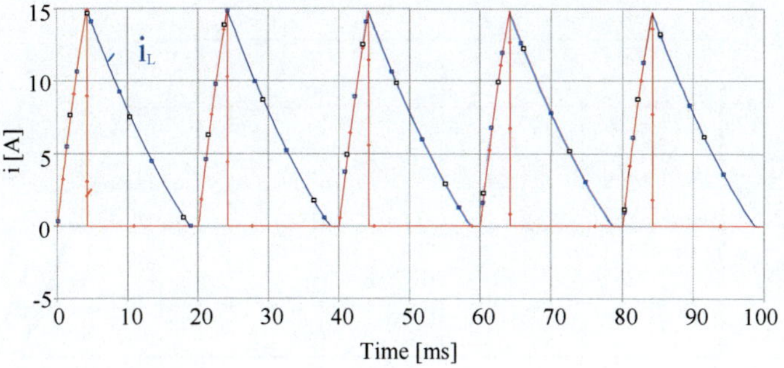

Fig. 6.10 Currents in the generator and the coil

During the energy absorption state, the coil draws energy from the braking Faraday generator and then delivers it to the supercapacitor in a state of energy transfer. In this way the supercapacitor can acquire a constant amount of energy per switching cycle. The MOSFET switches, $SW1$ and $SW2$, are synchronized for the first state and the turn-off time as well as switching frequency are adjusted to the ratio of the inductance L and total series resistances of R_L and $R = R_{in} + R_C$, where R_{in} and R_C denote the wire resistance and the series resistance of the supercapacitor, respectively.

Typically, the sum of the resistances R_L and R_{in} is greater than the equivalent series resistance R_C of the supercapacitor, which allows during simulations to neglect the influences of different characteristics of the supercapacitors provided by various manufacturers.

In the case of a Urban Driving Cycle test for the voltage waveform produced by the braking Faraday generator (shown in Fig. 6.11 as V_{in}) the switching frequency of 4 kHz was applied (Fig. 6.8) and 300 F supercapacitor was charged. The current waveforms of the braking Faraday generator and inductor are shown in Fig. 6.10. The inductor current rises to a peak value of 15A during the energy absorption stage. Then the current decays to zero and energy is transferred to the supercapacitor.

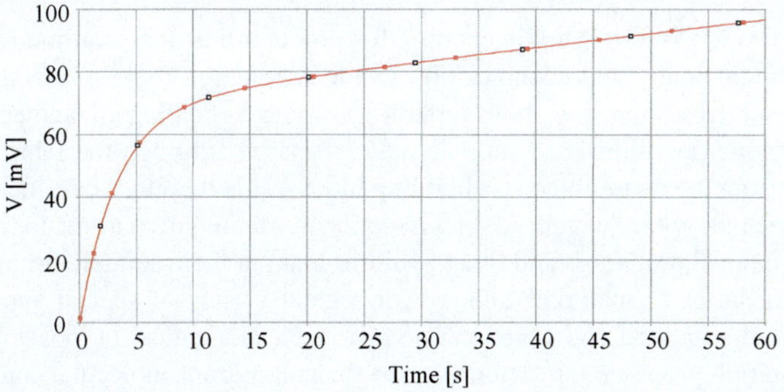

Fig. 6.11 Supercapacitor voltage

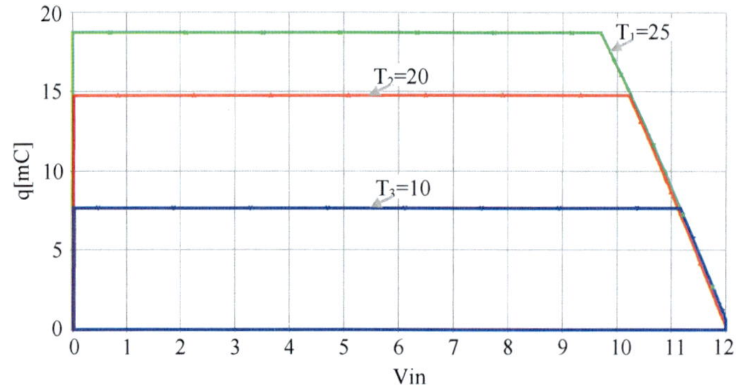

Fig. 6.12 One-period energy loops for the generator

Actually, the voltage of a supercapacitor (Fig. 6.11) increases slowly during charging and varies over a wide range then can effectively smooth the car power supply fluctuation caused by periodic acceleration and deceleration.

The obtained energy from periodic braking of the car depends on the period of repetition of braking. This is reflected in Fig. 6.12 by one-period energy loops for an example of three runs. They determine the energies $W_{T1} = 87.00$ mJ, $W_{T2} = 150.90$ mJ and $W_{T3} = 180.40$ mJ, respectively. The loop represented by the green line determining the highest energy corresponds to the longest time of single braking, what is also confirmed by experiments.

6.3 The Energy Approach to Electrochemical Corrosion Studies of Nano-copper Coatings

6.3.1 Introduction

Electrodeposition is a versatile technique for producing nanocrystalline materials. It is a technologically and economically viable production route to metals, alloys and metal matrix composites, both in bulk form and as coatings. Nanotechnology offers a unique possibility to manipulate the properties through the fabrication of materials using the nano-objects as building blocks. It is therefore considered as an enabling technology by which existing materials, virtually all man-made materials, can acquire novel properties and functionalities making them suitable for numerous novel applications as materials tailored for a great variety of special applications varying from structural and functional to advanced biomedical purposes [39–41]. While, materials properties are determined by their atomic and molecular constituents and structure, their functionalities emerge when the microstructure of these early ensembles is in the nanometer regime. The properties and functionalities of these ensembles may be different as their size grows from the nano- to micron-regimes and bulk structures. It is now evident that the nanotechnology has the potential for revolutionizing the ways in which materials and products are created and the range as well as the nature of functionalities that can be accessed. Improvements in environmental impact can be achieved utilizing nanostructure particulates in coatings and eliminating the requirement of toxic solvents.

Nanocrystalline copper and surface composite electrodeposits with grain sizes smaller than 100 nm have attracted extensive interest due to their improved electrochemical, mechanical and physical properties. Properties of nano-structured electrodeposits such as hardness, wear resistance, electrical resistivity and corrosion protection are strongly grain size dependent.

Improvements in environmental impact can be achieved utilizing nanostructure particulates in coating and eliminating the requirement of toxic solvents. Corrosion behavior of nanocrystalline copper and composite electrodeposits has been assessed by several techniques in various environments [42].

Recently, several methods have been proposed to evaluate the susceptibility of various metals coatings to electrochemical corrosion. This research direction has quite important significance since no single technique can be employed in all practical situations. The development of numerous techniques for the study of the mechanisms of corrosion and corrosion protection has, on one side, brought quite satisfactory results and, at the same time, there have been a lot of critical remarks on the interpretation of the differences between various physical models [39–44]. Even simple models are the cause of some controversy in the literature [45, 46]. Among the various electrochemical methods suitable for the measurements of the corrosion rate of metals, the harmonic current analysis method is quite useful because in the method the corrosion current can be obtained without the use of Tafel constants at corrosion potential. Significant advances in the study of the mechanisms of corrosion and corrosion protection have been achieved since the development of the impedance measurements. The measurements have been extended to the study of mass transport and stress corrosion cracking, as well as to the noise analysis used to study the localized corrosion, stress corrosion cracking and blistering of coatings [47–53]. In particular, the impedance measurements have been adopted to separate the different elementary steps involved in overall process, and receiving a better knowledge of mechanisms and progress of corrosion, as well as of possible means for the corrosion protection [54–61].

In this section we extend the results of [47–49, 52, 53] for the corrosion measurements and analyze them in the context of the activation energy and mass transfer effects.

6.3.2 *The One-Period Energy Approach*

It is well known that the electrochemical corrosion takes place when two different metals come into contact with a conductive liquid—usually impure water or soil moisture—resulting in a flow of direct current electricity. The current always flows away from the anodic metal (anode), and the anode is corroded. The current flows through the electrolyte to the cathodic metal (cathode), but the cathode is not corroded. The potential that causes the current to flow is always due to some kind of difference between the anode and the cathode, such as a difference in the two metals, concentration of the conductive liquid, a difference in temperatures, a difference in the amount of oxygen present, or some other difference in operation conditions.

A uniform attack of corrosion process environment is normally characterized by chemical and electrochemical reactions that proceeded uniformly over the whole exposed surface and over a large area of the sample. The products of corrosion may be dissolved species or solid corrosion products. In either case, the energy of the system is lowered as the metal converts to a lower-energy form. The change in the internal energy of the material is the driving force for the corrosion process and is a subject of thermodynamic principles.

Early identification of signs of corrosion especially in hard to reach locations where visual inspection is impossible, helps to prevent or slow down the onset of serious corrosion. The same approach works for evaluating protective coatings under real-world conditions. While excellent electrochemical kinetic models have been available for decades, each one is system specific and almost all apply only to dilute solutions and were developed to address specific questions. Thus, a more general model of corrosion rate in mixed solvent process environments is indispensable to predict component activities and transport properties of the bulk (aqueous, no aqueous or mixed) phase that is in contact with the metal. In a number of respects the impedance spectroscopy is the method of choice for characterizing the electrical behavior of corrosion systems in which the overall system behavior is determined by a number of strongly coupled processes, each proceeding at a different rate. However, such type of corrosion investigations can be carried out in the conditions of stability, causality and linearity of the tested electrochemical system. However, in a general case electrochemical systems are non-linear. For the above mentioned insufficiencies and limitations of the up-to-date methods being in using for identifications of corrosion processes an elaboration of a more accurate approach to the subject problems is indispensable. When being developed it must exhibit abilities for prediction the stability of multicomponent alloys as well as thermodynamic equilibriums, transport properties and rates of general corrosion as a result of existing electrochemical and chemical processes. An accomplishment of these goals is possible by an expansion and modifications of existing thermodynamic and kinetic modeling tools [49, 59–67].

In this work an attempt is put on description of a new method which takes into account the activation energy and mass transfer effects. It is based on non-sinusoidal periodic excitations of the corrosion cell and integrations of the measured response waveforms. This is done through the use of time-periodic signals of suitable amplitudes, and it is recommended not to introduce a DC potential difference across the system, which can induce further electrochemical activity. Using the small periodic non-sinusoidal excitation, and assuming that the system's time response is also periodic we can evaluate energy absorbed during the corrosion process [40–42, 55, 62]. The block diagram of the experimental setup is shown in Fig. 6.13. The wave

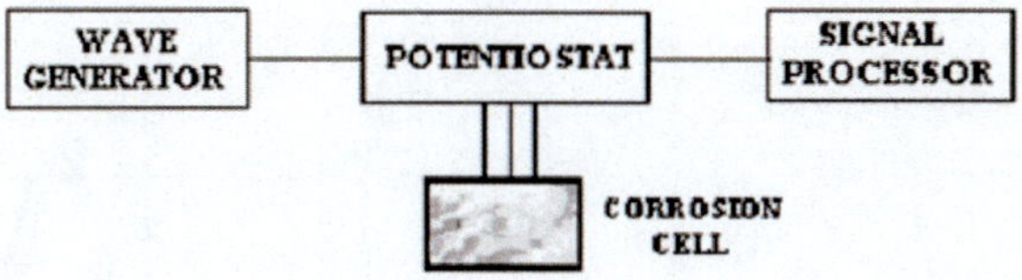

Fig. 6.13 Block diagram of the experimental setup

generator can supply a controlled amplitude time-periodic voltage wave of specific frequency and form. Such excitation being potentially static is applied to the corrosion cell using an elaborated electrochemical system and the current response of the corrosion cell is analyzed by a dynamic signal analyzer. An integrator of a measured signal is also applied. The samples of electrochemically produced copper thin layers and investigated in the corrosion cell are subject to an appropriate corrosion environment (usually 0.5 M NaCl solution). The important novelty of the present approach is exhibited by employing the energy state variables (time functions), to determine the susceptibility of a given material to corrosion. The main feature of this method is the complete elimination of the frequency analysis what leads to significant simplifications in the corrosion rate measurement process. The second important feature is that that the method is based on the analysis of appropriate loops on the energy phase plane leading to an easy estimation of the corrosion process through the evaluation of the loop's area. The presented methodology leads to easily interpretable physical results, with robust properties.

Thus to identify the effect of active species that are responsible for corrosion of the investigated samples we can construct an equivalent electric model of the corrosion process basing on knowledge of the physical properties and the electrochemical processes that occur as corrosion results. The general scheme of such a situation is shown in Fig. 6.14a. The steady state energy $W(\Delta t)$ delivered by the source $v(t)$ to the corrosion system during the time interval $\Delta t = nT$ (T is the period, n $\gg$ 0 is a positive integer) is expressed by

$$W(\Delta t) = n \cdot W_T \tag{6.45}$$

where W_T denotes the one-period energy delivered by the source. Thus, in the periodic state it is sufficient to evaluate W_T and by multiplying it by n obtained is the energy absorbed by a corrosion element during the given time interval Δt.

The derivation of the corresponding expression for W_T leads to

$$\begin{aligned} W_T &= \int_0^T v(t)i(t)dt = \int_0^T v(t)\frac{d}{dt}\left(\int i(\tau)d\tau\right)dt \\ &= \int_{q(0)}^{q(T)} v(t)dq(t) = \int_{\psi(0)}^{\psi(T)} i(t)d\psi(t) \end{aligned} \tag{6.46}$$

where

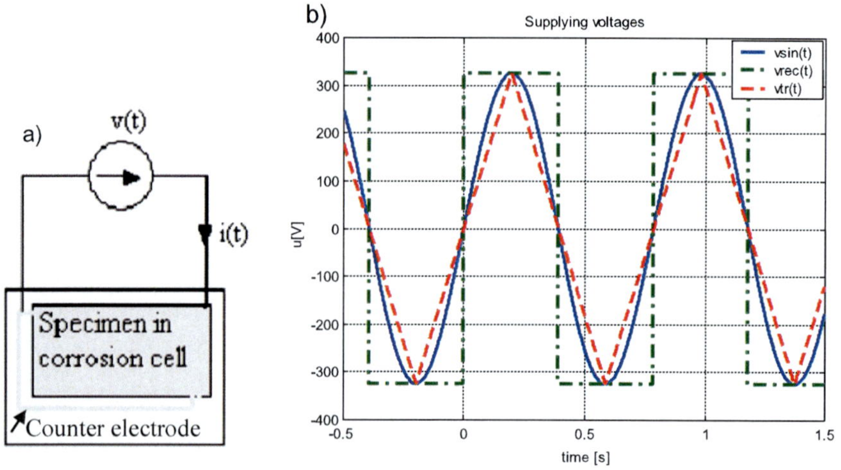

Fig. 6.14 Corrosion species in the periodic steady states: **a** electric system scheme, **b** waveforms of supplying voltage

$$q(t) = \int i(t)dt \quad \text{and} \quad \psi(t) = \int v(t)dt \tag{6.47}$$

denote the electric charge and quantity representing the integral of the source voltage $v(t)$, respectively. It follows from (6.46) and (6.47) that the area enclosed by a one-period loop on the energy phase plane with coordinates $(v(t), q(t))$ or, equivalently, $(\psi(t), i(t))$ determines the one-period energy W_T delivered to the controlled element operating under periodic non-harmonic conditions [63–68]. The above approach can also be applied when a given system operates at a permanent sinusoidal input excitation as well as in a non-linear condition.

It should be emphasized that it does no matter which energy phase plane we choose to determine W_T. In some cases it is easier to evaluate the phase variable $q(t)$ and in other ones the phase variable $\psi(t)$ should be used. Thus, for any corrosion specimen operating under periodic non-sinusoidal conditions it is possible to directly determine the one-period energy W_T without using the power or the Fourier series approach [62, 67, 69–72].

The second important feature of the one period energy approach is an easy way of evaluating the area enclosed by the loop in the energy phase plane. In most applications we have the preliminary knowledge concerning the forms of the physical system response on the given input. This allows choosing a priori the most appropriate energy phase plane calculations. Moreover, the results of calculations can be easily interpreted and have robust properties. The usefulness of the proposed technique has been examined for microcrystalline and nanocrystalline copper surface layers deposited on a polycrystalline substrate by the electrocrystallization method. The quantitative results obtained from the measurements of the one-period energy loops are applied to control the corrosion resistances of the micro- and nano-copper thin-layer coatings. Several experiments performed on real testing specimens verified the efficiency of the method in electrochemical corrosion analysis of many practical systems.

To illustrate the main feature of the one-period energy approach a series of experiments have been performed and analyzed. In all cases a three-electrode 'flat-cell' arrangement was used exposing 1cm^2 of working electrode made from polycrystalline technical copper plates. A large area platinum foil and a saturated calomel electrode were used as the auxiliary and reference electrodes, respectively. Quiescent 0.5 M NaCl ($pH = 6$) solution was used for electrochemical corrosion testing. Three different voltages supplied by the wave generator were taken into consideration successively, namely, clean sinusoidal $v_{sin}(t)$, rectangular $v_{rec}(t)$ and triangular $v_{tr}(t)$ waveforms changing in time with the same period T, and exhibiting the same magnitude values $|V_{\max}| = 10$ mV. The graphs of the supplying voltage waveforms are shown in Fig. 6.14b. The processes were monitored by measuring the current $i(t)$ during 30 min intervals. Each of these voltage excites in the corrosion cell corresponding currents $i_{\sin}$(t), $i_{\text{rec}}(t)$, and $i_{tr}(t)$, respectively, exhibiting variations in time more or less different with respect to supplying voltages. The excited currents were measured and then suitably processed in the context of the activation energy and mass transfer effects. Plots of the current waveforms corresponding to particular voltages shown in Fig. 6.14b are presented in Fig. 6.15a. It is easily seen that all currents as time functions take similar forms but their magnitudes differ importantly. Taking into account the input voltages shown in Fig. 6.14b we can measure the one-period energy absorbed by the samples on energy-phase plane ($q(t)$, $v(t)$) in accord to expression (6.46). The corresponding one-period energy loops are presented in Fig. 6.15b.

Since both the input voltages and output charges contain many harmonics with different spectra, the one-period energy loops have different shapes and surfaces. It is possible to arrange the energies absorbed by the investigated samples into the following inequalities

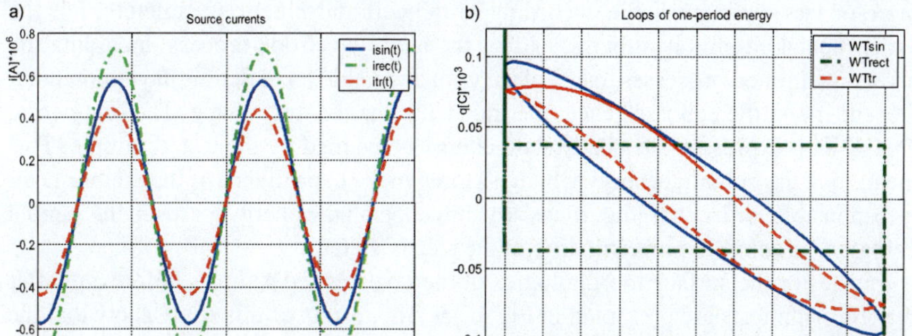

Fig. 6.15 Output values of the corroded samples: **a** current waveforms, **b** loops of one-period energy

$$W_{Trec} > W_{T\,\sin} > W_{Ttr} \tag{6.48}$$

This means that the input voltage of rectangular waveform exhibiting most rich spectrum of harmonic components makes better allowances for corrosion tests than other waveforms of the excitation voltage. It is easily seen from the Fourier series for the rectangular and triangular waves with period T, phase offset 0, and magnitude A which leads to

$$f_{rec}(t) = \frac{4A}{\pi} \sum_{n=1,2,5,\ldots}^{\infty} \frac{1}{n} \sin\left(\frac{2n\pi\, t}{T}\right),$$
$$f_{tr}(t) = \frac{8A}{\pi^2} \sum_{n=1,2,5,\ldots}^{\infty} \frac{(-1)^{(n-1)/2}}{n^2} \sin\left(\frac{2n\pi\, t}{T}\right) \tag{6.49}$$

It is worth noting that although both waveforms contain infinitely many harmonic components the rectangular one is much better convenient for corrosion studies because its all harmonic components have greater and positive magnitudes with respect to triangle wave. Thus supplying the corrosion cell by the source with rectangular waveform voltage we can eliminate inconveniences of the electrochemical impedance spectroscopy approaches. Moreover, such type of supplying voltage leads to a simple form of the one-period energy loops having the form of a rectangle on the $(q(t), v(t))$ plane.

The one-period energy absorbed by the specimens can be easily evaluated to be $W_{Trec} = 1.6\ \mu\text{J}$ so that during the whole test time $\Delta t = 30$ min we get $W(\Delta t) \approx 3.667$ mJ. In the case of two remaining supplying voltage waveforms the absorbed energies are appropriately smaller. Under corrosion cell conditions the applied voltage acts in the double layer immediately after the supplying source is connected. The resulting current response contains the different parts of current, like double-layer charging due to the voltage step, adsorption, and condensation more or less distinguishable according to the cell and kinetic parameters. For the experimental conditions this means that the entire electrode process, including the above-mentioned processes, takes place with a variable local double-layer potential. Dependent on the cell resistance, the transients change their shape. The one-period energy loop expresses the steady state energy consumed by the tested sample. This means that the potential energy which is stored in the capacitance of the double-layer is responsible for its charging. It has any influence on the energy losses in the sample being under corrosion investigation in the steady state.

Images of the surface morphologies of the investigated samples before and after the corrosion tests are presented in Fig. 6.16. The results of this study show that the corrosion tests performed with the voltage of rectangular waveform reveal much more details in the morphology of sample destroyed surfaces than the other waveforms of the supplying voltage. It is worth mentioning that, although computers can easily provide results up to four or more digits, care must be taken to ensure that the overall system remains linear. Moreover, the corrosion rate also depends on the material

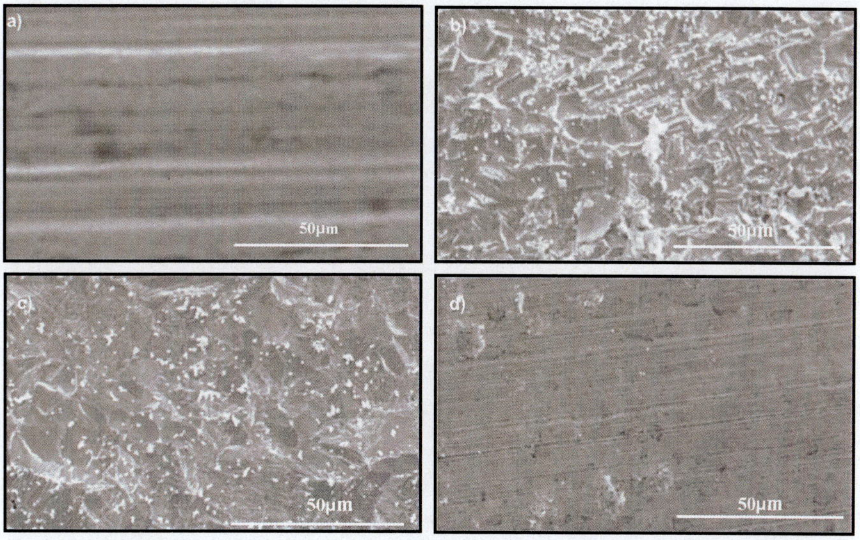

Fig. 6.16 SEM images of the copper surface morphology of the investigated samples: **a** before corrosion test, **b** after corrosion test with $v_{rec}(t)$, **c** after corrosion test with $v_{sin}(t)$, **d** after corrosion test with $v_{tr}(t)$

microstructure and such factors as the crystallographic texture, porosity, impurities and triple junctions.

6.3.3 Experiments

When the electrodes in corrosion cell are polarized by a periodic pulse voltage source, under stable conditions, the corrosion current also has a periodic waveform but, in a general case, its shape differs significantly from that of the excitation voltage. The voltage periodic waveform distorted from the sinusoid better describes the conditions that exist in practice. By perturbing a corroding system with a distorted periodic nonsinusoidal signal of low amplitude we do not need to perform a harmonic analysis. Moreover, electrochemical corrosion processes are non-linear in their nature. Polarization characteristic representing the complexity of electrochemical phenomena during the corrosion is given in Fig. 6.17a. The use of relatively high amplitude input nonsinusoidal signal and the determination of the total response signal which is excited at the output of the tested system allows its full corrosion characteristics. The suitability of the absorbed energy appears as a good base for the complex characterization of the tested system.

The approach presented in Sect. 6.3.2 can be applied even in a nonlinear case when a given system operates under permanent periodic conditions. Controlling the output of the wave generator at the voltage waveform shown in Fig. 6.16b we can measure

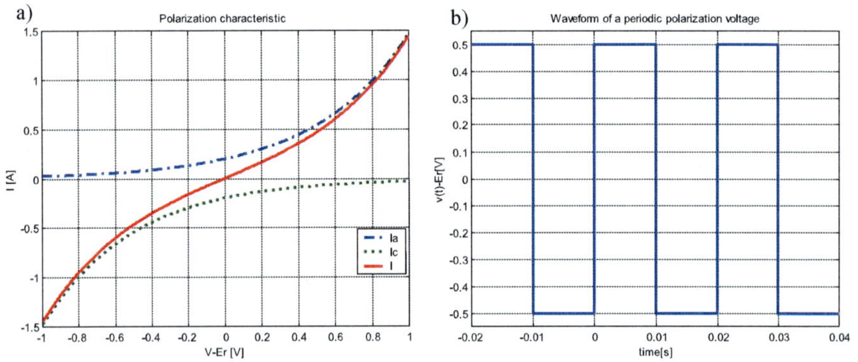

Fig. 6.17 **a** Typical polarization characteristic of electrochemical corrosion system, **b** waveform of the pulse voltage applied for corrosion tests

the one-period energy absorbed by the investigated sample. A direct comparison of the absorbed energies appears as a main factor determining the corrosion resistance of a given material. In this new approach the analysis of each harmonic component is completely eliminated and only the absorbed energy during the corrosion test determines the material quality.

The use of the one-period energy approach in evaluation of the micro- and nano-crystalline copper thin-layers and their interactions with various corrosive environments are illustrated by the SEM images shown in Fig. 6.18. The layers used in this study were deposited by the electrocrystallization method on polycrystalline copper substrates. Pure copper thin-layer electrodes with effective area of 1 cm^2 were used as the working electrodes. The corrosion tests have been carried out by using the set of apparatus shown in Fig. 6.13. Taking into account the voltage input shown in Fig. 6.17b the one-period energies absorbed by the samples have been measured. In order to do a comparison of obtained results the corrosion test measurements were repeated with sinusoidal voltage of the same amplitude and frequency as that rectangular one. In each case the working electrode was polished with 1/0, 2/0, 3/0 and 4/0 emery papers degreased with trichloroethylene. The solutions were made using analar grade chemicals with triple distilled water.

The quantitative results obtained from the measurements of the one-period energy loops were used for controlling the corrosion resistance of the micro- and nano-copper thin layer coatings. The obtained loops of one-period energy for non-sinusoidal periodic voltage excitation as well as sinusoidal one are presented in Fig. 6.19.

The one-period energy approach is based on the supplying of the corrosion cell by the voltage source of appropriate waveform. In such a situation when the voltage amplitude is constant the energy absorbed by the investigated sample is inverse proportional to its electrical resistance. Nano-crystalline copper layers exhibit much higher resistance to the current flow because the developed grain boundaries which may occupy as much as 50% of the material volume contain some additives originated from the bath, as for instance, trace quantities of sulfur. This results in much lower

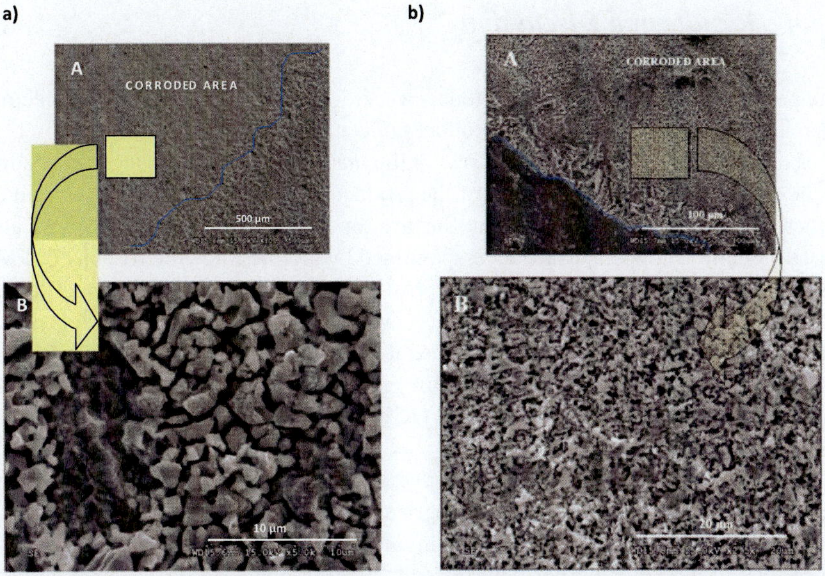

Fig. 6.18 SEM images of surface structure of the copper layers: **a** microcrystalline—**A** before and after corrosion, **B** enlarged fragment of the corroded area, **b** nanocrystalline—**A** before and after corrosion, **B** enlarged fragment of the corroded area

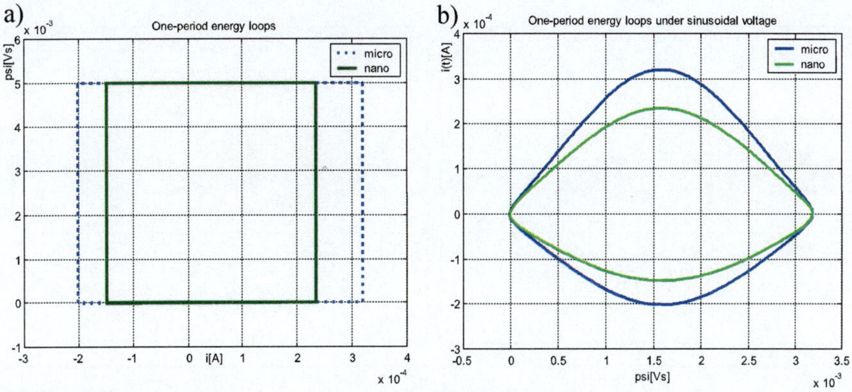

Fig. 6.19 One-period energy loops for copper thin layers obtained with input voltage of waveform: **a** rectangular, **b** sinusoidal

internal energy of the nano-crystalline corroded sample and then absorbed energy is smaller in respect to the micro-crystalline copper layers.

6.3.4 Results and Discussion

The performed corrosion investigations have showed that by the method of electrochemical reduction it is possible to produce copper layers with micro- and nano-meter sizes of grains. The decrease of the size of the material grains influences the change of the morphology and topography of the produced copper surface layers as well as it increases the rate of the corrosion. Surface topographies of technical copper and microcrystalline copper surface layer examined by atomic force microscopy (AFM) are presented in Fig. 6.20a and b, respectively. Surface roughness of copper layers was also examined but quantitative results can appear to be quite different depending on the scale which one investigates it. Illustration results relating to geometric profiles of the investigated materials are shown in Fig. 6.20c and d. The surface roughness was calculated as root mean square (*rms*) of heights by STM analyzer with different scales. The corresponding values are: $R_{aT} = 30$ nm and $R_{a\mu} = 82$ nm.

These results offer the important information for the process design of copper layer manufacturing in three-dimensional integration applications. This is due because the surface roughness is one of the most important properties of copper thin surface layers that affects many other behaviors of them like adhesion, friction, electrical properties, reflection or scattering and adsorption and etc.

We have also performed a series of experiments with using electrochemical impedance spectroscopy (EIS) and potentiodynamical investigations in quite equal corrosive conditions and obtained results are comparable with these acquired by the

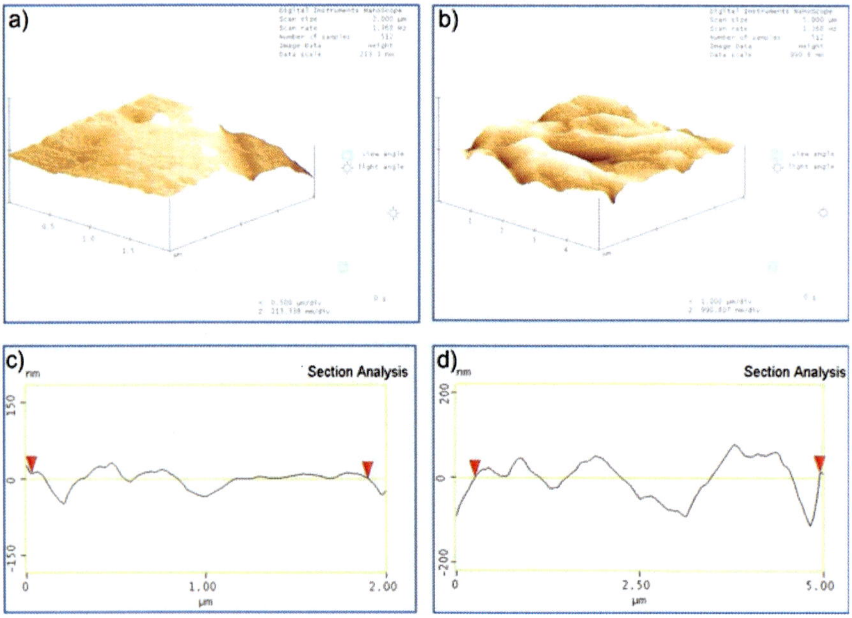

Fig. 6.20 AFM surface topography and geometric profiles: **a**, **c** technical copper, **b**, **d** microcrystalline surface copper layer

one-period energy technique. Height fluctuations and surface roughnesses of nano- and micro-crystalline copper thin-layer electrodeposits on polycrystalline substrate have been investigated and illustration results are shown in Fig. 6.20. Variation of roughness against copper layer thickness exhibited an exponential growth and rises of thickness cause an increase of roughness and then start to saturate.

Although *rms* values $R_{an} = 200$ nm and $R_{a\mu} = 1230$ nm of the nano-crystalline and microcrystalline copper layers, respectively, differ by the rate equal six there is no direct physical relationship between surface roughness and coating parameters. Moreover, the corrosion resistance of nanocrystalline copper surface layers is smaller than that of microcrystalline copper layers. This phenomenon can be understood as follows: in the nanocrystalline layer grain boundaries are the regions of few atomic distances in thickness that isolate grains of different orientations and more molecules of additive substance contained in the bath during the layer electrodeposition cover the copper grains and the inhibition effect is worse. Atoms located in the grain boundary regions have higher energies. These regions are also prone to easier diffusion and segregation of foreign atoms and impurities. In result, it is easier for corrosive chloride ions to attack the copper grains through the interspaces between the tail groups of the adsorbed molecules and damage layers on their surfaces.

It is well known that there are several ways for analyzing randomly rough surfaces. We often need a statistical approach to determine some set of representative quantities. It is worth noticing that the AFM data are usually collected as line scans along the x and y coordinates that are concatenated together to form the two-dimensional image. The surface area can be estimated by the following approach. Let f_i for i = 1, 2, 3, 4 be values of the surface heights in four neighboring points (pixel centers), and g_x and g_y pixel dimensions along corresponding axes. We assume that the sampling interval is the same in both the x and y direction. Moreover, we also assume that the surface height at a given point x, y can be described by a random function ξ(x, y) that has given statistical properties. If an additional point is placed in the centre of the rectangle which corresponds to the common corner of the four pixels (using the mean value of the pixels), four triangles are formed and the rough surface area can be approximated by summing their areas. This leads to the following formulas for the area of one triangle

$$S_{mn} = \frac{g_x g_y}{4}\sqrt{1 + \left(\frac{f_m - f_n}{g_x}\right)^2 + \left(\frac{f_m + f_n - 2\overline{f}}{g_y}\right)^2}, \quad m, n = 1, 2, 3, 4, m \neq n \tag{6.50}$$

where $\overline{f}$ denotes the value of the surface height in the pixel center.

The surface area of one pixel (k) is determined by

$$S_{(k)} = S_{(k)1,2} + S_{(k)2,3} + S_{(k)3,4} + S_{(k)4,1} \tag{6.51}$$

By counting in the area that lies inside each pixel, the total area is defined also for grains and masked areas.

The geometric size of the area is determined by $L_x \times L_y$ where L_x and L_y are the sizes of the sample along x and y axes, respectively. The rate η_S of surface roughness is defined by

$$\eta_S = \frac{\sum_{k=1}^{N} S_{(k)}}{L_x \times L_y} \tag{6.52}$$

where N denotes the number of pixels taken into account.

The performed calculations using Digital Instrument Nanoscope gives the following results: $\eta_{ST} = 1.064$ for the technical copper and $\eta_{S\mu} = 1.273$ for the micro-crystalline copper surface layer.

A determination of the quantitative correlation between absorbed energies and the surface roughness value is not a direct problem because it is dependent not only on surface roughness but also on the material structure and its chemical compositions (contaminations) (Fig. 6.21).

The quantitative results obtained from the measurements of the one-period energy loops are applied to control the corrosion resistances of the micro- and nano-copper thin-layer coatings. They were reported to the same conditions of the corrosion measurements but materials of these particular specimens of copper surface layers were characterized by different anodic and cathodic transfer coefficients. Fitting to experimental data the one-period energies absorbed by investigated samples can be

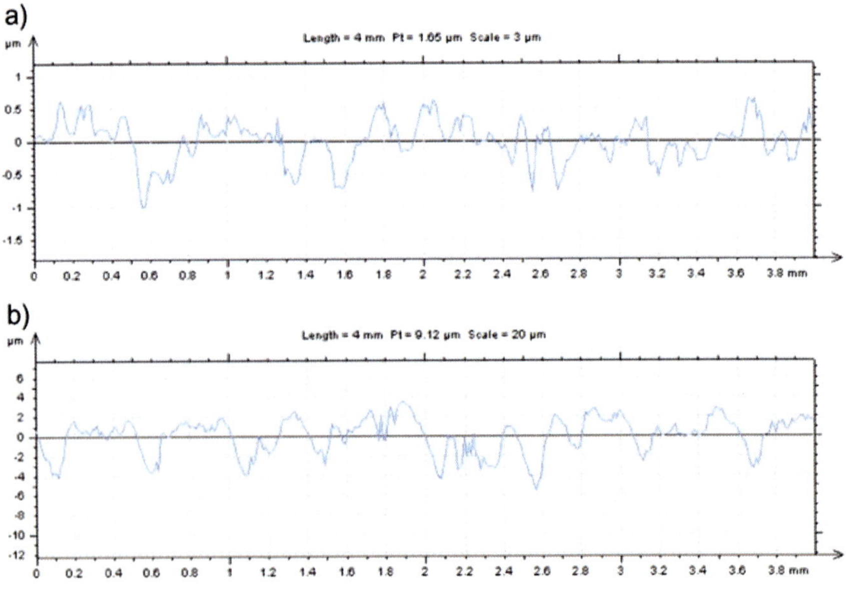

Fig. 6.21 Surface roughness of copper thin-layers: **a** nanocrystalline, **b** microcrystalline

a fairly straightforward task. It requires some knowledge of the corrosion cell being studied, its mechanism and a basic understanding of the behavior of cell elements. A good starting point for the development of the model is to compute the area of the one-period energy loop on the energy phase-plane ($q(t)$, $v(t)$) or equivalently ($\psi(t)$, $i(t)$). The corresponding surface integral denoted by W_T expresses the one-period energy absorbed by a given sample during the corrosion test. This method has the advantage that the prior knowledge of Tafel constants is not necessary and the calculation of the corrosion rate is accomplished within a very short time interval. Since both the input and output signals contain infinitely many harmonics, the one-period energy loops have different shapes that those obtained in the case of sinusoidal excitations. The method works very well both in linear as well nonlinear conditions. The applied methodology leads to easily interpretable physical results attesting the worst corrosion resistance of the nanocrystalline copper surface layers with respect to microcrystalline copper layers even they were produced by the adequate electrodeposition processes under appropriate conditions. All studied materials were exposed to the same corrosive environment. The decrease of the size of the material grains influences the change of the morphology and topography of the produced copper surface layers as well as it increases the velocity of the layer corrosion.

Different morphologies of the nanocrystalline copper and microcrystalline copper were observed in the SEM images presented in Fig. 6.18. The pitting corrosion failure on the surface of the microcrystalline copper is significant because of the smaller active intercrystalline area. This form of corrosion, however, is of great concern from the technical standpoint, because the life of equipment can be accurately estimated on the basis of the above comparatively simple tests. When localized corrosion phenomena occur, the characteristics of the fluctuations change dramatically, and these characteristics can be used to identify the most probable mode of attack. From a plant corrosion control perspective, this is important, since the technique provides an early warning of incipient localized corrosion [57].

Therefore, some concentrated active anode spots yielded serious pitting corrosion. Pitting results in holes in the metal. These holes may be small or large in diameter, but in most cases they are relatively small. Pitting is one of the most destructive and insidious forms of corrosion. It causes equipment to fail because of perforation with only a small percent weight loss of the entire structure. It is often difficult to detect pits because of their small size and because the pits are often covered with corrosion products. Pitting is particularly vicious because it is a localized and intense form of corrosion, and failures often occur with extreme suddenness. It is believed that during the initiation or early growth stages of a pit, conditions are rather unstable. Many new pits become inactive after a few minutes. It appears that pitting, though quite similar to crevice corrosion, deserves special consideration since it is a self-initiating form of crevice corrosion, which does not require a crevice. It is also known that all systems that show pitting attack are particularly susceptible to crevice corrosion, but the reverse is not always true: many systems that show crevice attack do not suffer pitting on freely exposed surfaces. Stable passivation does not occur on the surface of conventional microcrystalline copper.

The results of this study show that the corrosion pits, likely formed on grain boundaries, impurities, triple junctions and surface defects, are significant on a microcrystalline surface and the uniform attack is present on nanocrystalline structured specimen, respectively. The above mentioned facts are illustrated by the one-period energy loops presented in Fig. 6.18. The energy absorbed by the tested specimens when supplied by rectangular voltages can be easily evaluated and they are: $\mathrm{W_{Tmicro}} = 2.6437\ \mu\mathrm{J}$, and $\mathrm{W_{Tnano}} = 1.9013\ \mu\mathrm{J}$, respectively. Analogous measurements performed for one-period energies absorbed by investigated samples when sinusoidal input voltages were applied give: $W_{Tsinmicro} = 0.8243\ \mu\mathrm{J}$, and $W_{Tsinnano} = 0.6008\ \mu\mathrm{J}$. A simple comparison of the determined energies leads to conclusion that during the corrosion tests the nanocrystalline copper layer absorbs about thirty percent less energy than the microcrystalline copper layer. It can be also seen that the nanocrystalline copper layers exhibit worse corrosion resistance than the microcrystalline cooper layers, both being exposed to the same corrosive environment. This can be generally attributed to the unique microstructure of the nanocrystalline materials in which the grain boundaries may occupy as much as 50% of the material volume.

To estimate the corrosion rate, which is equal to

$$\chi = \frac{dm}{dt} = \frac{Ai_a}{zF} \tag{6.53}$$

where m is the weight loss of the metal, A is the atomic weight, i_a denotes the anodic current of the bath, z is the valence of the electrolyte and F is the Faraday constant, we must measure the time-evolution of i_a. The plot of mass loss evolution of nanocrystalline copper thin-layers during the anodic time-interval of the corrosion test is presented in Fig. 6.22 where the instantaneous values of anodic current in significantly reduced scale are also shown. Taking into account (6.53) it is easily state that the changes in corrosion rate χ are proportional to that of the anodic current. The plots shown in Fig. 6.22 exhibit the dependence of the copper thin-layer corrosion on parabolic law of the mass loss with time run, namely

$$m^2 = k \cdot t \tag{6.54}$$

where k is a constant depending on microstructure of the corroded nanocrystalline copper layer and its thickness. When the layer is thicker than the mass loss is less. This appears as a fundamental aspect which governs the corrosion processes in nanocrystalline copper thin-layers produced by electrochemical method.

The presented considerations and measurements of the non-linear system representing corrosion processes of micro- and nano-crystalline copper surface thin-layers produced by electrochemical method are based on one-period energy approach. We have shown that the nanocrystalline copper layers exhibit inferior corrosion resistance compared to the microcrystalline copper layers even when both type of layers were produced by the same electrochemical process. The decrease of the size dimension of the material grains influences on the change of the morphology and topography

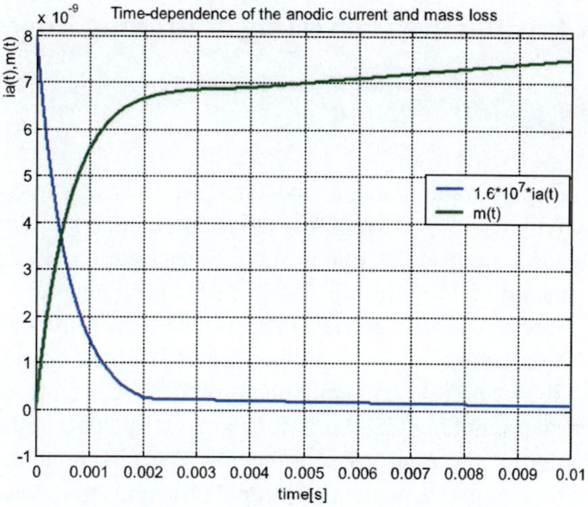

Fig. 6.22 Time dependences of the anodic current and mass loss of nanocrystalline copper thin-layers during the corrosion test

of the produced copper surface layers as well as increases the velocity of the layer corrosion.

The one-period energy loops such that shown in Fig. 6.19 can provide new information which could not be obtained previously with traditional techniques such as measurement of the open circuit potential, the polarization resistance, recording of polarization curves and/or the electrochemical impedance spectroscopy [39–43]. It is evident that applying the above tools we can accurately predict all the factors that are responsible for corrosion. The corrosion behavior of nanocrystalline copper surface thin-electrodeposits in this investigation suggests that interactions of crystallites are modified by the presence of numerous grain boundaries and other structural defects that likely provide preferential attack sites upon exposure to a corrosive environment. By perturbing a corroding system with a distorted nonsinusoidal periodic voltage of suitable amplitude we do not need to perform a harmonic analysis of the current response [73–76].

The main feature of the present approach is the complete elimination of the harmonic analysis which in turn simplifies the appliances used to measure the corrosion rate. The method is based on the time-domain data only and there is no need to use the second domain of analysis such as, for example, the frequency domain used in the EIS method [39, 41, 55]. The use of the typical mathematical tools, such as the Fourier series and Fourier transforms [56] is avoided. This is very important result in this respect that copper surface layers manufacturers must continue to rise to the challenge by designing copper products with attributes that meet the more stringent requirements of the future.

6.4 Effective Harvesting of Braking Energy in Electric Cars

6.4.1 Introduction

Transport systems are now becoming the basis for integrating vital spheres of human activity in cities and regions, and for their analysis, planning, design and operational functioning in-depth diagnosis of transportation system elements are essential. The transport sector is responsible for 21% of greenhouse gas emissions throughout the European Union countries. In accord with the European Union directive, the average CO_2 emissions of 95 g/km for all cars, with all passengers set, were fixed to reach in 2020, and 80 g/km is planned for 2030. Currently it is slightly below 120 g/km. Further reducing CO_2 emissions to 80% by 2050 will require 95% of the decarbonisation of the road transport sector. The recommended standards can be achieved only through adequate financing of strategies to increase the use of electric vehicles and the implementation of pilot electromagnetic mobility programs in cities. This requires the introduction of new technologies, mainly by putting into practice the general transport systems based on electrical vehicles and promoting the intensification of activities aimed at improving the economic situation and strengthening the protection of the environment in all European states [77–79]. Taking into account the above imperatives, presently many cars are designed to use only electricity as motive energy, which ensures high efficiency energy conversion and its transfer between different types of secondary energy sources. This is very important issue because of the need to take care of minimizing greenhouse gas emissions and the negative impact of urban transport on the environment. Moreover, the electric car makes possible to vary easily the driving speed and change load during traffic in the city and in non-urban areas [78].

It is worth to emphasize that in recent years great technical progress has been made in the areas of system structures and key component developments for electric vehicles, but there is still a performance gap between electric vehicles and conventional vehicles with respect to driving range, energy saving and power car efficiency as well as passengers safety. However, the electric car has already become a technology that in the world market has successfully busted through the bastions so far dominated by vehicles driven by conventional combustion engines [80]. The global design of replacing an internal combustion engine into an electric motor can be compared in a very simplistic way with the past realization of the project of electrification of the railways when the electric locomotive turned the steam locomotive. However, the energy significance of the implementation of an electric car for road transport is much more effective than the electrification of railways. This is due, in part, to the fact that the Well-to-Wheel (WtW) energetic efficiency [81], counted from the extraction of energy resources to use of energy by the vehicle while driving on a road does not exceed 30% for combustion engine vehicles while in the case of an electric motor car already in the near future can be expected at 60%. Moreover, the WtW efficiency of small electric cars is more than 2.5 times than that of equivalent petrol cars. The energy is also saved by the fact that the electric car engine does not work when the car is standing and also because it can regenerates the braking energy and accumulates it in the battery, supercapacitor, and flywheel or in pneumatic accumulator. This gives particularly large additional effects when moving in the city with very often stops at traffic lights or with traffic jams and that means significant lowering of costs in exploitations of electric cars [82]. It is now recognized that the tendency for electrification of road transport is at a stage of significant intensification and that this is an irreversible process and is particularly geared towards urban mobility. The justification for this statement is that every major automotive company has or is currently developing electrical models, and that a large number of countries have set significant plans to accelerate development and the deployment of electric vehicles. At the present stage of technical development, the use of electric vehicles is expected primarily for road transport in highly urbanized areas.

6.4.2 Energy Losses in Sub-systems of Electric Cars

Currently, the production and operation of electric vehicles using integrated and efficient energy sources is seen as an important factor in ensuring the realization of urban transport at a low emission of harmful carbon compounds and noise signals. In order to achieve maximum utilization of the available energy during the operation of electric vehicles a full informative system and monitoring of inside and outside parameters as well as their constant modifications can be designed to reach maximal as possible energy efficiency during the city traffic. The most important possibilities for increasing energy efficiency of electric vehicles can be reached, regarding energy savings accumulated in the vehicle itself and increasing the range of performances of the cars with given initial resources. Nowadays, a progress in improving the energy efficiency of electric cars can be achieved by installing on the vehicle board a kinetic energy recovery system harvesting the energy under braking and vibrations or shocks generated from the road disturbances. Dampers or shock absorbers are the mechanical devices that are designed to absorb the shocks and minimize the vehicle vibrations. They are also installed to maintain the contact between the vehicle tires and road surfaces [83–85]. The recovered energy is stored in such reservoirs as an electrochemical battery and a supercapacitor for later use under acceleration. Once the energy has been harvested, it is stored in the reservoirs and released when required. Supercapacitors that provide high power density increase the acceleration of vehicle as well as collecting all the energy from instant braking; therefore, improvements of the characteristics of power supply are made [86]. Figure 6.23 illustrates some of the sub-systems in the electric car that experience energy losses during the driving on a road. The illustrated losses are energy being converted in the brakes and the dampers. Most of these losses are heat dissipating during mechanical processes, such as the friction and vibrations damping. Note, that in urban driving cycles a significant amount of energy is consumed by braking [87, 88].

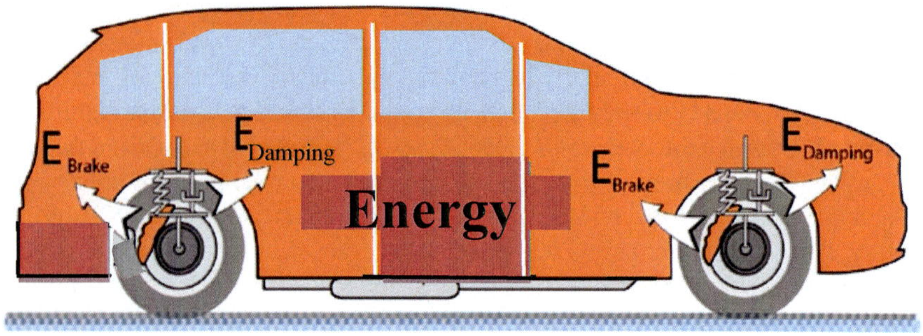

Fig. 6.23 An illustrative figure of some energy losses from sub-systems in a moving electric car

It is now expected that soon on the race tracks will be racing vehicles with electric motors, which emit low noise and are almost silent, and in a few moments will accelerate to over 300 km/h. Moreover, they will need less maintenance than complicated internal combustion race vehicles, that there are so far only in operation. The dissipation of kinetic energy during braking and damping by an electric car can be recovered advantageously by controlling the power electronics for total energy management on board the vehicle. Therefore, regenerative braking and damping is an efficient technology to improve the efficiency of electric vehicles. Research indicates that substantial energy savings are in fact achievable, from 8% to as much as 25% of the total energy use of the vehicle, depending on the driving cycle and its control strategy (Fig. 6.24).

One of the most effective methods for implementing energy harvesting is to convert kinetic energy produced by mechanical vibrations into useful electric energy which can be stored in supercapacitors and then used to power sensors and/or active systems and/or on board auxiliary electrical loads.

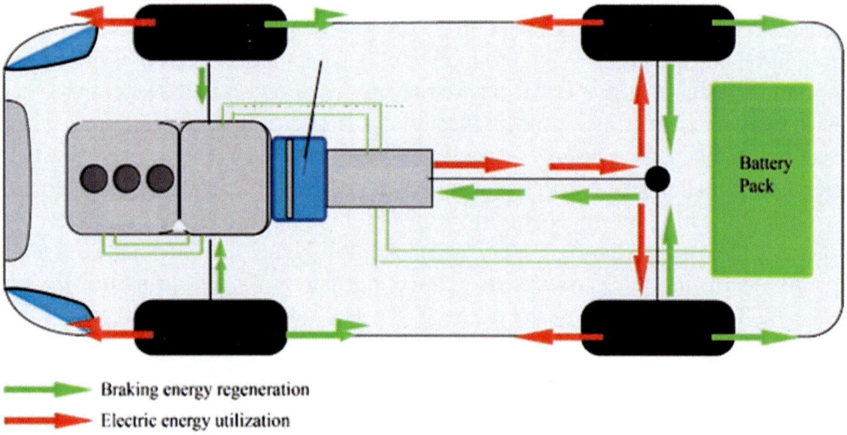

Fig. 6.24 Controlling the energy transfer in an electric car during traffic

6.4.3 Energy Regeneration in Sub-systems of Electric Cars

To improve the performance of electric vehicles, the regenerative braking and damping sub-systems can been developed. The classic brakes, for example, are installed to provide means of decelerating the vehicle, thus helping with speed alteration. This is usually done by applying a brake pad to a brake disc which in turn creates friction and slows down the wheel (Fig. 6.25a). In this instance the rotary kinetic motion of the wheel is turned into heat energy which heats the brake discs. In the case of the damping system it is designed to cushion the driver and passengers of the vehicle from the uneven road surface. When the damper moves up and down inside of the spring, kinetic energy is transformed into potential and thermal energies (Fig. 6.25b). However, most of this energy dissipates in the form of heat in the damper. It is obvious that all of these energy losses are unwanted. They are a byproduct of the fact that energy can only be transformed and not destroyed, thus making it obviously preferable to have as low losses as possible. From the analysis of research conducted so far in this field it follows that two types of kinetic energy regeneration, i.e. systems with chemical accumulation (Fig. 6.25c) and capacitive energy accumulation dominate (Fig. 6.25d) over others of lesser importance [89].

By combining these technologies, concepts and their improvements, it is possible to face towards energy-efficient vehicles which will greatly simplify our urban transport systems. It is well known that the braking performance strongly influences the safety and riding comfort of vehicles since various phenomena uncontrollable by the driver's pedal operations occur when breaking.

In new regenerative braking strategy design, the goal are not only the safety, reliability and easy driving but, above all, the most important factor appears to be the transformation of kinetic energy of the braking car into electric energy. In using regenerative braking, a significant change must be made to the architecture of the vehicle braking system. This can be done in a number of ways and two of which are presented and explained more thoroughly in the following subsection. It has to be emphasized that in this section the focus lies on a system that in some way assists the cars movement.

6.4.4 Modification of the Car Brake Sub-system

Vehicle speed and the driver's brake force command have large impacts on braking efficiency. An effective way to improve the energy efficiency of an electric vehicle is the ability to rebuild the braking system so that the kinetic energy of braking on each wheel of the car can be converted into electricity that must be stored in a battery or/and in a supercapacitor. Full use of the possibilities of this process ensures the application of the disk Faraday generator, whose principle of operation is illustrated in Fig. 6.26a. It utilizes a strongly compacted device formed by two cylindrical permanent magnets separated by a thin good conductive disk. The disk homopolar

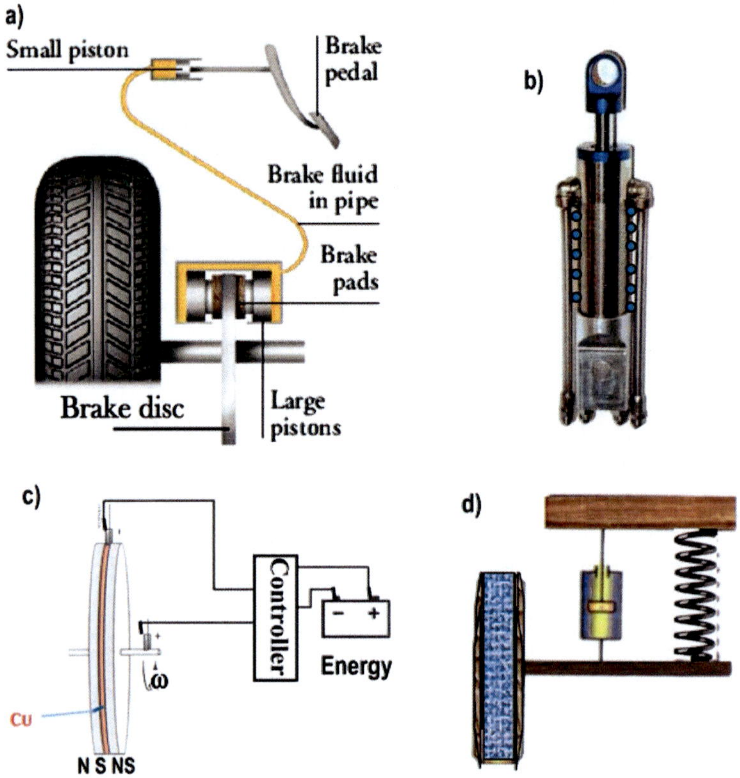

Fig. 6.25 Energy-harvesting sub-systems: **a** car brake with classic disc, **b** shock absorber, **c** modified disc charging an energy storage device, **d** shock absorber supplying a supercapacitor

generator distinguishes itself from other common generators in that no commutation or alternating of the magnetic poles is necessary for this machine in order to generate electric energy. When the car wheel rotates a voltage between brushes attached to the axle and the rim of the conducting disc is generated and directed appropriately from the driven compact and converts the kinetic energy to electrical energy that can be used for recharging the battery or supercapacitor [90–94].

To examine the effect of electromagnetic characteristics on the output of the disk Faraday generator, and to investigate the conditions that result in the maximum efficiency of the generator a comprehensive model that takes into account most of the experimental variables is taken into account and validated by considering limited conditions (Fig. 6.26b).

The value of the generated voltage can be derived by using Faraday's law of induction and the Lorentz force law, namely

$$V(t) = \int_{r}^{R} \overrightarrow{E_i} \cdot \overrightarrow{dr} = |\overrightarrow{B}|\omega \frac{R^2 - r^2}{2} \tag{6.55}$$

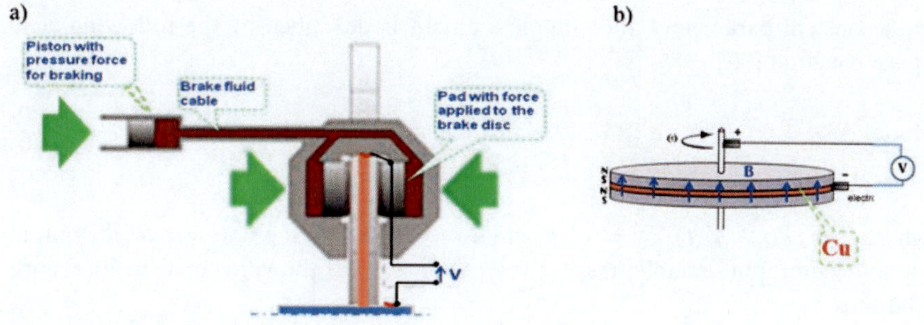

Fig. 6.26 Modified car brake: **a** with Faraday generator, **b** model for voltage evaluation

where $\overrightarrow{E_i}$, $\vec{r}$ and $\vec{B}$ denote the vectors of internal electric field intensity, of the radius from the axle and of the magnetic induction, respectively. The rotation velocity is denoted by ω and the external and internal radii of the conducting disk are denoted by R and r, respectively.

Assuming the radial direction of the current in the conducting disk, the resistance of a disk-shaped conductor can be expressed as:

$$R_t = \rho \int_r^R \frac{dl}{2\pi h l d\alpha} = \frac{\rho}{2\pi\pi} \int_r^R \frac{dl}{l} = \frac{\rho}{2h\pi} \ln\left(\frac{R}{r}\right) \quad (6.56)$$

where ρ and h denote the resistivity and thickness of the conducting disk.

The efficiency η of the generator can be defined as the total output energy divided by the input energy, namely

$$\eta = \frac{\int_0^T p(t)dt}{\frac{1}{2} J\omega_0^2} \quad (6.57)$$

where $p(t)$ and J denote the instantaneous power delivered by the generator to a load and the rotational inertia of the disk, respectively. The time of observation is denoted by T and ω_0 is the initial angular velocity of the disk.

In order to determine the power produced as a function of time, we can express $p(t)$, namely

$$p(t) = V(t)I(t) \quad (6.58)$$

where $V(t)$ is determined by (1) and $I(t)$ denotes the current delivered by the generator. The instantaneous values of the current depend on the disk resistance (6.56) and parameters of dynamic elements representing the load and an intermediate connecting network.

If we assume that a supercapacitor with capacity C_s represents the load and the connecting network exhibits the structure of an R_c, L_c, C_c standard two-port network with constant parameters, the complete circuit is described by the following state space equation [95]

$$\begin{aligned} \dot{\boldsymbol{x}} &= \boldsymbol{A}\boldsymbol{x} + \boldsymbol{B}\boldsymbol{u} \\ \boldsymbol{y} &= \boldsymbol{C}\boldsymbol{x} + \boldsymbol{D}\boldsymbol{u} \end{aligned} \tag{6.59}$$

where $\boldsymbol{x} = [I(t) \quad V(t)]'$, $\boldsymbol{y} = V_s(t)$ and $\boldsymbol{u} = V(t)$ denote the state vector, the output variable and input variable, respectively, with $V_s(t)$ as the voltage at the load port. Matrices

$$\boldsymbol{A} = \begin{bmatrix} -\frac{R}{L_c} & -\frac{1}{L_c} \\ \frac{1}{C} & 0 \end{bmatrix}, \quad \boldsymbol{B} = \begin{bmatrix} \frac{1}{L_c} \\ 0 \end{bmatrix}, \quad \boldsymbol{C} = \begin{bmatrix} 0 \; 1 \end{bmatrix}, \quad \boldsymbol{D} = [0] \tag{6.60}$$

depend on circuit element parameters with $R = R_t + R_c$ and $C = C_s + C_c$ as the equivalent resistance and capacitance of the complete circuit, respectively.

The state space equation (6.59) with (6.60) can be useful to design a suitable control of regenerative braking of the car. A limitation of this model is that that it does not accurately describe a situation when a current with a strong angular component is produced on the disk due to a high angular velocity, or with a high overall current. Thus the change in the magnetic flux along the path of the electrons creates eddy currents, which consequently resists the rotation of the disk.

To the above two governing equations must be added the description of the physical friction and the armature reaction which represent factors decelerating the rotation of the disk. The frictional torque $M(t)$ acting on the conducting disk can be modelled through the following equation

$$M(t) = \alpha \cdot sign(v(t)) + \beta v(t) \tag{6.61}$$

where $v(t)$ denotes the car velocity and α and β are the coefficients of the dry kinetic friction and of the viscous friction, respectively.

The armature reaction commonly referred as a back torque $H(t)$ that resists the rotation of the generator disk is simply due to the Lorentz force by the induced current and is expressed by

$$H(t) = \int_r^R BI(t)x \cdot dx \tag{6.62}$$

where x denotes a point on the disk.

It has to be noted also that the resulting back torque affects the magnitude of the induced current. Therefore, Eqs. (6.59)–(6.62) constitute the mathematical model of the car braking energy regeneration with the Faraday disk generator which can be used to designe and consequently update the parameters and the variables every time the car is run.

It should be underlined that the implementation of this sub-system into electric vehicles should not be much difficult since it seems possible to alter the dimensions of the braking system in order to make it fit in the current control board of passenger cars. It is important to note though that changing the component dimensions might affect imperceptibly the output power of the car. Lastly, the fact that this type of system can replace the braking solutions up-to-date installed in the cars could possibly lower the exploitation cost of the system.

6.4.5 Regenerative Damping

Today, the problem of energy becomes so important that all the attention of modern societies is turning towards clean and renewable energies (solar energy, wind energy, etc.) and their efficient consumption. This is why energy regeneration systems of damping in electric cars come also into play [89, 96].

A damper, which is also referred as a shock absorber represents a separate construction device in classic cars for damping excessive vibrations of the sprung and unmanned masses by changing the kinetic energy of mechanical vibration movement to heat energy dissipated to the surrounding atmosphere. This means that the damper absorb and dissipate excess energy of dynamic loads. The use of dampers is to achieve smooth movement of the car and eliminating the phenomenon of tearing off the road wheels from the road surface, at maximized comfort and grip. Depending on the damper destination its structure is always a compromise between driving comfort and grip.

The small amount of energy that is produced when the engine and passengers compartments of a vehicle vibrate while in motion is a possible source for energy harvesting. It is possible to convert vibration/kinetic energy to electric energy by using regenerative shock absorber effectively. The harvested energy from electric car shock absorbers will result in a much more economical vehicle performance and increased comfort of running [97].

Typically, the new linear shock absorber consists of a small magnetic tube with high flux intensity that slides inside a larger, hollow coil tube. Due to bumps and vibrations from normal driving, the sliding tube can produce an electric voltage. When installed in a medium-sized passenger car traveling at 100 km/h, the shock absorber can generate the energy at power of 100–400 watts under normal driving conditions, and up to 1600 W on particularly rough roads.

The harvested energy is then used to charge the battery and power the vehicle's electronics, which is typically 250–350 W with optional electronic systems turned off. This energy reduces the load on the vehicle's battery, which usually supplies the DC motor. In this way, the harvested damping energy could increase the battery's efficiency by 1–8%. As a side benefit, the shock absorber also creates a smoother ride due to the ability to adjust the suspension damping and implement self-powered vibration control.

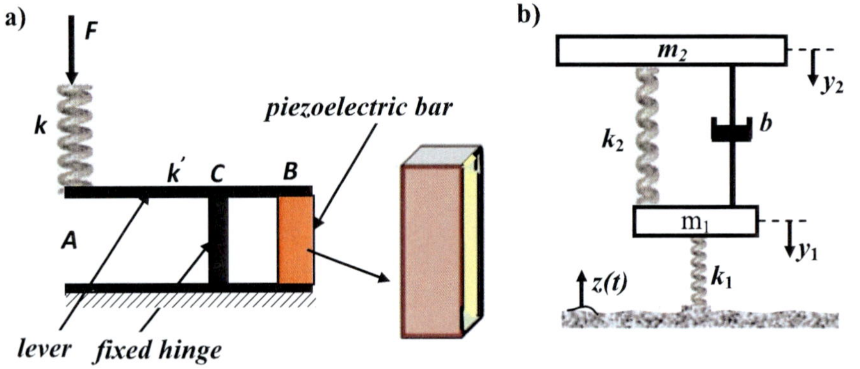

Fig. 6.27 Model of the dual-mass piezoelectric bar harvester: **a** piezoelectric bar transducer, **b** dual-mass piezoelectric energy harvester

Currently, the mostly available mechanisms suitable for vibrations-to-electric energy conversion are electromagnetic, electrostatic, and piezoelectric transducers. Among these three types of energy transducers, the piezoelectric transducers (Fig. 6.27a) are preferred because of their efficiency which is much higher than the two others. It was yet found out that the energy density of piezoelectric transducer is three times higher than the other two transducers.

To achieve a new effective design with piezoelectric technology of damping energy harvesting for driving vehicles, a dual-mass piezoelectric bar harvester can be developed for absorbing energy from vibrations and motions of a suspension system under excitations of the vehicle from road roughness.

According to the Newton second law of dynamics the governing differential equations of the dual-mass piezoelectric bar harvester system (Fig. 6.27b) are expressed as follows

$$\begin{aligned} &m_1\ddot{y}_1 + b(\dot{y}_1 - \dot{y}_2) + k_2(y_1 - y_2) + k_1(y_1 - z(t)) = 0 \\ &m_2\ddot{y}_2 + b(\dot{y}_2 - \dot{y}_1) + k_2(y_2 - y_1) = 0 \end{aligned} \tag{6.63}$$

where k_1, k_2 and b denote elastances and damping coefficient of the springs and damper, respectively. The displacements of the unsprung mass and sprung mass with respect to their respective equilibrium positions are denoted by y_1 and y_2, respectively. The form of the road surface in the transverse motion of the car is denoted by $z(t)$.

Taking into account the principle that the dissipation energy of a damper is equal to the electric energy generated by the piezoelectric bar harvester, the damping coefficient b can be expressed as

$$b = n^2 d^2 k_2^2 / (\pi^2 C f) \tag{6.64}$$

where n, d, C and f denote the ratio of the moment arms of the lever, the piezoelectric constant in the polling direction; the electrical capacity of the piezoelectric bar and the first natural vibration frequency of the car suspension.

Consequently, we can obtain the instantaneous displacements and velocities $\dot{y}_1$, $\dot{y}_2$ of the unsprung mass and sprung mass at their respective equilibrium positions. The relative displacements $y_{12} = y_1 - y_2$ and velocities $\dot{y}_{12} = \dot{y}_1 - \dot{y}_2$ of the unsprung mass and the sprung mass can also be determined. Then the generated charge $q(t)$, and voltage, $V(t)$, from the piezoelectric bar at time t can be expressed by relations

$$\begin{aligned} q(t) &= ndk_2 y_{21}(t), \\ V(t) &= \frac{q(t)}{C}, \\ I(t) &= ndk_2 \dot{y}_{21}(t). \end{aligned} \tag{6.65}$$

Thus, the instantaneous power $p(t)$ and energy $E(t)$ generated from 0 to t by the piezoelectric bar can be evaluated as

$$\begin{aligned} p(t) &= C^{-1}(ndk_2)^2 y_{21} \dot{y}_{21} \\ E(t) &= \frac{(ndk_2)^2}{C} \int_0^t y_{21} \dot{y}_{21} d\tau \end{aligned} \tag{6.66}$$

It is easily seen that the energy $E(t)$ increases with an increase in the velocity of vehicles and the class of road surface, an increase in the ratio of the moment arms of the lever, and with a decrease in the width of the piezoelectric bar. It can be expected that in practice four or more of the novel piezoelectric bar energy harvesters could be installed on a vehicle and provide more efficient energy harvesting as an auxiliary energy of electric cars. These offer the energy regeneration as much as possible from the damping kinetics and store it to be used for a useful purpose such as an auxiliary on board electric energy source.

6.4.6 Supercapacitor Characterization

The supercapacitor also known as ultracapacitor is a high power density energy storage device that can deliver high short-term discharging current and acquire burst of charging current. It can effectively absorb the regenerative energy during braking and provide extra-current for hill climbing. It should be emphasized that the supercapacitor does not have the drawbacks of electrochemical batteries like poor temperature coefficient, limited charging and discharging cycle, and critical charging current. It can be used in tandem with batteries for performance improvement of electric cars [96, 98].

In general, the supercapacitors have higher power densities than batteries, as well as sub-second response times. The capability of the present material technology to synthesize nanostructured electrodes with tailored, high-surface-area architectures offers the potential for storing multiple charges at a single site, increasing charge density. The addition of surface functionalities could also contribute to high and reproducible charge storage capabilities, as well as rapid charge–discharge functions. The design and manufacturing of new materials with tailored architectures optimized for effective capacitive charge storage will be catalyzed by new computational and analytical tools that can provide the needed foundation for the rational design of these multifunctional materials. These developments also provide the molecular-level insights required to establish the physical and chemical criteria for attaining higher voltages, higher ionic conductivity, and wide electrochemical and thermal stability in electrolytes. Electrical energy is stored between its plates in the electric field and as a result of this stored energy a voltage exists between the two plates. A charged ultra-capacitor can store this electrical energy even when removed from the voltage supply until it is needed acting as an energy storage device. The amount of energy stored is proportional to the capacitance C and the square of the voltage V across its terminals.

A supercapacitor can accept a wide of charging current so that precise current control is not necessary. The criterion for terminating charging is the maximum rated voltage of a supercapacitor. It is under repetitive charging with a quit short time constant and can be completely discharged within a few minutes or even just a couple of seconds, and then fully charged again within a short period.

The capacities of the supercapacitors coming up to 2700 F (Fig. 6.28a), combined with the short charging/discharging cycle time, are achievable at very high power density (unit power), up to 10 kW/kg, which exceeds the unit power of modern batteries. Supercapacitors are suitable for use in systems where low energy is transferred at high values of the power. They operate typically well in impulse state conditions. The permissible operating voltage is in the range 2–3 V. If a switching mode DC-DC converter with pulse width modulation (PWM) is in use as a charger then it operates in such a way that provides a maximum charging current for a short period and then reduces to zero until next charging cycle.

The second-order circuit model of a supercapacitor is shown in Fig. 6.28b. It consists of a series resistance and an inductance, and the leakage current is represented by a resistor in parallel with the capacitor. The series resistance ranges from a few milliohms to several tens milliohms. The inductance depends on the construction and can be ignored for low frequency operation. The leakage resistance can also be ignored for short-term operation. Actually, the leakage current of a supercapacitor with capacitance over 500 farads is less than 10 mA and its rated current is over a hundred of amperes.

6.4.7 *Simulations*

The present expansion of transportation, mostly in large towns, induces serious problems resulting from inability of municipal authorities to cope with situations on crowded roads and crossroads. There are more requirements put on modern vehicles, namely to achieve high accelerations during start-ups at crossroads and, last but not least, during overtaking which is closely related with not only the safety of car

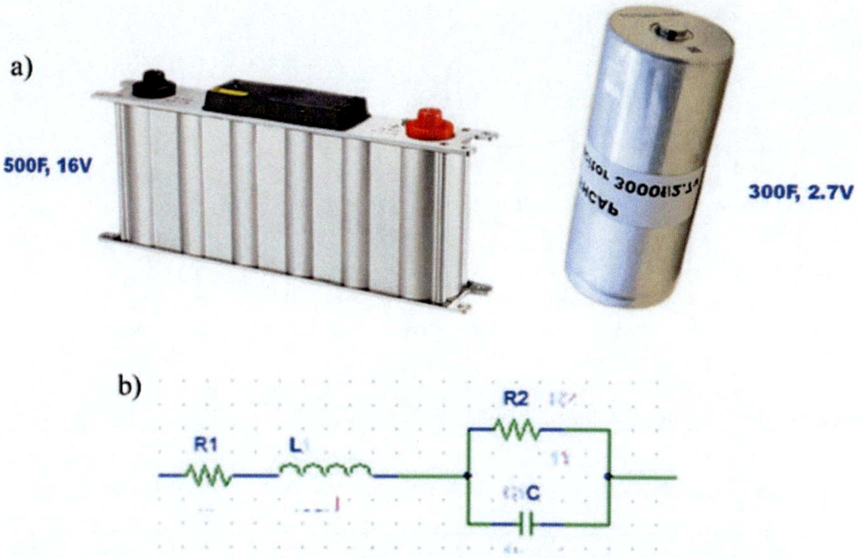

Fig. 6.28 Supercapacitors: **a** view of fabricated product, **b** equivalent circuit of a supercapacitor

passengers but also of pedestrians near the streets. One objective is to evaluate—on the basis of numeric simulations—an effective influence of the energy harvesting during car braking and converting via particular devices the kinetic energy of the car suspension into electric energy.

Two homologation tests were selected for the city traffic simulation: UDC (Urban Driving Cycle) showed in Fig. 6.29 and the Stop and Go special test to simulate driving in traffic congestion which is presented in Fig. 6.30. Observe that this test is characterized by frequent acceleration and braking, and the vehicle's average speed is 5.8 km/h. Moreover, the test Stop and Go is characterized by stronger dynamic properties than the UDC test.

To enhance the efficiency of the energy harvesters their mathematical and computational models have been developed and experienced. It has to be emphasized that the theory, modelling, and simulation can effectively complement experimental efforts and can provide insight into mechanisms, predict trends, identify new materials, and guide experiments.

For a given Faraday generator installed in the car wheel's disc brake, the relationship between the speed and the generated voltage can be derived applying the relation (6.55) and the relationship between the circular and linear speeds $v(t) = r_w \omega(t)$ with r_w as radius of a wheel tire. Thus we have

$$V(t) = \int_r^R \overrightarrow{E_i} \cdot \overrightarrow{dr} = \left|\vec{B}\right| \frac{R^2 - r^2}{2r} v(t) \tag{6.67}$$

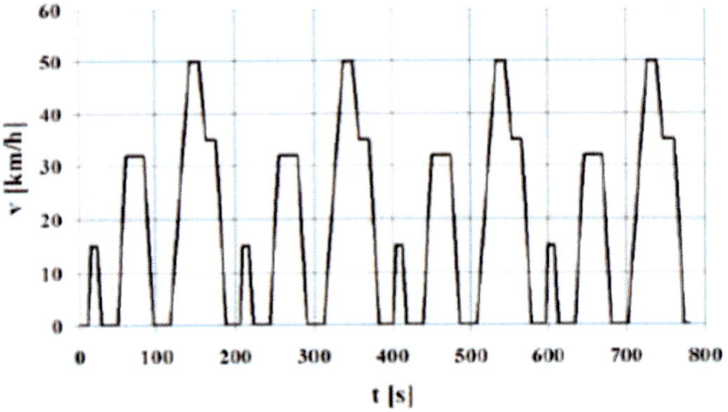

Fig. 6.29 Time-varying car velocity according to UDC test

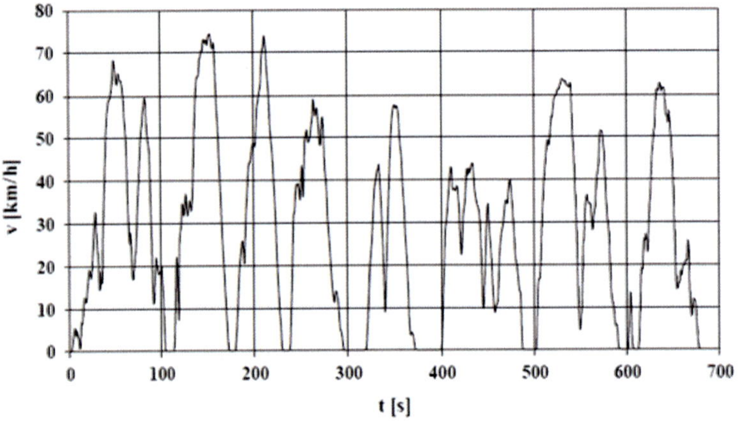

Fig. 6.30 Time-varying car velocity according to Stop and Go test

Since in practice r ≪ R, then with a good precision we can convert expression (6.39) into the following form

$$V(t) = |\overrightarrow{B}|\frac{R^2}{2r_w}v(t) = Av(t) \tag{6.68}$$

where the structural constant A is defined by the expression

$$A = |\overrightarrow{B}|\frac{R^2}{2r_w} \tag{6.69}$$

Taking $|\overrightarrow{B}| = 1.6T$, $R = 15$ cm and $r_w = 1.25$ cm yields $A = 1.2$ Vs/m.

The full potential of supercapacitors as electric energy storage devices was checked out by using the model system representing a supercapacitor charged with the aid of a Faraday generator attached to a car wheel. The circuit mapping the process of the supercapacitor charging is shown Fig. 6.31, which consists of only five major components. Energy operates at fixed frequency and duty cycle with V_{in} as voltage generated by braking. Although the circuit does not consist of any current transducer or closed loop control circuit, the input current is limited because the peak current of the inductor is limited by a fixed turn-on time.

When the synchronous switches, $S1$ and $S2$, are turned on, the inductor current can be expressed by:

$$V_{in} = Ldi_{on}/dt + R_L i_{on} + 2V_S \tag{6.70}$$

where V_S is the voltage drop across a synchronous switch. The values of V_S and R_L are small, and the voltage drops are negligible in this stage.

Then

$$V_{in} = Ldi_{on}/dt \tag{6.71}$$

If the turn-on time is short, (6.71) can be approximated by a linear equation. Thus, the peak inductor current during T_{on} is given by

$$I_{peak} = V_{in}T_{on}/L \tag{6.72}$$

During the energy transfer state, the synchronous switches $S1$ and $S2$, are turned off. The inductor current can be expressed by:

$$L\frac{di_{off}}{dt} + (R_L + R_w + R_C)i_{off} + \frac{1}{C}\int_0^t i_{off}dt + V_{C0} = -2V_d \tag{6.73}$$

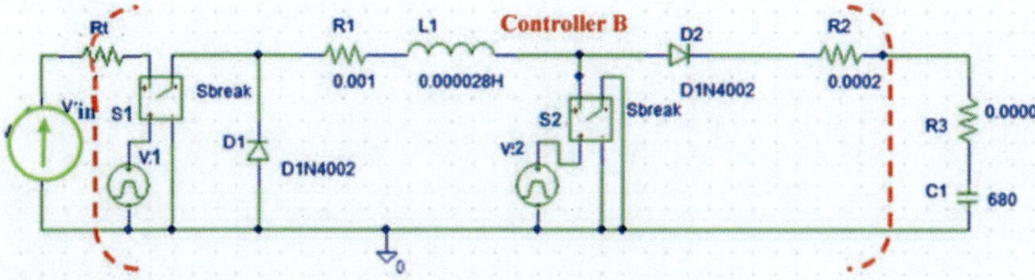

Fig. 6.31 Charging the supercapacitor 680 F

where V_{C0} is the previous voltage of the supercapacitor, and V_d is the voltage drop of each diode. As an extreme condition, V_{C0} is zero when the supercapacitor is completely discharged.

The inductor current of this stage can be expressed as

$$i_{off}(t) = \left(I_{peak} - \frac{2V_d + V_{CO}}{L}\right)\left(\frac{ae^{at} - be^{bt}}{a - b}\right) \tag{6.74}$$

where

$$a = -\frac{R}{2L} - \frac{1}{2}\sqrt{\left(\frac{R}{L}\right)^2 - \frac{4}{LC}}, \quad b = -\frac{R}{2L} + \frac{1}{2}\sqrt{\left(\frac{R}{L}\right)^2 - \frac{4}{LC}},$$

denote the roots of the characteristic equation of (6.73) with $R = R_L + R_W + R_C$.

When there is no initial charge stored in the supercapacitor, then V_{C0} is zero. To ensure the circuit operation at boundary condition, the inductor current i_{off} should reduce to zero at the end of energy transference stage. Substituting $V_{C0} = 0$ and (6.68) into (6.70), we have

$$V_{in}T_{on}(ae^{at} - be^{bt}) = 2V_d(e^{at} - e^{bt}) \tag{6.75}$$

Solving (6.49) with respect to T_{on} yields (Fig. 6.32)

$$T_{on} = \frac{2V_d(e^{at} - e^{bt})}{V_{in}(ae^{at} - be^{bt})} \tag{6.76}$$

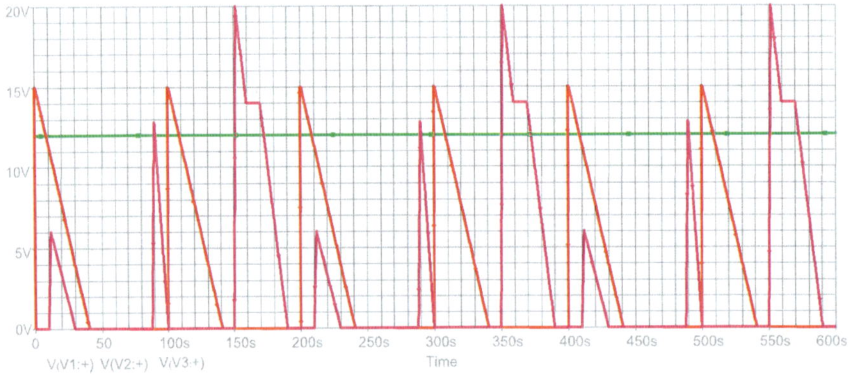

Fig. 6.32 Generated voltages taken as V_{in} into simulations

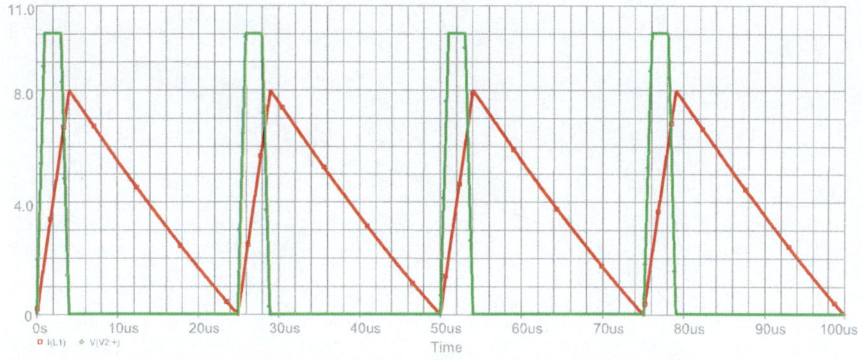

Fig. 6.33 Current in the coil and switching voltage for charging the supercapacitor at $V_{in} = V_1$

There are two states of operation per switching cycle, namely energy absorption and energy transference states, respectively. During the first state, the inductor absorbs energy from the supplying source. Then, it delivers the energy to the supercapacitor at the second state and the supercapacitor can acquire constant amount of energy per switching cycle (Fig. 6.33).

By testing the appropriate cases of city driving, we can not only change the voltage V_{in} generated at the input in the model circuit, but also control the driver's switches and thus obtain information on the optimal parameters of the practical system. Taking the wire resistance R_W less than 0.5 Ω it is possible to minimize the conduction loss. The turn-off time and switching frequency is determined by the ratio of the inductance L and total series resistances of R_L, R_W and R_C. However, the total resistance of R_L and R_W can be taken as greater the equivalent series resistance of the supercapacitor, and the influences of different characteristics of the supercapacitors provided by various manufacturers are negligible.

Numerical experiments with regard to vehicle braking indicate that the highest voltage on the supercapacitor is generated at both high speed and deceleration time. This is evident in the Stop and Go test when the supercapacitor is charged fastest and at the higher voltage (Fig. 6.36) when comparing to the remaining test cases (Figs. 6.34 and 6.35). The obtained result is very significant when we take into consideration that it has been achieved only due to a better controlling of the regenerative energy by means of an appropriately chosen switching frequency and values of the coil inductance, without any change of the unipolar disk generator dimensions or maximum value of ride velocities.

6.4.8 Summary and Conclusions

To realize the full potential of the Faraday disk generator in connection with supercapacitors as electrical energy recovering and storage devices, new materials and

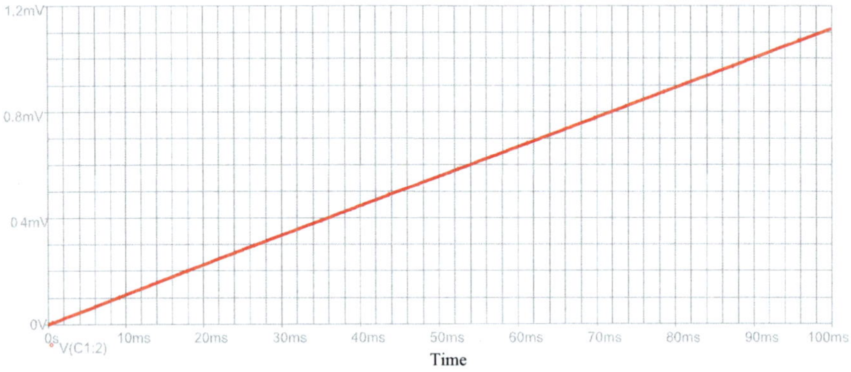

Fig. 6.34 Supercapacitor voltage versus time when $V_{in} = V_1$

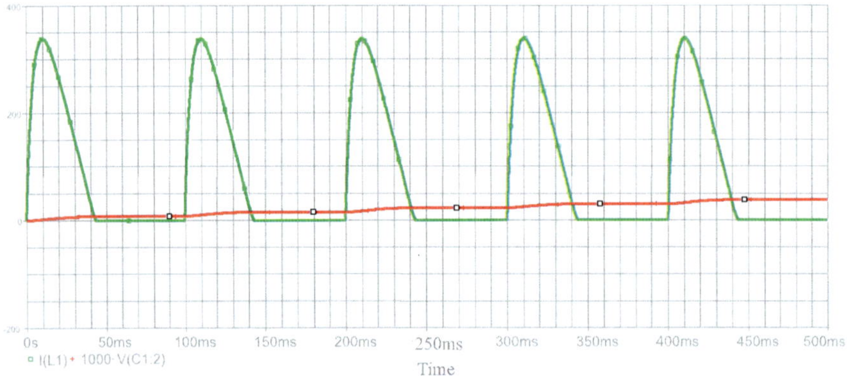

Fig. 6.35 Current in the coil and supercapacitor voltage versus time when $V_{in} = V_2$

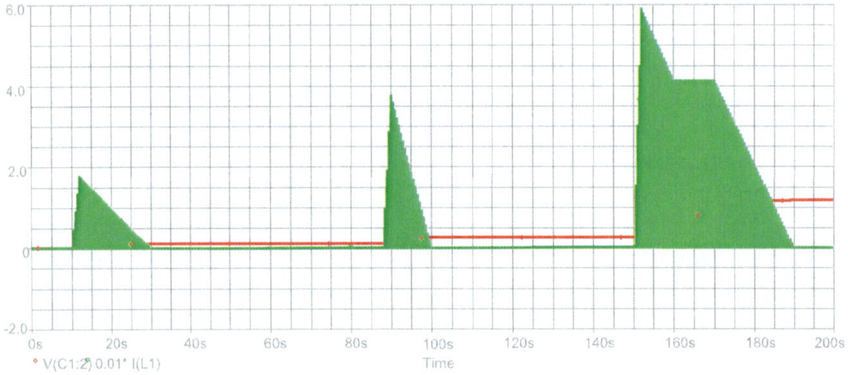

Fig. 6.36 Current in the coil and supercapacitor voltage versus time when $V_{in} = V_3$

chemical processes can improve the efficiency of energy harvesting in electric cars by increasing both the recovered energy and its storage densities. The performance of energy storage systems is limited by the efficiency of the constituent materials—including active materials, conductors, and inert additives. Advances in nanoscience and nanotechnology offer particularly exciting possibilities for the development of revolutionary three dimensional architectures that simultaneously optimize components and energy harvesting capacities. Theory, modelling, and simulation can effectively help experimental design and efforts and can provide better insight into mechanisms, predict trends, identify new materials, and guide experiments. The discovery of such new nanostructured materials as graphene and carbon nanotubes has opened opportunities to develop novel tailored system components [99, 100].

Electric cars, even if powered by electricity generated by coal powered plants are much more efficient than combustion engine based vehicles and as a result release less CO_2 emissions and are noiselessly. Between the main benefits resulting from the spread of electric vehicles in cities is a "shift" of harmful air pollutants from municipal areas to rural regions where the exposure of the living organisms to the detriment of biological state is lower. It is worth underline, especially in urban driving conditions, that the noise level generated by electric vehicles is so low that it often can even pose a danger to pedestrians in the streets.

The supercapacitors play important roles in the development of electric vehicle industry. A new charging circuit with minimal number of components count has been presented. It provides high reducing in the production cost as well as circuit failure rate. Applying the developed equations optimal inductance and switching frequency of the PWM controller can be determined. The charger can fully charge up a 680F supercapacitor over a dozen minutes. The circuit does not consist of complicated closed loop control and is free of stability problem. It favors the application of supercapacitor-battery combination system for electric cars.

6.5 Electromechanical System for Charging Batteries of Electric Cars

6.5.1 Introduction

Presently, the problem of energy becomes so important that all the attention of modern societies and entire industries are turning towards clean, renewable energy (solar energy, wind energy, hydrogen fuel cell, etc.) designed to use only electricity as motive power, which reduces greenhouse gas emissions and noise level close to zero. It should be emphasized that electric drive vehicles are ones of the most advanced vehicles at the moment taking into account an important decrease of environment contamination. Therefore, the current priorities for the development of electric drives in vehicles are primarily ecological. Electric vehicles (EVs) in the field of use do not emit any pollution (in fact, of course, they emit solid particles, derived from the pair of tribological parts in the vehicle, the cooperation of the wheels with the road surface and dust from the road). EVs are also characterized by much lower noise emissions than those with combustion engines. Moreover, an important advantage of using electric cars but the ability to protect raw materials for the production of liquid and gaseous fuels because there are many types of primary energy carriers that can be used to power EVs such as, for instant, renewed energy sources. In the case of vehicles with electric drives there is a possibility of "moving" places of emission sources from the most dangerous areas—from the centers of large urban agglomerations—to the location of power plants. However, the electric car has already become a technology that in the world market has successfully busted through the bastions so far dominated by vehicles driven by conventional combustion engines [34]. From a technical point of view it should be appreciated that in today's vehicles are the weakest element of electric energy accumulators. This is by no means a surprise, as it is well-known that energy storage is a particularly difficult problem—so far unsurpassed use is the use of liquid fuels [101]. It is worth to emphasize that in recent years great technical progress has been made in the areas of system structures and key component developments for EVs, but there is still a performance gap between electric vehicles and conventional vehicles with respect to driving range, energy saving and power car efficiency as well as passengers safety. For these, among others, reasons today major research efforts in the wide world are invested in the development of high energy density batteries with minimum electric service requirements (ESR). Note also that modern EVs have full information system that is under constant modifications and does monitoring of inside and outside parameters in order to achieve maximum energy savings. Development of new components, improved connections and electric engine control algorithms allow increase of efficiency of power appliances, therefore electric engine itself, to the possibly maximum theoretical limits. This cautious approaching to EVs is not in any way contradictory to the dynamic development of electric drives not only in small passenger cars but also in buses, lorries and two-wheel vehicles [102]. The condition for wide dissemination of EVs is not only the availability of charging stations, but also the short charging time of the car's battery. It is essential that charging stations can be available at both the home and the workplace, at shopping centers, offices, parks, recreation centers and all other places that the driver uses. So far installed battery chargers are built on the basis of power converters, which are fundamental flaws negative impact on the supplying power network, as strongly deformed charge current (Fig. 6.37) and this happens their complex structure and significant failure rate [34].

Recent advances in electric-drive technologies enabled commercialization of plug-in electric vehicles (PEVs), which integrate a charging infrastructure or other

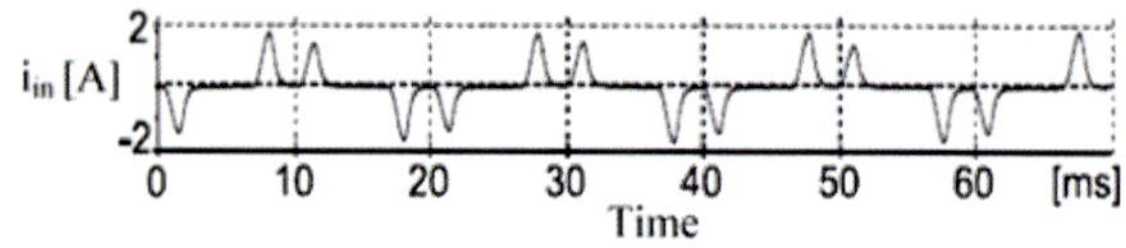

Fig. 6.37 Non-sinusoidal current absorbed from the supplying power network

power source with batteries, regenerative braking, and an electric motor to boost fuel economy.

It is worth mentioning that the complexity and cost of the charging system is primarily dependent on the type of battery and the recharge time. Hence, it is important to develop a cheaper and more accessible charger for electric vehicle applications using renewable energy sources which address the problems of pollution and depleting non-renewable sources of energy.

Of the many types of batteries suitable for use in EVs should be distinguished lithium-ion (Li-ion) batteries that have worked so far as a source of energy used in cell phones and laptops [99]. These batteries last longest in operations if they are frequently charged; fully discharging them will degrade their capacity relatively quickly. A Li-ion battery is unique, as it is charged from a fixed voltage source that is current limited. Such a case of battery charging is usually referred to as constant voltage charging (C-V charger). The essential properties of such systems are shown in Sect. 6.5.2.

As a candidate device for recharging batteries can be considered the Faraday disk generator also called a homopolar machine or an acyclic machine that has preserved a nimbus of mystery because it works on the principle of the Lorentz force which keeps a life of its own in engineering since this law is not directly contained in Maxwell's equations, being the basis of electrodynamics [103]. The principle of its operation and properties are presented in Sect. 6.5.3. However, the Lorentz force is nothing else than the relativistic transformation law which Maxwell's equations are based on, so nothing cryptic is contained in the principle of homopolar induction [90]. A precise explanation of the operation of such type device is possible in the context of Einstein-Cartan-Evans (ECE) theory [104]. Section 6.5.4 presents considerations related to the mathematical model of the system under examination. Therefore we give a deeper description of it in the next section and report on simulations which corroborate that a Faraday's disk generator appears as an effective solution for the implementation of a good and universal EVs battery charger. The results of simulation calculations using the MATLAB program package are presented in Sect. 6.5.4. Discussion of simulation's results and conclusions are included in Sect. 6.5.5.

6.5.2 *Principle of Batteries Charging with the Faraday Disk Generator*

All-electric vehicles are powered only by one or more electric motors. They are powered with electricity by plugging into the electrical power grid and store it in batteries. They consume no petroleum-based fuel while driving and produce no tailpipe emissions. The present days electrical power grids were not designed for this new type of load, which corresponds to the batteries charging systems of PEVs, therefore the impact caused by the proliferation of EVs cannot be neglected. The challenge is to accommodate the electrical power grids, as early as possible, as "smarter" as possible, and the most environmentally friendly as possible. Smart grids appear to achieve these objectives, which are not characterized as a single technology or device, but rather as a near perspective of a distributed electrical system, supported by such reference technologies as power electronics devices, integrated communications, energy storage systems, and advanced controlling infrastructures [94, 105].

PEVs are increasingly popular and the main energy storage systems of these vehicles are the electrochemical batteries. The time required to recharge EVs batteries depends on the total amount of energy that can be stored in the battery pack, and the voltage and current (i.e., power) available from the battery charger.

However, taking into account nowadays limits of those still most used technologies of the energy storage, the EVs have limited travel range. This results in the need to recharge the battery during the long travel route, which can be implemented on the appropriate charging stations. An interactivity between the vehicles and the electric power grid is expected to be one of the key technologies in the future of the smart grids and batteries charging systems in the context that the energy stored in EVs' batteries may be suitable for providing regulation services, spinning reserves and peak power demand. This seemingly impossible task can only be accomplished with the help of engineering simulation. As is well known the engineering simulation is a time-tested tool for accelerating technology development. Computer simulation is conducted in a virtual environment and is significantly faster than physical prototyping and testing, expediting a new product's time-to-market. With computer simulation design parameters can be virtually tested with precision, speed and cost economy. Such an approach is far less expensive than physical prototyping and testing, and can cut costs by an order of magnitude. Moreover, it provides deep insights into the underlying physics involved in the construction and operation of a product, helping solve quality issues upfront [106].

6.5.3 Battery Property and Modelling

Nowadays automotive industry considers Li-ion batteries as the preferred choice over the conventional batteries like Ni-MH, Ni-Cd and lead acid batteries due to its high single cell voltage, high energy density and long service life. One of the major challenges faced by the current EVs is the overall driving range, which is much lower compared to the classic internal combustion engine vehicles. Adding to the problem is a frequent lack of a battery management system that can estimate and predict the actual remaining power of a battery i.e., to predict the residual driving range. Therefore preventing EVs from running out of charge or leaving the passengers stranded appears presently as the main concern [107].

Studies about the impact of residential EVs batteries charging systems based on power electronic transducers showed that, for the use of the EVs, the total harmonic distortion (THD) of the grid current presents a variation from 3% (at the beginning of charging, with a unitary power factor) to 28.11% (at the end of the charging, with a power factor equal to 0.96). So, it is clear that the simultaneous use of a great number of EVs batteries charging systems connected to the electrical distribution grid can cause a significant degradation of the electrical power quality.

Taking into account the above unsatisfactory results and the previous experiences with operations of the car battery charging pilot stations, it becomes essential to develop effective solutions in the concerned field. This requires the introduction of new technologies, mainly by putting into practice the charging systems based on application of homopolar induction phenomenon and developing effective circuit models of Li-ion batteries [107].

In general, the car batteries charging systems can be of two types: public chargers and residential chargers. Public chargers are an optimal solution to charge the batteries of the vehicles using energy from several sources of energy (such as from wind or sun) and can be deployed at strategic places around a town or city, like for example, at companies, public buildings and shopping centers parking lots. On the other hand, residential chargers are operating with delivering low power, in an efficient way, since in general they are used to make a complete charge of the batteries during long periods of time (as, for instance, slow charging during the night). It is worth noticing that at the present time, the uncontrolled way is the most common charging procedure.

In order to safeguard the performance and to extend the battery's useful life and at the same time to predict the residual range of the electric vehicle are required better control and management strategies for battery capacity. One of the parameters directly involved in its identification is state of charge (SoC) which describes the battery's remaining capacity status and is critically important for accurate measurement of safe operations of EV battery.

The relationship between the battery factors and its running characteristics can be suitably described by an equivalent circuit model which captures the electrochemical physics and exhibits the advantage of simplicity and could be easily implemented using only electrical components. In a general case the equivalent circuit model includes an nth order resistor–capacitor (RC) sub-circuit with a series connected open circuit voltage (E), making it easy to incorporate into the system model of an electric vehicle with more precision and lower computational cost.

An equivalent circuit model for Li-ion battery with the E source and a series–parallel RC network is shown in Fig. 6.38. Essential properties of the battery are expressed by parameters: R_1—the ohmic resistance of the battery which describes the electrolyte resistance and connection resistance of the battery, R_2—the polarization resistance, C—polarization capacitance, and V denotes the terminal voltage of the battery.

The established model presents a good reflection of the dynamic and static characteristics of lithium-ion batteries, and on the other hand its complexity is also appropriate, easy to implement in engineering, and it is easy to implement parameter identification with high precision. Having determined the model element parameters of the lithium-ion battery we can connect it with the charger model and perform simulations in the Matlab/Simulink program of the whole system. Therefore, in the

Fig. 6.38 Equivalent circuit model for Li-ion battery

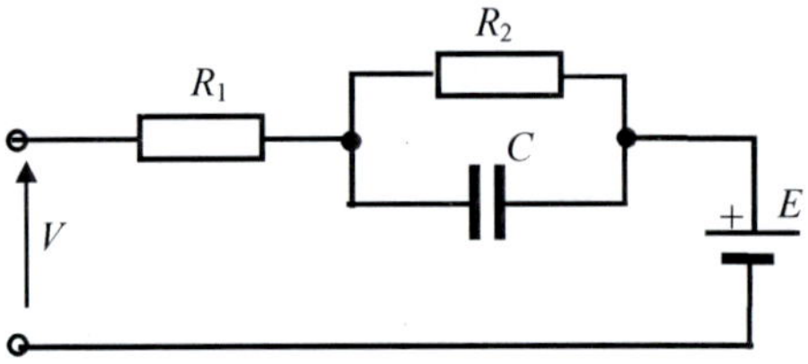

case of testing the Li-ion battery charging process, the developed model could be the preferred choice.

6.5.4 *Electro-magnetic-mechanical Model of Faraday Disk Generators*

The Faraday disk generator is a DC electrical generator comprising an electrically conductive disk rotating in a plane perpendicular to a uniform static magnetic field (Fig. 6.39a). A potential difference E is created between the centre of the disk and the rim, the electrical polarity depending on the direction of rotation and the orientation of the field (Fig. 6.39b). In contrast to classic DC generators, there is no effect of polarity change because there is no any effect of current rectifying. For this reason, the Faraday disk generators are placed among homopolar machines. They are unusual in that they can source tremendous electric current, because the homopolar generator can be made to have very low internal resistance. This is not strange since in the scientific community the homopolar electric generation is considered as naturally extra-efficient [108].

The voltage E generated in the rotating conducting shield is fundamentally due to the Lorentz force on the electrons in the moving conductor placed in a magnetic field.

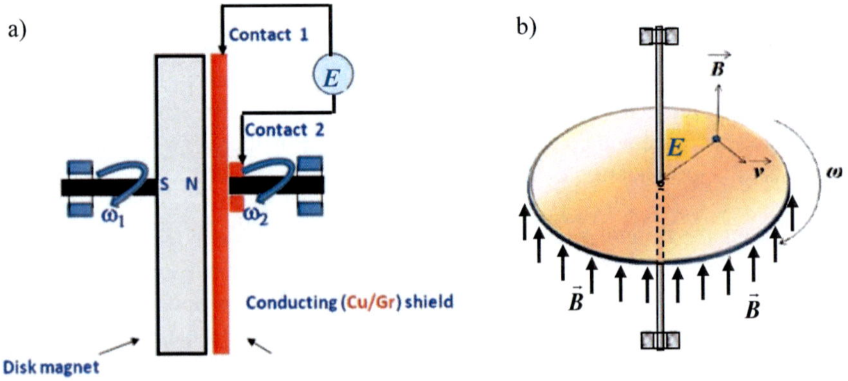

Fig. 6.39 Principle of the Faraday disk generator: **a** structure, **b** generated voltage

Table 6.1 Conditions for rotations of a magnetic disk and a conductive shield

Magnet	Shield	Voltage
ω1 = 0	ω2 = 0	$E = 0$
ω1 = 0	ω2 > 0	$E > 0$
ω1 > 0	ω2 > 0	$E > 0$
ω1 > 0	ω2 = 0	$E = 0$

The electrons q move with the initial angular velocity ω of the conducting shield and in the presence of a magnetic field $\vec{B}$ [109].

The dynamics of the Faraday disk homopolar generator can be explained in a straightforward manner assuming that the established field near the permanent magnet can be considered to be stationary relative to any motion of the magnetic disk that creates it. Thus the established field can be considered to be a static field and consequently no invocation of relativity physics is necessary to explain the actions of the Faraday disk generator. Only the relative motion between the conductive shields being in rotation through a zero velocity magnetic field that is created by the disk made of permanent magnet is necessary. Possible cases are detailed in Table 6.1 and indicate the generation of voltage between the center and the circumference of the conductive shield only when it rotates relative to the magnetic field. The rotation of the magnet does not matter because the field it creates is stationary. Having the stationary field with zero velocity relative to the motion of the magnetic disk that creates it explains why there is no voltage generated in a Faraday homopolar generator if the shield is held motionless and the magnet disk is rotated [78, 110].

The Lorentz force F_L acting on an electron q can be expressed as

$$\overrightarrow{F_L} = q\left(\vec{v} \times \vec{B}\right) \tag{6.77}$$

where q, $\vec{v}$ and $\vec{B}$ denote electron's charge, linear velocity and magnetic field density, respectively.

Now, applying usually the Faraday's law of induction in terms of the Lorentz force yields the voltage E generated between the centre and the rim of the conducting shield, namely

$$E = \int_{r_1}^{r_2} \frac{\overrightarrow{F_L}}{q} \cdot \overrightarrow{dr} = \int_{r_1}^{r_2} (\vec{v} \times \vec{B}) \cdot \overrightarrow{dr} = \frac{1}{2}\omega B(r_2^2 - r_1^2) \tag{6.78}$$

where B, ω, r_1 and r_2 denote the magnetic field density, angular velocity, internal and external radii of the conducting shield.

Therefore, it is clear that the voltage on the generator's terminals is greater when greater are: the rotation and the external radius of the shield, the magnetic induction B produced in the shield and the radius of an annular permanent magnet.

It is also seen that the output voltage E increases as the difference of the square of the inner and outer radius length increases, while only in a linear fashion as a direct function of the angular velocity as well as of the magnetic field density [111].

Thus, it is possible to design a generator for commercial use of large diameter yielding relatively substantial voltage output at a fairly low rotation rate. This means that the use of the Faraday disk generator as a charger for EVs batteries would be ideal since the size of the disk magnet would not be limited by gravitational force stressing the magnets due to their sheer weight consideration.

In order to calculate the resistance for the conductive shield, we start with the assumption that the direction of the current I is radial. Then the resistance of a shield-shaped conductor can be calculated by considering an infinitesimal fraction of a circle (an arc) as a thin wire, of which the cross sectional area A is dependent on r. The material of the rotating shield is homogeneous, isotropic, and electrically linear with the material constants ρ, ε_o, μ_o. The dimensions are appropriately chosen for given applications. Therefore, the resistance R_s of a shield-shaped conductor can be expressed as

$$R_s = \rho \int_{r_1}^{r_2} \frac{dr}{\int_0^{2\pi} hr d\alpha} = \frac{\rho}{2\pi h} \int_{r_1}^{r_2} \frac{dr}{r} = \frac{\rho}{2h\pi} \ln\left(\frac{r_2}{r_1}\right) \tag{6.79}$$

where ρ and h denote the material resistivity and thickness of the shield.

Note that this model does not include electrical energy loss due to eddy currents because we test only relatively low angular velocities of the conductive shield. It may be argued that increasing the radius of the conductive shield will increase the resistance proportionally, but this is not the case since the resistance is proportional to the product of resistivity and natural logarithm from the ratio of radii of the shield and this is then divided by the thickness of the shield multiplied by 2π. As the resistance of the disk varies both with the radius of the shield and the thickness according to Eq. (6.79), it is easily seen that the expected behavior of the resistance is hard to predict. Thus, it is worth mentioning that when the resistance will actually go down, then an increase of the output current must take place. Convenient results in this direction can be achieved by producing a shield of conductive material of low resistivity, such as copper/graphene composite [112].

From the expression (6.78) it follows that the magnetic field induced in the generator exerts a directly proportional effect on the voltage generated in the conductive disk. Neodymium magnets can be chosen because they are strong and may be easily coated by copper/graphene composites [81]. The diameter of the magnet is chosen to be slightly larger than the diameter of the conductive shield. This allows a large enough space for the brush to maintain contact with the circumference of the conductive shield while avoiding the larger moment of inertia caused from a larger diameter magnet. Typically, the disk magnet and the conductor disk are strongly bonded to each other and are held on the same shaft that drives their rotation. Such a compact can be located in a sealed and suitable plastic cylinder with a vacuum to minimize resistance during rotation. Using the above concept, a point will be reached where only a small portion of drive energy will be required to balance bearing friction and brushes [113].

6.5.5 Structure of the Charging System

Although there have been up-to-date many experimental approaches to find an efficient homopolar generator, very few attempts have been conducted to develop an electro-magnetic-mechanical model of a Faraday disk generator. In this section, we focus on modeling the electro-magnetic behavior of a Faraday disk generator depending on various system characteristics. We intend to investigate the conditions where the maximum efficiency can be obtained when basing on simulation results. This research is significant in that it suggests a general strategy in building a Faraday disk generator under the numerical and experimental constraints. To solve these problems a new and effective battery-charger system is developed in this section. Its structure is shown in Fig. 6.40.

In this design the single disk with a static magnetic field is replaced by a twin disk. A one conventional permanent magnet for one disk and an electromagnetic coil for the other are used in this special design. The current produced by the machine flows through the electromagnet and provides a positive feed-back effect. If the current in the coil is higher, than higher is the magnetic field and higher is the current again. A charged battery is connected between the shaft and the rim of the conducting shield by means of brush contacts, and when the magnet and shield rotate a current is observed to flow. The conduction electrons in the shield take up its rotation, and they experience a radial magnetic force which drives them towards the brush and hence round the circuit. On the other hand the external circuit containing the battery is not in motion, and its electrons do not experience a magnetic force in the direction of the circuit. Thus the force driving the electrons round the circuit originates in the rotating shield. The electromagnetic torque exerted on the conductive shield due to the Lorentz force is

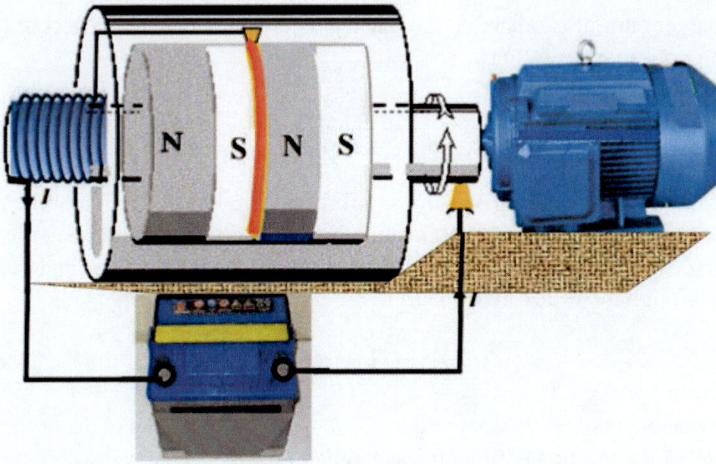

Fig. 6.40 Structure of a battery-charger system

$$M_s = -\frac{1}{2} I B(r_2^2 - r_1^2) \tag{6.80}$$

The minus sign in formula (6.80) means that the Lorentz force acts on the shaft in the direction opposite to the direction of its rotation. Hence, for the drive of the Faraday generator shaft, an electric motor is used that provides a torque equal to an absolute value of (6.80) but with the opposite sign [114].

6.5.6 *State-Space Equations*

The dynamics of a Faraday disk generator charger for EVs batteries can be explained in a straightforward manner if the state-space equations governing the whole system are considered taking into account simultaneously all processes running in the system. According to the Kirchhoff voltage law, the sum of voltages in the system of Fig. 6.40 gives

$$L\frac{dI}{dt} + R_1 I + U + E(I) = \frac{1}{2}\omega(B + B_s)(r_2^2 - r_1^2) \tag{6.81}$$

where I, U, ω denote state variables, and L is the coil inductance, R_1 denotes the sum of the coil resistance, conducting shield resistance and series resistance of the battery. The feed-back magnetic induction B_s is determined by

$$B_s = \frac{\mu N I}{l} \tag{6.82}$$

where N and l denote the turn number and length of the coil, respectively.

The charger current is related to the battery polarization voltage in accord to the Kirchhoff current law as follows

$$C\frac{dU}{dt} + \frac{U}{R_2} = I \tag{6.83}$$

To the above two governing equations must be added the description of the physical friction and the armature reaction which represent factors decelerating the rotation of the disk. The frictional torque $M_f(t)$ acting on the conductive shield can be modelled through the following equation

$$M_f(t) = \alpha \cdot sign(v(t)) + \beta v(t) \tag{6.84}$$

where $v(t)$ denotes the car velocity and α and β are the coefficients of the dry kinetic friction and of the viscous friction, respectively.

Taking into account the Newton dynamic equation for the rotational movement yields

$$J\frac{d\omega}{dt} + M_f(t) - M_s = M \tag{6.85}$$

where J is the sum of inertia moments of the Faraday disk generator and the drive motor with respect to the system shaft and M the driving torque.

Considering Eqs. (6.81)–(6.85) together and subjecting them to appropriate transformations, we can present the system of equations described in the form of equation in the state space

$$\begin{aligned} \frac{d}{dt}\boldsymbol{x} &= \boldsymbol{f}(\boldsymbol{x}, M) \\ y &= h(\boldsymbol{x}, M) \end{aligned} \tag{6.86}$$

where $\boldsymbol{x} = [I, U, \omega]$ means the state vector, $\boldsymbol{f}(\boldsymbol{x}, M)$ is the state evolution function, y—the response variable and $h(\boldsymbol{x}, M)$—read-out map.

The state evolution function takes the form

$$\boldsymbol{f}(\boldsymbol{x}, M) = \begin{bmatrix} -\frac{R_1}{L}x_1 - \frac{1}{L}x_2 + \frac{a}{L}x_3 + \frac{b}{L}x_1x_3 - \frac{1}{L}E(x_1) \\ -\frac{1}{C}x_1 - \frac{1}{CR_2}x_2 \\ -\frac{\alpha}{J}sign(r_2x_3) - \frac{\beta r_2}{J}x_3 + \frac{1}{J}M \end{bmatrix} \tag{6.87}$$

with $a = \frac{1}{2}B(r_2^2 - r_1^2)$ and $b = \frac{1}{2l}\mu N(r_2^2 - r_1^2)$. As the system response we take $y = E$ and the read-out map is determined as

$$h(\boldsymbol{x}, M) = E(x_1) \tag{6.88}$$

The state-space equations (6.86) accompanied by (6.87) and (6.88) contain quite a few parameters and to make their set manageable, we may choose the simplest ones compatible with the physical meaning of the distribution of the quantities entering into the problem. A detailed analysis of the appearance of various regimes with a change in all parameters is beyond the scope of this study, because our goal is to demonstrate the compatibility of the simplest Faraday disk generator models with the needs of effective battery charging for electric cars. The determined mathematical model describing the operation of the tested system appears as the basis for the implementation of appropriate computer simulations, which allow determining the optimum parameters of the components [81].

6.5.7 Computer Simulations

Computer modeling and simulation methods are important elements in the domain of design and operation of battery charging systems. Several reasons justify simulation tests, namely: the cost of building a charger can be minimized based on the results of previous ones simulation, analysis of its operation can be performed with minimal risk, dynamic analysis can be done without the need to build a prototype, simulation analysis can be made in the system design phase for a fraction of the construction cost. Because the main goal of the conducted research was not simulation taking into account all the smallest details, but the global behavior of the charging system of the electric battery, it was adopted its model with the form of state-space equations (6.86), which for this research are accurate enough. For simulating the time behavior of the state variables we assume that the rigid shaft in the system is driven by an external electric motor with a constant torque M. To solve the state-space equations (6.86) the numerical procedure *ode*23 from MATLAB program package has been used for various sets of system element parameters. Computer simulations have been implemented for many sets of parameters of the elements forming the system and after selecting the obtained results corresponding to the ones most needs of the practice, the optimal set was established. In the completed computer simulations, the following parameters have been adopted: $B = 1.5$ T, $r_2 = 55$ cm, $r_1 = 2.50$ cm, $L = 0.2$ H, $C = 0.75$ F, $R_1 = 2.3\ \Omega$, $R_2 = 10^9\ \Omega$, $\alpha = 0.5$ Nm, $\beta = 0.5$ Ns, $J = 0.5$ Nms, $\mu = 4\pi 10^{-5}$ H/m, $l = 5$ cm, $N = 250$, $M = 4.8$ Nm. The voltage at the terminals of the

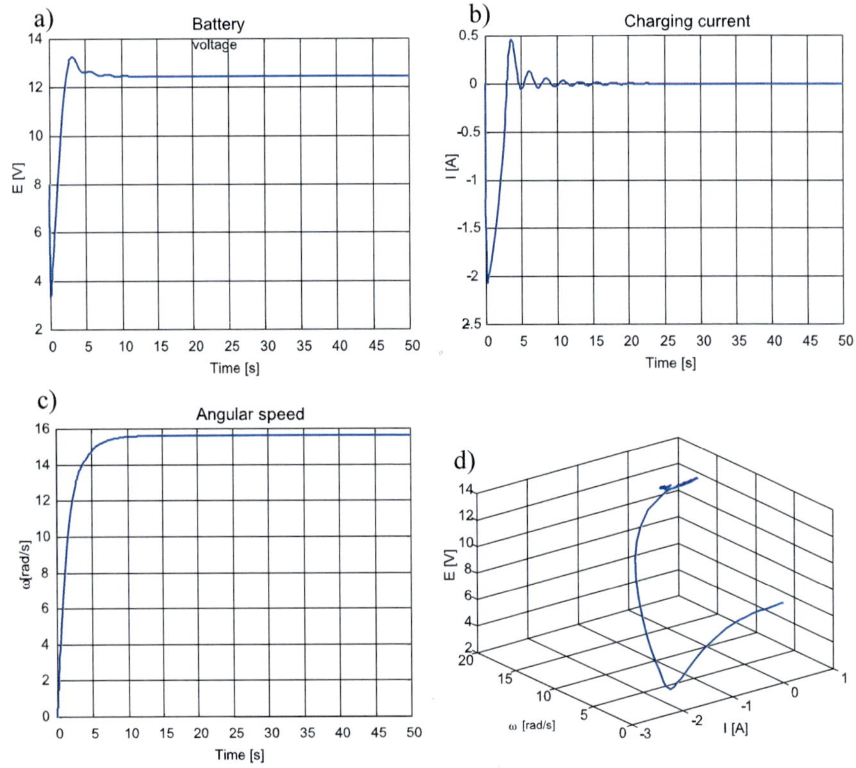

Fig. 6.41 State-variables variations for selected system element parameters: **a** battery voltage, **b** charging current, **c** rotation speed of the shaft, **d** 3D plot

charged battery was modeled by means of the relationship: $E(x1) = a_0 + a_1 x_1 + a_3 x_1^3$ with $a_0 = 8$ V, $a_1 = 0.75$ Ω, $a_3 = 0.5$ Ω/A^2. Corresponding results of performed simulations are shown in Fig. 6.41. The battery voltage and current waveforms in the system supplied by the Faraday disk generator exhibit the preferred shapes. Initial very fast current change in the battery, with the forms quite similar to a short pulse and subsequent fast-damped sinusoidal oscillations with small amplitudes are very easy to see. Operation of the battery in these conditions can lead to its very rapid charging without any damages. The results obtained from the performed simulation can allow the potential user to observe the benefits of using a Faraday disk generator for effective charging of the electric vehicle battery, as well as evaluate benefits for the battery itself, and thus the state of its charge, more even its vitality and limiting peak battery current values that can be accepted.

6.5.8 Discussion and Summary

The time required to recharge electric vehicle batteries from the Faraday disk generator depends on the total amount of energy that can be stored in the battery pack, and waveforms of the voltage and current (i.e., power) available from the battery charger. It is possible to create a high-efficiency generator by reducing the amount of time it takes for the rotational energy to be converted into electrical energy. This can be accomplished by strengthening the magnetic field and by decreasing the resistance of the conductive shield by making it from copper/graphene composite. The Faraday disk generator appears as a new development in the battery recharging operation that decreases the time required to recharge electric vehicle battery to as little as 1.0–1.5 min without damaging it. When this technology will be fully deployed electric charging stations similar to gas stations will allow the electric vehicle operator to quickly recharge the battery pack. This new charger technology, coupled with advanced Li-ion batteries with a range of 300 km between recharging, will allow the electric vehicle operator the same freedom of the road currently enjoyed by today's operators of gasoline-powered vehicles. It is possible to design a generator for commercial use of large diameter yielding relatively substantial voltage output at a fairly low rotation rate. This means that the use of the Faraday disk generator as a charger for EVs batteries would be ideal since the size of the disk magnet would not be limited by gravitational force stressing the magnets due to their sheer weight considerations. The results obtained from the performed simulation can allow the potential user to observe the benefits of using a Faraday disk generator for effective charging of the electric vehicle battery, as well as, evaluate benefits for the battery itself, and thus the state of its charge, more even its vitality and limiting peak battery current values that can be accepted.

6.6 Modeling of Energy Processes in Wheel-Rail Contacts Operating Under Influence of Periodic Discontinuous Forces

6.6.1 Introduction

In the analysis of wheel-rail contact several modelling methods can be applied and in the course of the last decade new demands have emerged. They are mainly connected with the reduction of costs of maintenance, technical diagnostics of the track and of railway vehicles, and elimination of negative effects on the environment. Many of the works have been related to the increase in train speed of railway vehicles without affecting safety and comfort [115–120]. Recently, the advanced technology is confronted increasingly with damping problems that do not address to issues of scaling. The contact zone (roughly 1 cm^2) between a railway wheel and rail is small compared with their overall dimensions and its shape depends not only on the rail and wheel geometry but also on how the wheel meets the rail influence. Thus, studying the complex motions of railway vehicles could give significant new insight into the defect properties of materials—a research domain where relatively little first-principles progress has been made. A hard problem arises when the wheel-rail contact is subject to an action of time-discontinuous forces. It is well known that in the contact zone between railway wheel and rail the surfaces and bulk material must be strong enough to resist the normal (vertical) forces introduced by heavy loads and the dynamic response induced by track and wheel irregularities. Thus, the interactions of the surface and the volume of a solid rail are important. It is not possible to materials grow without dislocations and/or other disturbances to crystalline order, such as vacancies, interstitials, or substitution impurities. In the case of polycrystalline materials, the memory features of hysteresis may be important according to the methods of their fabrication. Long before defects escalate to the point of incipient failure, they still influence vibrations. In its simplest form the equations of dynamic motion of a wheel-rail system are given by a non-linear second order system of ordinary differential equations (ODEs) of moderate dimension that may be solved numerically by standard methods [118–128]. The goal of this Section is to assist the subject progress towards a healthier balance between these extreme factors. The main attention is focused on modelling of the energy absorbed by the rail rested on an elastic sleeper and completely rigid foundation. To determine the susceptibility of a given contact to the strength induced by the rail roll, an approach employing the energy state variables (time functions) is established. Then, the established model of the wheel-rail contact dynamics has been applied to that same roll plane but with taking into account a nonlinear characteristic of the sleeper with respect to the ground. We conclude that under operations of periodic discontinuous forces the energy absorbed in contact can be measured by loops of one-period energy in the energy phase plane. Several numerical examples are included.

6.6.2 The Problem Description

The contact between wheel and rail is the basic constitutive element of railway vehicle dynamics. For its modelling two aspects must to be considered: (1) the geometric or kinematical relations of the wheel-rail contact and (2) the contact mechanical relations for the calculation of the contact forces. Strictly connected to contact mechanics is the problem of formation of corrugations on the rail treat, which is now in the centre of attention of many research teams. The knowledge of railway vehicle dynamics allows us to predict with confidence what the values of contact stress, tangential creep forces and creep age in the wheel/rail contact patch are for a wide range of different conditions. This gives a valuable insight into the influence of the many different factors that affect the incidence of rolling contact fatigue in rails due to the combination of these parameters [129–138]. At present, many scientific centres in the world are involved in research focused on different problems of railway vehicle dynamics. Increasing operational speeds and comfort demands require focusing on both riding qualities of railway vehicles, transport safety and comfort. Several factors have a negative influence on ride comfort, and emphasize the necessity of more advanced suspensions. As the challenges of higher speed and higher loads with very high levels of safety require ever more innovative engineering solutions, better understanding of the technical issues and use of new computer based tools is required [138–146]. Usually, railway vehicles operating in modern countries use wheel sets comprising two wheels fixed to a common axle (Fig. 6.42). Wheels are rigidly connected with the axle and roll in the direction which they are heading for. As both wheels are rotating at the same speed, the contact forces ultimately are repartitioned symmetrically on both wheels. These forces are a major cause of the rails wear. A completely linear multi-body formalism must be taken into considerations and the kinematical nonlinearity can be replaced by quasi-linearization. Strictly connected to contact mechanics is the problem of formation of corrugations on the rail treat, which is now in the centre of attention of many research teams [132, 147]. The development of rolling contact fatigue in rails depends on the interplay between crack growth, which is governed by the contact stress and the tangential force at the contact patch. Moreover, the wear depends on the tangential force and the creep age at the contact patch (Fig. 6.43). These parameters are dependent on a large number of inter-dependent factors, in particular:

- Vehicle Configuration: wheelbase, axle load, wheel diameter,
- Suspension design: in particular primary yaw stiffness,
- Track geometric quality,
- Wheel profiles: nominal profile and state of wear,
- Rail profiles: nominal profile and state of wear,
- Wheel/rail friction,
- Cant deficiency (depending on speed, radius and cant),
- Traction and braking forces,
- Wheel and rail material properties.

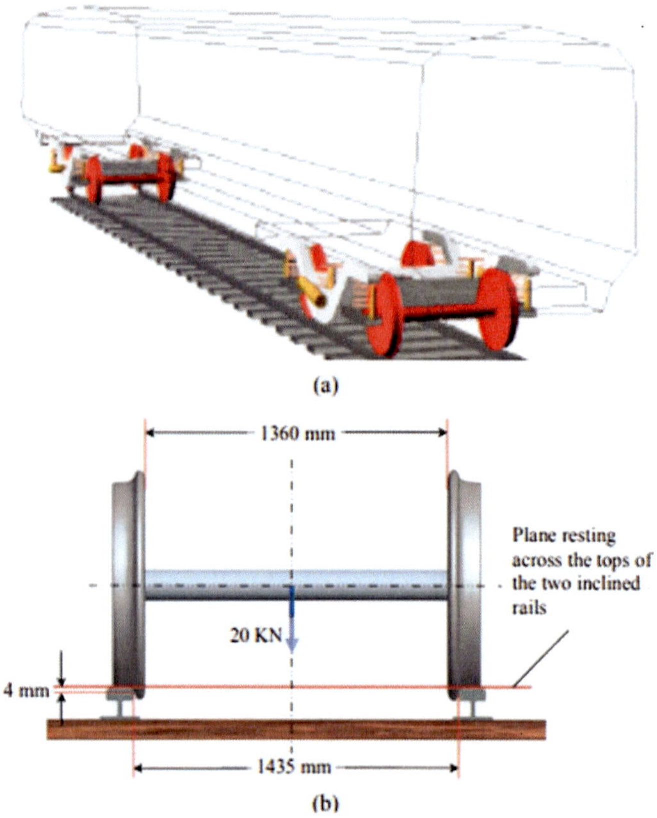

Fig. 6.42 Schematic views: **a** train vehicle; **b** wheels, axle, rail and sleeper

Fig. 6.43 Spalling from high contact stresses of the: **a** rail; **b** wheel

The estimates have showed that in the case of the wheel-rail system, the elastic correlation length far exceeds the length of the contact region (about 1 cm). Multibody dynamics and wheel-rail contact models make detailed analysis possible not only for ride and handling, but also for determining their effects on rail non-uniformities. Experimental studies have revealed that estimations of both the size of the contact region and pressure distribution in the contact depend on the accuracy of which the surface micro-relief was determined. The size and shape of the contact zone where the railway wheel meets the rail can be calculated with different techniques. The surfaces of railway wheels and rails, as many other technical surfaces, have micro-heterogeneities at many scale levels; by experimental studies it has been established that these are self-similar surfaces in the wide range of surface scales and can be referred to the class of fractal ones [120, 121]. This suggests that both the size of the contact region and pressure distribution in the contact depend on the accuracy with which the surface micro-relief can be determined. Through mathematical analysis it is possible to build a deep and functional understanding of the wheel-rail interface, suspension and suspension component behaviour, simulation and testing of mechanical systems interaction with surrounding infrastructure, noise and vibration.

6.6.3 Dynamic Model of Wheel-Rail Contacts

Issues related to the phenomena occurring in the contact between the wagon wheel and the railway rail was initially signalled in Chap. 3. A wide range of problems caused by the movement of a train at a constant speed was shown there. An in-depth approach to these issues will be presented in this subsection. The contact between wheel and rail is the constitutive element of railway vehicle dynamics. The geometric or kinematical relations of the wheel-rail contact and the contact mechanical relations for the calculation of the contact forces are to be dealt with very careful attention. This can be done within a completely linear multi-body formalism taking into account kinematical nonlinearities by quasi-linearization. From the short-term dynamic calculations a periodic non-harmonic motion can be obtained in terms of the generalized displacements. For the subsequent calculation of stress only typical contact characteristics are of interest. As is well known the potential and success of vehicle-track dynamic simulations very much depends on how well the system is mathematically modelled and fed with pertinent input data. We use a simplified dynamic model for the wheel set and rail contact. The wheel set is running on straight track and the wheel set and track are considered as rigid bodies. Any contact stiffness is included. However, the contact stiffness is not the only one elasticity to be taken into account, and the track stiffness itself can be used to smooth out the load variation. A large part of the wheel-rail contact modelling leads to the load transfer when flanging, and more generally when there is a jump between two contact points on the profiles. We assume that the wheel set is rigid and in the rail model the discrete sleepers under the rail are resting on completely rigid foundation. The interface between the wheel and the rail is a small horizontal contact patch. The contact pressure on this small surface is closer to a stress concentration than in the rest of the bodies. The centre of this surface is also the application point of tangential forces (traction and braking F_x, guiding or parasite forces F_y, see Fig. 6.44). The knowledge of these forces is necessary to determine the general wheel set equilibrium and its dynamic behaviour. For the overall model we need to consider all specific phenomena occurring in the system during the train movement on the rail. To have a large set of information on the contact, as a function of the vertical relative displacement d_z the study of a single wheel-rail pair is enough. The first modelling of the flange contact has been presented to consider it as an elastic spring whose value comes from the track and the rail beam deformation. For these purposes we consider a scheme presented in Fig. 6.45. The rail is considered as Timoshenko beam which can be divided into small segments so that a sleeper is assigned to every segment. The sleeper has both a mass and the pad between rail and sleeper and the ballast are replaced by springs and dampers. As a natural improvement a more complex model is also considered. It takes into account the roll effect of the other wheel-rail pair of the wheel set. The lateral displacement and the yaw angle can be considered as two small displacements relative to the track and they do not exceed approximately $\pm$ 1%. Then we take into account the fact that the wheel sets are connected to the chassis of the bogies by a set of joints, which control the relative motion between wheels and chassis, and by a set of springs and dampers. In turn, the bogie chassis is connected to a support beam to which one of the car shell connecting shaft is attached. The connection between chassis and support beam is also done by a set of kinematic joints, springs and dampers, designated by secondary suspension [124, 139]. Each train car body is generally mounted in the top of two bogies. The attachment between the car body and each bogie is done by a shaft rigidly connected to the car body that is inserted in a bushing joint located in the support beam of the bogie. Moreover, the rail remains in its original 1in20 or 1in40 inclined position. Measured values showed a maximum of about 0.3° and 0.95° respectively for the UK and German sites. In light of this reason it is assumed for the developments that follow that the motion of each wheel set tracks exactly the geometry of the railway and lateral inclinations can simply be ignored during numerical simulation. The rail structure is particularly subjected to dynamic load, which is induced by moving wheels of the vehicles. This dynamics action can excite vibrations of the track, contribute to bad track conditions and consequently lower the comfort of travelling. The dynamic analysis of the rail structure subjected to a moving load is a very complex problem. Such models can be used for the simulation of real actions of vehicles and detailed response of the structure can be obtained. When an elastic body, such as a wheel, is pressed against another elastic body, such as a rail, so that a normal load is transmitted and a contact area is formed. As the elastic deformation in the vicinity of the contact area is small its effect on the stress response cannot be neglected. Then, assuming that in the vicinity of the contact patch the curvatures of the wheel and rail are constant, the imprinted contact patch is small compared with the radii of curvature and the dimensions of the wheel and rails, the contacting bodies can be represented by elastic half-spaces and their shape can be approximated by quadratic surfaces. Usually it can be assumed that the material properties of wheel and rail are the same and in this case it can be shown that the tangential tractions do not affect the normal pressures acting between the bodies. Then with these assumptions, for the case where the wheels and rails are smooth, the dimensions of the contact area can be obtained from the theory of Kalker described in [45]. Thus, we can consider the stationary rolling contact of elastic bodies and suppose that material properties of wheel and rail are the same, i.e. the bodies in contact are quasi-identical. In this case the contact problem translates to the normal action between the wheel and rail. It means that the normal pressure at a point of the contact patch is proportionate to the interpenetration of the contact bodies at the point. In this system, inertial, stiffness and damping properties vary piecewise-continuously with respect to the spatial location. A continuously vibrating system may be approximately modelled by an appropriate set of lumped masses properly interconnected using discrete spring and damper elements (Fig. 6.46). An immediate advantage resulting from this lumped-parameter representation is that the system governing equations become ordinary differential equations. A linear model is usually justified on the basis of small corresponding deflections. For each of such segments as that shown in Fig. 6.46 a system of equations can be formulated as follows

$$\begin{bmatrix} M_r & 0 \\ 0 & M_S \end{bmatrix}\begin{bmatrix} \ddot{z} \\ \ddot{z}_S \end{bmatrix} + \begin{bmatrix} b_r & -b_r \\ -b_r & b_r + b_S \end{bmatrix}\begin{bmatrix} \dot{z} \\ \dot{z}_S \end{bmatrix} + \begin{bmatrix} k_r & -k_r \\ -k_r & k_r + k_s \end{bmatrix}\begin{bmatrix} z \\ z_s \end{bmatrix} = \begin{bmatrix} F(t) \\ 0 \end{bmatrix} \tag{6.89}$$

The system vibrations are excited by the loading force $F(t)$ exhibiting variations in time shown in Fig. 6.47. It is worth noticing that the short duration pulses correspond

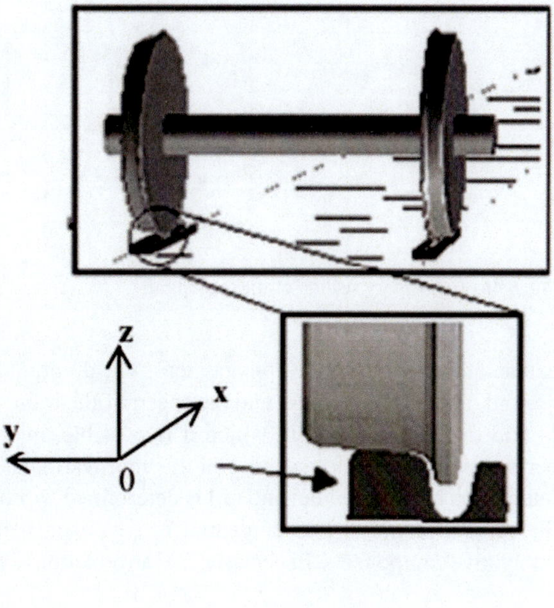

Fig. 6.44 Rail, wheel and contact frames

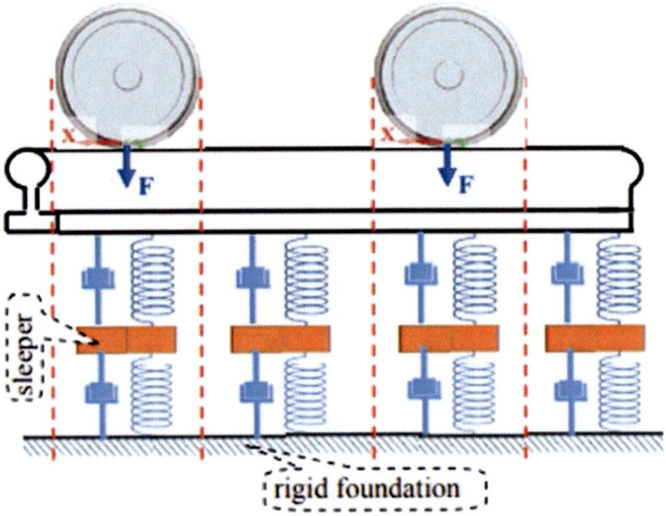

Fig. 6.45 Subdivision of the wheel-track set into segments

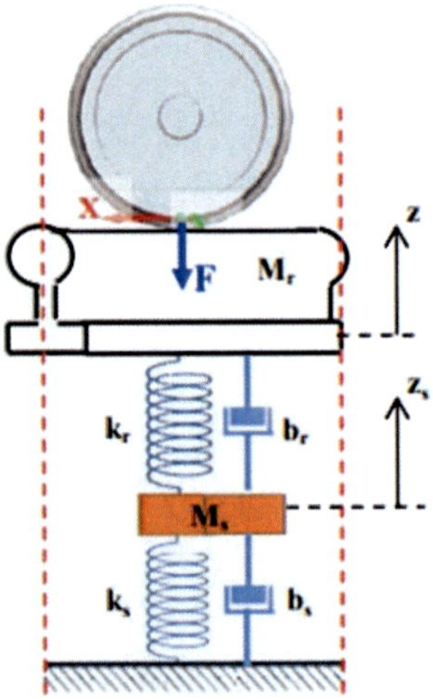

Fig. 6.46 Elementary section of the loaded rail

to small dimensions of the contact areas. The duration and intensities of these pulses depend on the vehicle mass and the speed of the train. Accordingly, we are required to find the proper frame in which it is possible, in an easy way, to determine the periodic non-harmonic response of this multivariable linear dynamical system. The interaction between wheel and rail is determined by normal (along a line connecting the centers of mass) and tangential forces, each being the sum of potential and dissipative components. Especially, the introduction of geometric tools like hysteresis

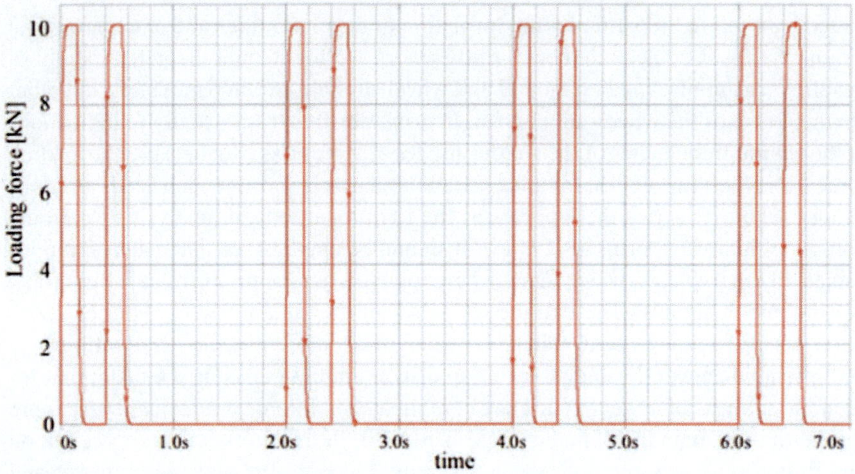

Fig. 6.47 Variations in time of the loading force

loops on energy phase plane greatly advances the theory and enables the proper generalization of many fundamental concepts known for computer aided geometric designs to the world of periodic non-harmonic waveforms. The main problem in the hysteresis loop method is to specify the energy absorbed by the rail during its periodic loading. According to the issue posed in the work, the calculations were performed using a model based on the deformation theory of plasticity with unloading by the elastic law.

6.6.4 *Exact Periodic Solutions*

There are several methods available to analyze (6.89). The most of up-to-date used approaches of these is an eigenvalue calculation on the matrices that represent the equations of vibration (6.89). In the time-domain a periodic solution of (6.89) can always be found by integrating it after the transient responses die out [134, 135]. Such an approach, known as the "brute force method", is a rather time-consuming task and computationally expensive, particularly for the slowly varying systems. The frequency domain approach, known as the harmonic balance technique (HBT), is an iterative method which matches the frequency components (harmonics) of a set of variables defined for the two sides of (6.89). Although the HBT avoids the computationally expensive process of numerical integration of (6.89), its serious drawback is the large number of unknown variables that must be determined. The model (6.89) has a two-fold purpose: first, it can account all the facts discovered experimentally, and second, it can predict the system behavior under various conditions of operations. Therefore, it is important to "explicitly" analyze the role of the material structure, in particular the formation of discontinuities and crack growth at the interfaces of contact elements. In this work we focus on a theoretical study of the effect of impulse load on the rail and wheel strength, strain parameters and other characteristic of the wheel-rail contact. In principle, this leads usually to indications of the natural frequencies of the various modes of vibrations. But because the loading force $F(t)$ is periodic non-harmonic excitation such a classic approach needs an application of Fourier series that appears as the traditional tool for analysis of periodic non-sinusoidal waveforms [141–144]. It should be emphasized that a discontinuous signal, like the square wave, cannot be expressed as a sum, even an infinite one, of continuous signals. The extraneous peaks in the square wave's Fourier series never disappear; they occur whenever the signal is discontinuous, and will always be present whenever the signal has jump discontinuities. Thus, it is evident that for accurate analysis of large systems and complicated harmonic producing terms, more formal time-domain mathematical tool is needed. In this work an attempt is put on description of a new method which takes into account the activation energy and effects of loading force transfers. It is related to non-sinusoidal periodic excitations of the wheel-rail contact and integrations leading to the identification of response waveforms.

Taking into account the above requirements and insufficiencies of the methods based on Fourier series, which are up-to-date most commonly used for studies of periodic non-harmonic states of linear systems we propose in the sequel new method for obtaining, in closed form, the response of any linear system corresponding to piecewise-continuous periodic non-harmonic forcing terms. This newly involved approach is based on the exact solution of (6.89), segment concatenation and periodizer functions. In our approach, the solution is exact, and by means of suitable unification of its piecewise representation, we can get with ease the exact expressions for its time derivatives. The method presented here depends on a "saw-tooth waveform" and a scheme used for unified representation of composite periodic nonharmonic waveforms. We discuss properties of linear systems with periodic non-harmonic excitations and develop a systematic Fourier series-less method for their studies. From this basis, more advanced theoretical results are developed. The main feature of this method is the complete elimination of the frequency analysis what leads to significant simplifications in the process analysis. The second important feature is that that method is based on appropriate loops on the energy phase plane leading to an easy estimation of the energy delivery to the wheel-rail contact process through the evaluation of the loop's area. Taking into account the periodic supplying force $F(t)$ with pulses within each period we can represent it by more suitable form shown in Fig. 6.48. Thus $T = T_1 + T_2 + T_3 + T_4$. The supplying force $F(t) = F(t + T)$ with two pulses within each period can be represented as follows

$$F(t) = F(t + T) = \begin{cases} A \text{ for } 0 \le t \le T_1 \\ 0 \text{ for } T_1 \le t \le T_2 \\ A \text{ for } T_2 \le t \le T_3 \\ 0 \text{ for } T_3 \le t \le T \end{cases} \tag{6.90}$$

where A and T_k, with $k = 1, 2, 3$, denote the magnitude and moments, respectively, describing the pulses in the supplying force. The pulsed supplying force (6.90) can be represented by more convenient formula when introducing the unit Heaviside functions $H_k(t, T_k)$, $k = 1, 2, 3$ which are shifted at the portions of period with respect to the origin point $t = 0$, namely

$$F(t) = A + \sum_{k=1}^{3} H(t, T_k)\left[(-1)^k A\right] \tag{6.91}$$

A steady-state periodic solution of (6.89) depends on system eigenfrequencies

$$s_1 = s_2^* = -\alpha_1 + j\omega_1 \quad \text{and} \quad s_3 = s_4^* = -\alpha_3 + j\omega_3 \tag{6.92}$$

where $j = \sqrt{-1}$ and the real and imaginary components of the system eigenfrequencies are determined by system parameters. It is easily seen from (6.89) and (6.91) that the resulting steady state coordinate $z(t)$ can be expressed as follows

$$z(t) = z(t+T) = z_1(t) + \sum_{k=1}^{3} H(t, T_k)\left[z_{k+1}(t) - z_k(t)\right] \tag{6.93}$$

where $z_k(t)$, $k = 1, 2, 3, 4$ denote the coordinate components in the successive parts, respectively, of the supplying force period. When $t \in (T_k, T_{k+1})$ and the supplying force is equal to A or 0 with $k = 1, 2, 3, 4$ we obtain

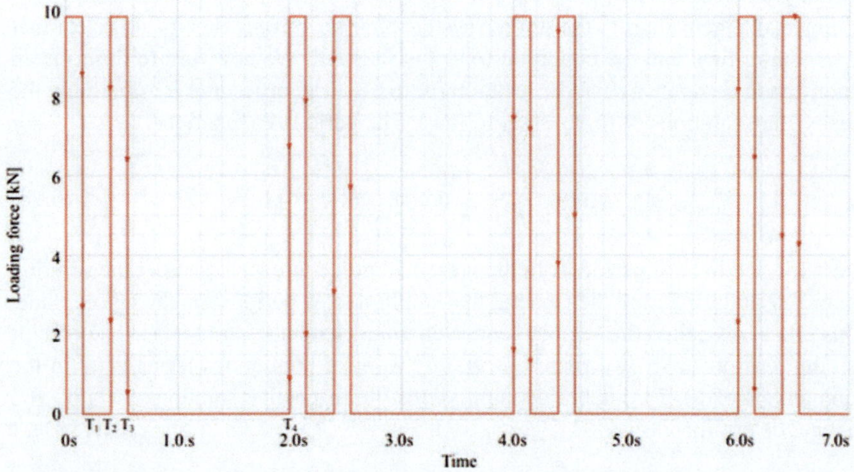

Fig. 6.48 Loading force with ideal pulses within the period

$$z_k(t) = \sum_{m=1}^{4} B_{m,k} e^{s_m t} + Z_{f,k} \tag{6.94}$$

where $Z_{f,k}$ denotes the steady state term forced by F_k and B_m, k, m = 1, 2, 3, 4, are constants to be determined from continuity and periodicity conditions fulfilled by respective components of the system whole response. The steady state forced term $Z_{f,k}$ follows from (6.89) when successively and respectively substituting F_k. In result we get

$$Z_{f,k} = \frac{k_r + k_s}{k_r k_s} F_k \tag{6.95}$$

To determine integration constants $B_{m,k}$ we take into account the continuity and periodicity conditions which must be fulfilled by the coordinates z(t) and $z_s(t)$ [121]. For the sake of compactness of presentation we transform (6.89) into standard state variable equation

$$\dot{\boldsymbol{x}}(t) = \boldsymbol{A}_1 \boldsymbol{x}(t) + \boldsymbol{g}(t) \tag{6.96}$$

where vectors $\boldsymbol{x}(t)$ and $\boldsymbol{g}(t)$ as well as constant element matrix $\boldsymbol{A}_1$ have appropriate dimensions. Here it is worth noticing that $\boldsymbol{A}_1$ must be non-critical with respect to T, i.e. the relation

$$\det\left(\boldsymbol{I} - e^{\boldsymbol{A}_1 T}\right) \neq 0 \tag{6.97}$$

must be fulfilled.

Since all eigenvalues of the studied system have negative real parts, it is clear that $\boldsymbol{A}_1$ is non-critical and the condition (6.97) is satisfied. We now turn to describe an algorithm to compute a periodic solution for (6.96). The procedure for obtaining the steady state solutions of (6.96) is as follows. The periodizer function

$$p(t) = \frac{T}{2} - \frac{T}{\pi} \arctan\left(\cot\left(\frac{\pi}{T} t\right)\right) \tag{6.98}$$

is established to have period T. In the presented procedure we construct the solution for one period $[0, T)$ and then extend that solution to be periodic on the whole t line. This process is called the segment concatenation. Generalizing (6.91) to the case of forcing term in (6.96) we represent $\boldsymbol{g}(t)$ by a set of continuous terms $\boldsymbol{g}_k(t)$ in the subintervals of $[0, T)$. Such a division yields the respective solutions $x_k(t)$ of $\boldsymbol{x}(t)$ that are determined as

$$\boldsymbol{x}_k(t) = e^{\boldsymbol{A}_1 t} \boldsymbol{J}_k + \boldsymbol{X}_{f,k} \tag{6.99}$$

for $k = 1, 2, 3, 4$, where $\boldsymbol{X}_{f,k}$ is a forced steady state (index f) solution of (6.96) and has the same waveform shape as that of $\boldsymbol{g}_k$(t) in the corresponding subinterval, and

$\boldsymbol{J}_k$ is a constant vector to be determined. We can find vectors $\boldsymbol{J}_k$ by analyzing the periodicity condition for the total solution as follow

$$\boldsymbol{J}_1 + \boldsymbol{x}_{f,1} = e^{\mathbf{A}_1 T} \boldsymbol{J}_4 \tag{6.100}$$

and from the continuity conditions it follows that

$$e^{\mathbf{A}_1 T_k} \boldsymbol{J}_k + \mathbf{X}_{f,k} = e^{\mathbf{A}_1 T_k} \boldsymbol{J}_{k+1} + \boldsymbol{X}_{f,k+1} \tag{6.101}$$

for $k = 1, 2, 3$.

Because the period $[0, T)$ is divided into subintervals $[t_k, t_{k+1})$, $k = 0, 1, 2, 3$, then (6.100), (6.101) yield the block matrix equation

$$\begin{bmatrix} \boldsymbol{I} & 0 & 0 & -e^{\mathbf{A}_1 T} \\ e^{\mathbf{A}_1 T_2} & -e^{\mathbf{A}_1 T_2} & 0 & 0 \\ 0 & e^{\mathbf{A}_1 T_2} & -e^{\mathbf{A}_1 T_2} & 0 \\ 0 & 0 & e^{\mathbf{A}_1 T_3} & -e^{\mathbf{A}_1 T_3} \end{bmatrix} \begin{bmatrix} \boldsymbol{J}_1 \\ \boldsymbol{J}_2 \\ \boldsymbol{J}_3 \\ \boldsymbol{J}_4 \end{bmatrix} = \begin{bmatrix} -\boldsymbol{X}_{f,1} \\ -\boldsymbol{X}_{f,1} \\ \boldsymbol{X}_{f,3} \\ -\boldsymbol{X}_{f,3} \end{bmatrix} \tag{6.102}$$

From the Kronecker-Weierstrass form [146] it follows that the system (6.102) has a unique solution $\{\boldsymbol{J}_1, \ldots, \boldsymbol{J}_4\}$ for any $T, T_1, \ldots, T_3$, and $\boldsymbol{A}_1$. Using the periodizer function $p(t)$ we sew on continuous solutions in all subintervals during the concatenation process yielding periodic steady state solutions. Because the solutions are exact, the typical drawbacks of classical methods, such as the Gibbs effect [116], are avoided.

6.6.5 Application of One-Period Energy Approach

The approach presented in the previous sections can be applied to compute energy delivered by the independent supplying force of the load elements in the wheel-track contact. Under periodic conditions the delivered energy is measured by one period energy loop on the energy phase plane. This new concept has been discussed in more details in the recent papers [143–145]. Then the total energy, W_T delivered by supplying force $F(t) = F(t + T)$ in one period equals

$$W_T = \int_0^T F(t)v(t)\mathrm{d}t = \int_{z(0)}^{z(T)} F(t)\mathrm{d}z(t) \tag{6.103}$$

where $v(t) = \dot{z}(t)$ denotes the velocity of the rigid rail oscillations. The above integral is of the Riemann–Stieltjes type. The solutions obtained by using the periodizer and concatenation procedure can be easily used in (6.103) to find W_T. Two following numerical examples illustrate the concatenation procedure in the considered context.

Example 1 Consider a particular case of the system shown in Fig. 6.46 with forcing term $F(t)$ characterized by $A = 110$ kN, $T_1 = 0.15$ s, $T_2 = T_3 = T = 2$ s and $M = 110$ kg. The rigid sleeper is directly incorporated to the rigid foundation. We take into account the remaining parameters: $b_r = 632$ Ns/m and $k_r = 2.5$ kN/m. Performing computations in accord to the above presented algorithm yields the one-period energy loop that is presented in Fig. 6.49. It can be easily found that the energy delivered by the loading force $F(t)$ equals $W_T = 110\ \text{kN} \cdot 0.6\ \text{cm} = 0.66$ kJ.

For the comparison purposes we have considered the same as the above system but with $F(t) = 110\sin(20\pi t)$ kN. The result of computation is presented in Fig. 6.50. Note, that sinusoidal loading force of the same magnitude excites vibrations exhibiting much smaller magnitude and delivers smaller one-period energy than in the case of pulsed force. One-period energy loop involved by sinusoidal loading force is shown in Fig. 6.50. The one-period energy was computed by using the Matlab function *quad.m* to yield $W_{Tsin} = 0.124$ kJ.

Example 2 Let us consider the system shown in Fig. 6.46 with an elastic sleeper that is connected with the rigid foundation with a spring and a damper. The data given in Example 1 are supplemented with the sleeper mass $M_s = 156$ kg and remaining parameters: $b_s = 520$ Ns/m and $k_s = 1.5$ kN/m. For the above taken data we have determined the one-period energy loop that is shown in Fig. 6.51. The computed one-period energy equals $W_{T2} = 1.485$ kJ. It has to be noted that for the loading force exhibiting two identical pulses within the period the magnitude of the rail vibrations is much greater and the one-period energy delivered to the system is more than five times greater with respect to the case of one pulse within the period. If the loading force F(t) takes sinusoidal form with the same magnitude and frequency $f = 10$ Hz then the one period energy delivered to the system has the form shown in Fig. 6.52. The area of this loop equals $W_{T\sin2} = 1.125$ kJ. Because the frequency range of interest is very limited in the involved model of the wheel-rail contact a

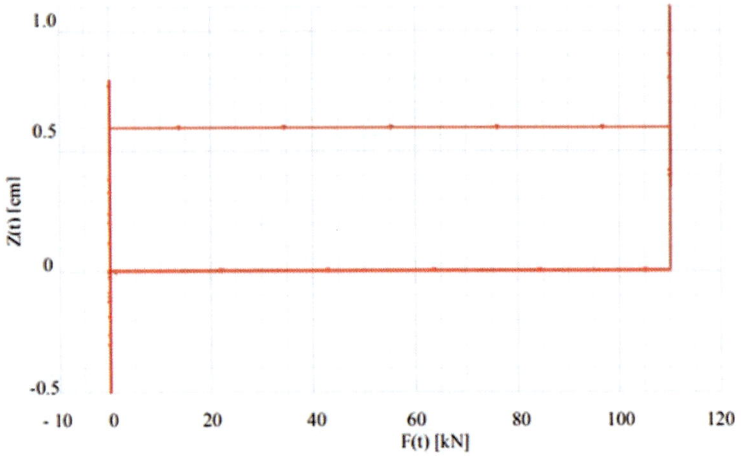

Fig. 6.49 One-period energy loop for loading force with one pulse within the period and rigid sleeper

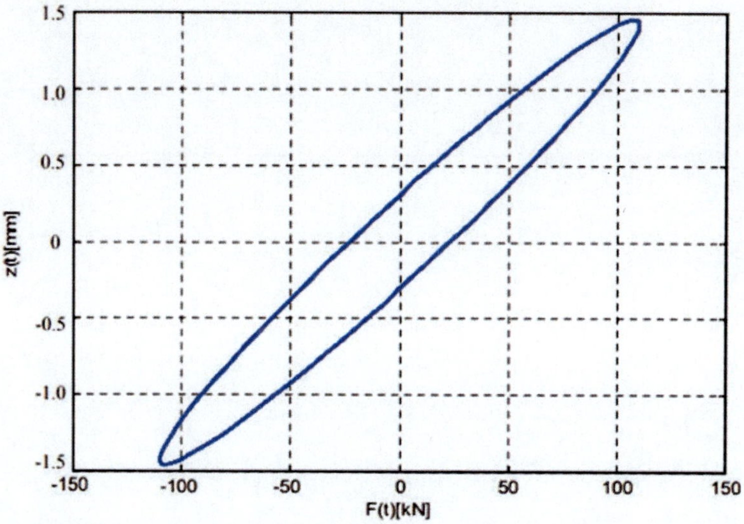

Fig. 6.50 One-period energy loop for sinusoidal loading force

linear spring in parallel with a linear viscous damper (dashpot) is sufficient here. For larger frequency ranges this model can be easily transformed into appropriate one that provides strong frequency dependence, giving a very significant stiffness and damping at high frequencies. The presented methodology can be applied to predict the durability of wheel rail systems subject to wear and crack growth. Under the action of the cyclic load obtained from the contact calculation the growth of the crack can be predicted. All eigendamping properties of rubber elements, as well as other parasitic damping must be considered in the model parameters for correct modeling. Some unfavorable conditions that are particularly detrimental for ride comfort are:

- low damping or even instability of the car body mode initiated through the coupling of the self-excited sinusoidal wheel movement with rail and sleeper eigenmodes;
- resonance from the eigenmode of the vehicle components with the periodic excited loading force. To study the influence of vehicle speed, we have to consider the excitation by track unevenness. This problem is not considered in this subsection for the sake of presentation compactness. In the next subsection influences of nonlinear characteristics of the elastic support will be considered.

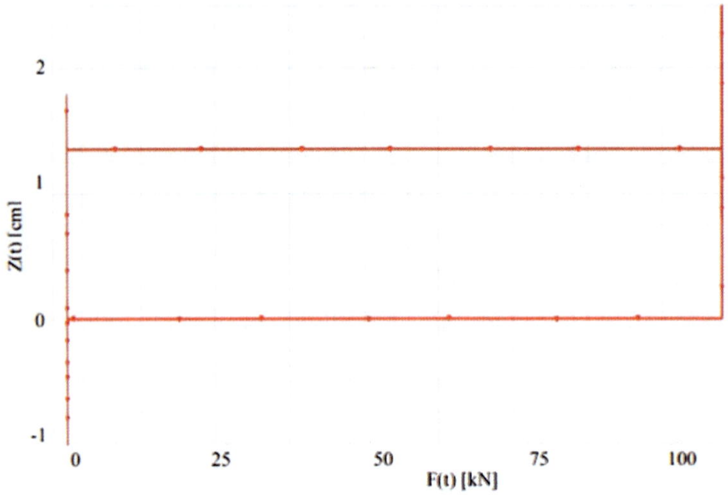

Fig. 6.51 One-period energy loop for loading force with two pulses within the period and elastic sleeper

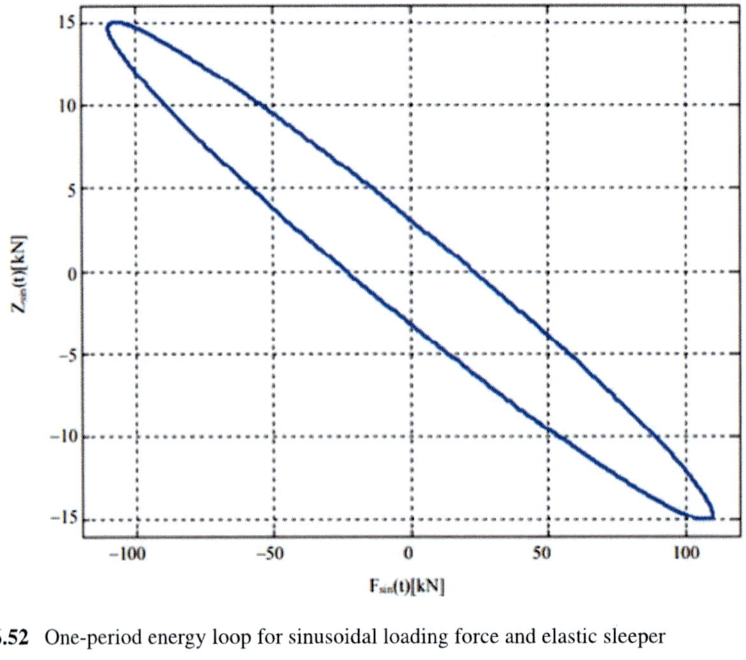

Fig. 6.52 One-period energy loop for sinusoidal loading force and elastic sleeper

6.6.6 Sleeper Nonlinear Characteristics

In order to check the efficiency of the proposed method which completely eliminates the Fourier series approach as well as to exhibit its advantages in applications the above established model of the wheel-rail contact dynamics has been applied to that same roll plane but with taking into account a nonlinear characteristic of the sleeper with respect the ground. It is represented by the relation

$$f_2(z_s) = k_2 \cdot \left(\alpha \cdot z_s^2 + \beta \cdot z_s^3\right) \tag{6.104}$$

where constant parameters k_2, α and β can be considered as bifurcating values. For $\alpha = 1.75$, $\beta = -1$ the plot of the relative value $f_2/k2$ is shown in Fig. 6.56. In our approach, periodic solutions were determined by appropriate modifications of standard numerical solver used for solutions of nonlinear ordinary differential equations [23]. They appear as computationally not expensive alternatives to the traditional harmonic balance approach and lead to quite satisfactory results. For a set of system parameters and the forcing term shown in Fig. 6.48 and nonlinearity presented in Fig. 6.53 the calculated time variations of the wheel-rail contact deformations are depicted in Fig. 6.54. Then using the alternative expression for the one-period energy,

$$W_T = \int_0^T F(t)v(t)\mathrm{d}t = \int_{P(0)}^{P(T)} v(t)\mathrm{d}P(t) \tag{6.105}$$

where $P(t) = \int_0^t F(\tau)d\tau$ denotes the force impulse, we can determine the energy absorbed by the rail during one-period of the applied force. The corresponding loop of one-period energy is presented in Fig. 6.55. Applying suitable numerical procedure we can evaluate the loop surface which determines the energy absorbed by the rail during one-period of the force action. It takes value of 3.0 kJ. Observe that the nonlinearity influences the energy absorbed by the rail more than two times. Moreover, parameter estimation is an important problem, because many the contacts of the wheel and rail in an important way. However, one can notice that the nonlinearity increases the absorbed energy in the contact. This is mainly due to the fact that, in these formulations, the motion of the wheel is assumed to travel with a constant forward velocity.

Parameters simply cannot be measured physically with good accuracy, especially in real-time applications. One of the advantages of the one-period energy approach used in this study is that when the number of nonlinear elements is increased, the form of the input force impulse remains unchanged and results of computer simulations can be used directly to estimate the energy absorbed by the rail. It can be noticed that when using the minimum number of input force pulses required to perform the estimation of the absorbed energy, i.e. a number of excitation points equal to the number of wheels in one side of the bogie, it leads to decreasing the number of particular areas in the resulting loop of one-period energy and adding extra information into an algorithm for whole loop surface evaluations. Although the proposed technique has

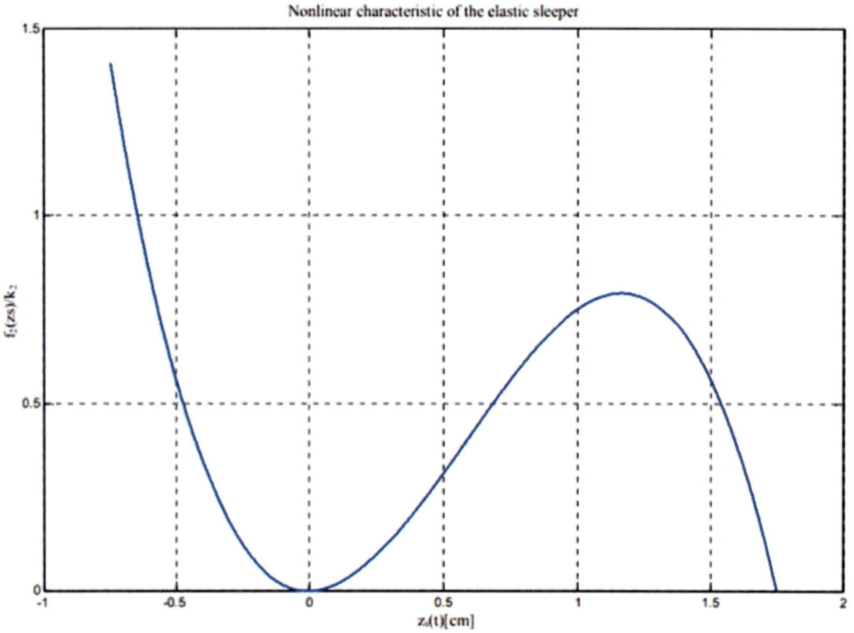

Fig. 6.53 Nonlinear characteristic of the sleeper stiffness

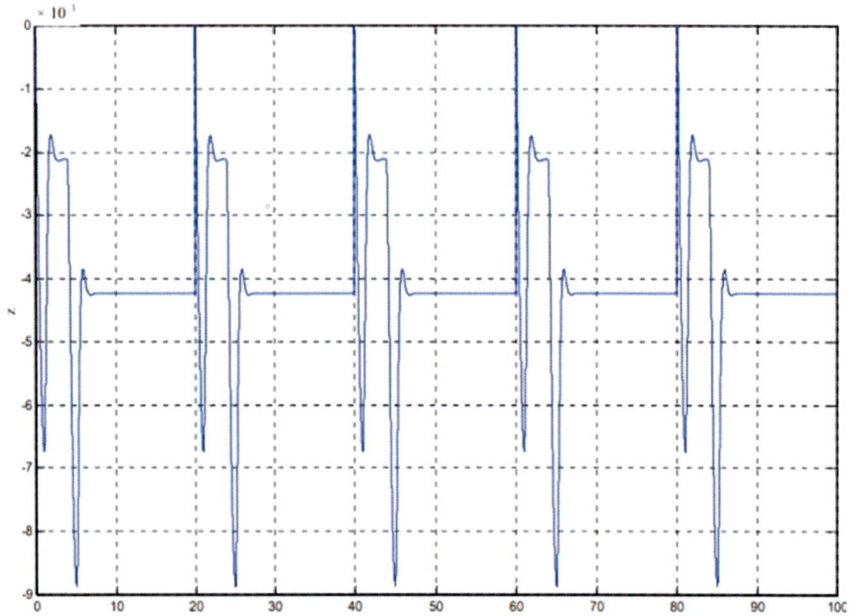

Fig. 6.54 Deformation of the wheel-rail contact

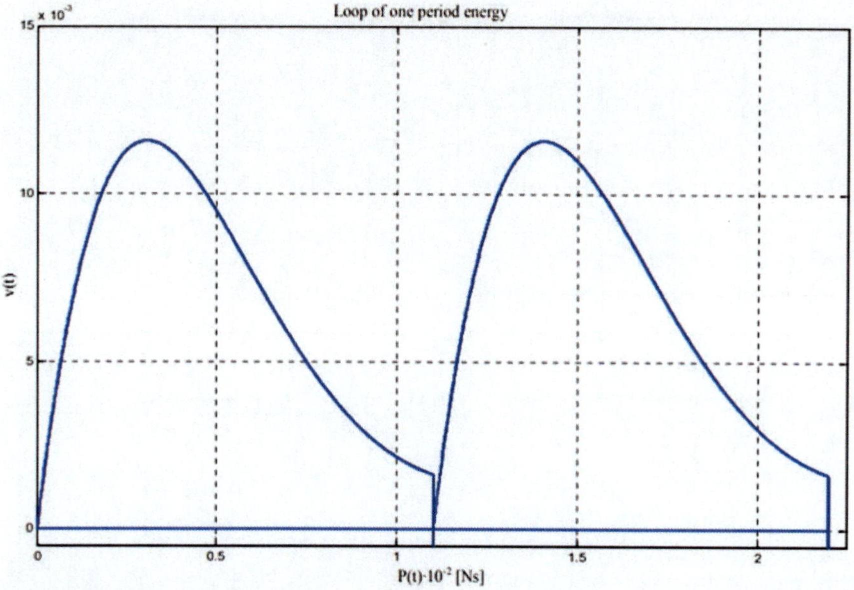

Fig. 6.55 Loop of one-period energy

been applied in order to identify parameters of a real mechanical system its use to solve analogous problems in other fields of human activities is recommended. The problem may find applications in multi-body dynamics as well, including dynamic vibration absorbers, while other applications include bridges, aircraft structures, and turbo-machinery blades.

6.6.7 Discussion and Summary

The introduction of damping elements into the model might well result in a change in the computed critical speed. The elementary analysis given in this chapter gives a good idea of the complexities involved in the stability analysis for rail vehicles but also shows that there are some similarities between the stability problems of railcars and other vehicles. The dynamics of the situation when the wheel set is not exactly centered and has a small yaw angle with respect to the rails will be subsequently studied. The interaction between steel wheels and steel rails is actually quite complex but for a linear stability analysis, a simplified treatment is adequate. When there are lateral and longitudinal forces between the wheel and the rail, the contact point on the wheel will exhibit rather small apparent relative velocities with respect to the rail in the lateral and longitudinal directions. The existence of a creepage does not imply a sliding velocity between the wheel and the rail. The introduction of damping elements into the model might well result in a change in the computed critical speed.

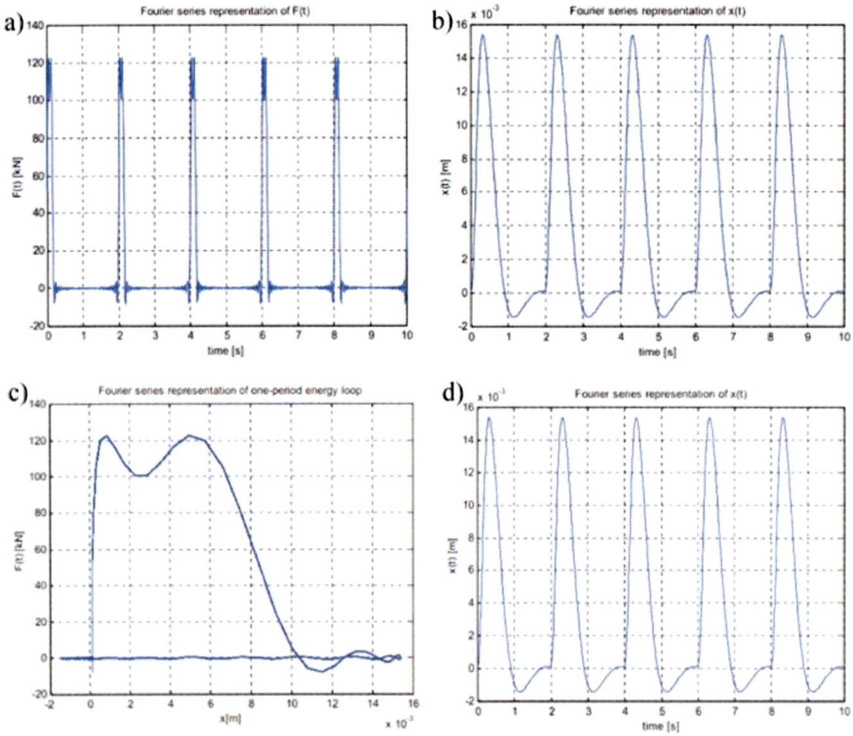

Fig. 6.56 Results of the Fourier series computations: **a** forcing term exhibiting Gibbs effect, **b** variations in time of the system response; **c** Fourier series representation of the one-period energy loop, **d** forcing term by Fourier series

It seems that the existence of a creepage does not imply a sliding velocity of the wheel and the rail. The formation of nonelliptical contact patches for real profile combinations and the so-called transient contact problem, which is important for high frequency vibration of wheel and rail, may give an attempt leading to a general view of the problems of contact mechanics concerning complete aspects of wheel-rail contact. The results may predict how many cycles are required for the crack to grow to a critical size and the crack growth rate. The methods and tools presented can be applied to provide critical design information for engineers responsible for railway system durability. Theoretical models describing the interaction between a railway wheel and the track were used for studying the wheel and track vibrations. The main emphasis has been put on one-period energy delivered to the rail by the loading force exerted by the railway vehicle. This is in response to modern technological demands where the couplings and interactions between various components of the model is a central mechanism in controlling the system. Accordingly, the proper frame in which it is possible, in an easy way, to determine the periodic nonharmonic response of multivariable linear dynamical systems was presented. The load transfer has been calculated on the basis of the elastic deformation in the neighborhood of each contact. In practice, the dynamic response of vehicles also has a significant influence on the rolling contact fatigue behavior. It has been shown that the time domain representation of a system by means of the concatenation procedure, as opposed to the frequency domain representation by means of the system transfer function, has become more advantageous approach to the exhibition of system dynamics in periodic non-harmonic states. Especially, the introduction of geometric tools like hysteresis loops on energy phase plane greatly advances the theory and enables the proper generalization of many fundamental concepts known for computer aided geometric designs to the world of periodic non-harmonic waveforms. The fatigue damage and rail corrugation can be treated as a further development of this research. For the comparative purposes we have considered the case presented in Example 1 (see Sect. 6.6.5) by up-to-date widely used approach based on the Fourier series. In this case, the periodic forcing term $F(t)$, which exhibits jump discontinuities, takes the form presented in Fig. 6.56a. The 25th partial sum of the Fourier series has large oscillations near the jump, which decrease the maximum of the partial sum above of the function itself. The overshoot does not die out as the number of frequency components increases, but approaches a finite limit. This is one of causes of insufficiency of the Fourier approach. As a result, the Fourier series computations give the variations in time of the system response $x(t)$ which is shown in Fig. 6.56b. Using the Fourier series approximations of both $F(t)$ and $x(t)$ we can present the one-period energy loop on the energy phase plane in the form presented in Fig. 6.56c and d. Thus, it is clear that there are important discrepancies with respect to exact results presented in Sect. 6.6.5. Hysteresis energy loops are viable alternatives to active, reactive, distortion and apparent powers. The proposed method exhibits several advantages. Simulations using one-period energy loops are exact and much faster than Fourier series simulations [147–158].

References

1. Huang, K., He, W., Zhao, R.: A two-step sensing circuit for the hysteresis loop selector based resistive non-volatile memory arrays. IEEE Trans. Circuit Syst. II: Express Briefs **65**(1), 101–105 (2018)
2. Jin, P., Wang, G., Ho-Ching Iu, H., Fernando, T.: A locally-active memristor and its application in chaotic circuit. IEEE Trans. Circuits Syst. II: Express Briefs **PP**(1), 1–5 (2017)
3. Chandıa, K.J., Bologna, M., Tellini, B.: Multiple scale approach to dynamics of an LC circuit with a charge-controlled memristor. IEEE Trans. Circuits Syst. II: Express Briefs **65**(1), 120–124 (2018)
4. Marszalek, W., Trzaska, Z.: Dynamical models of electric arcs and memristors: the common properties. IEEE Trans. Plasma Sci. **45**(2), 259–265 (2017)
5. Strukov, D.B., Snider, G.S., Stewart, D.R., Williams, R.S.: The missing memristor found. Nature **453**, 80–83 (2008)
6. Chua, L.O.: Memristor—the missing circuit element. IEEE Trans. Circuit Theory **18**(8), 507–519 (1971)
7. Marszalek, W., Trzaska, Z.W.: Memristive circuits with steady state mixed-mode oscillations. Electron. Lett. **50**, 1275–1277 (2014)

8. Chua, L.O.: Nonlinear circuit foundations for nanodevices, Part I: The four-element torus. Proc. IEEE **91**(11), 1830–1859 (2003)
9. Biolek, D., Biolek, Z., Biolkova, V.: Pinched hysteretic loops of ideal memristors, memcapacitors and meminductors must be 'self-crossing.' Electron. Lett. **47**(25), 1385–1387 (2011)
10. Zhang, H., Abhyankar, S., Constantinescu, E., Anitescu, M.: Discrete adjoint sensitivity analysis of hybrid dynamical systems with switching. IEEE Trans. Circuits Syst. I: Reg. Pap. **64**(5), 1247–1259 (2017)
11. Pershin, Y.V., Di Ventra, M.: Memory effects in complex materials and nanoscale systems. Adv. Phys. **60**, 145–227 (2011)
12. Borghetti, J., Snider, G.S., Kuekes, Ph.J., Yang, J.J., Stewart, D.R., Williams, R.S.: 'Memristive' switches enable 'stateful' logic operations via material implication. Nature **464**, 873–876 (2010)
13. Hsieh, Y.-C., Lin, Z.-R., Chen, M.-C., Hsieh, H.-C., Liu, Y.-C., Chiu, H.-J.: High-efficiency wireless power transfer system for electric vehicle applications. IEEE Trans. Circuits Syst. II: Express Briefs **64**(8), 942–946 (2017)
14. Czarnecki, L.S.: Power properties of electrical circuits and their miss interpretations by the instantaneous reactive power p-q theory. In: Proc. XII Int. Sem. Theor. Electr. Techn. (ISTET'2003), vol. 12, no. 1, pp. 261–267 (2003)
15. Willems, J.L., Ferrero, A.: Is there a relationship between non-active currents and fluctuations in the transmitted power? ETEP **8**(4), 265–270 (1998)
16. Wheeden, R.L., Zygmund, A.: Measure and Integral: An Introduction to Real Analysis. Marcel Dekker, New York (1977)
17. Dragomir, S.S.: Accurate approximations for the Riemann–Stieltjes integral via theory of inequalities. J. Math. Inequal. **3**(4), 663–681 (2009)
18. Wiliams, R.S.: How we found the missing memristor. In: IEEE Spectrum, NA, pp. 29–35 (2008)
19. Gerard, C., Gonze, D., Goldbeter, A.: Dependence of the period on the rate of protein degradation in minimal models for circadian oscillations. Philos. Trans. Royal Soc. A **367**, 4665–4683 (2009)
20. Briand, D., Yeatman, E., Roundy, S. (eds.): Micro Energy Harvesting. Wiley-VCH, Hoboken (2015)
21. Lin, D., Hui, S.Y.R., Chua, L.O.: Gas discharge lamps are volatile memristors. IEEE Trans. Circuits Syst. I: Reg. Pap. **61**, 2066–2073 (2014)
22. Nelson, M.E.: Electrophysiological models. In: Koslow, S., Subramaniam, S. (eds.) Databasing the Brain: From Data to Knowledge. Wiley, New York (2004)
23. Marszalek, W.: Bifurcations and Newtonian properties of Chua's circuits with memristors. DeVry Univ. J. Sch. Res. **2**(2), 13–21 (2015)
24. Vasconcellos, M.T., Luna, J.H.M., Mateus, J.C., Portugal, S.A.: Harmonic distortion as an influence quantity on reactive static electrical energy meters. J. Phys.: Conf. Ser. **575**, 1–11 (2015)
25. Barton, J.P., Infield, D.G.: Energy storage and its use with intermittent renewable energy. IEEE Trans. Energy Convers. **19**(2), 441–448 (2014)
26. Sun, Y., Yuan, Z., Liu, Y., Li, X., Wang, Y., Wei, Q., Wang, Y., Narayanan, V., Yang, H.: Maximum energy efficiency tracking circuits for converter-less energy harvesting sensor nodes. IEEE Trans. Circuits Syst.—II: Express Briefs **64**(6), 670–674 (2017)
27. Machado, J.E., Griñó, R., Barabanov, N., Ortega, R., Polyak, B.: On existence of equilibria of multi-port linear AC networks with constant-power loads. IEEE Trans. Circuits Syst. I: Reg. Pap. **64**(10), 2772–2782 (2017)
28. Emanuel, E.: Power Definitions and the Physical Mechanism of Power Flow. Wiley, London (2010)
29. Aiello, M., Cataliotti, A., Nuccio, S.: A PC-based instrument for harmonics and interharmonics measurement in power supply systems. Measurement **35**(4), 371–380 (2004)

30. Peretto, L., Willems, J.L., Emanuel, A.E.: The effect of the integration interval on the measurement accuracy of rms values and powers in systems with nonsinusoidal waveforms. In: Proceedings of the 7th International Workshop on Power Definitions and Measurements under Non-sinusoidal Conditions, Cagliari, Italy, July 2006
31. Chua, L., Kang, S.: Memristive devices and systems. Proc. IEEE **64**, 209–223 (1976)
32. Biolek, Z., Biolek, D., Biolkova, V.: Analytical computation of the area of pinched hysteresis loops of ideal MEM-elements. Radioengineering **22**(1), 32–35 (2013)
33. Czarnecki, L.S.: On some misinterpretations of the instantaneous reactive power p-q theory. IEEE Trans. Power Electron. **19**(3), 828–836 (2004)
34. Wolf, E.L.: Nanophysik und Nanotechnologie. Wiley-VCH, Weinheim (2015)
35. Biernacki, M., Pidek, H., Ryll-Nardzewski, C.: Sur une inégalité entre des integrales definies, Ann. Univ. Mariae Curie-Sklodowska **A4**, 1–4 (1950)
36. Mitrinovic, D.S., Pecaric, J., Fink, A.M.: Classical and New Inequalities in Analysis. Springer, New York (1993)
37. Trzaska, Z.: Effective harvesting of breaking energy in electric cars. KONES **93**(3), 77–85 (2017)
38. Priya, S., Inman, D. (eds.): Energy Harvesting Technology. Springer, New York (2008)
39. Briand, D., Yeatma, E., Roundy, S. (eds.): Micro Energy Harvesting. Wiley-VCH, Weinheim (2015)
40. Brett, C., Brett, A.: Electrochemistry. OUP, Oxford (1993)
41. Bard, A.J., Faulkner, L.R.: Electrochemical Methods; Fundamentals and Applications. Wiley, New York (2000)
42. Barsoukov, E., Macdonald, J.R. (eds.): Impedance Spectroscopy; Theory, Experiment, and Applications, 2nd edn. Wiley, New York (2005)
43. Kelly, R.G., Scully, J.R., Shoesmith, D.W., Buchheit, R.G.: Electrochemical Techniques in Corrosion Science and Engineering. Marcel Dekker, New York (2002)
44. Cha, C.S.: Introduction to Kinetics of Electrode Processes, 3rd edn. Science Press, Beijing (2002)
45. Welch, Ch.M., Simm, A.O., Compton, R.G.: Electroanalysis **18**, 965–970 (2006). http://physchem.ox.ac.uk/~rgc/publications/files/articles/2006/as22%282006%29567.pdf
46. Tao, D., Chen, G.L., Parekh, B.K.: J. Appl. Electrochem. **37**, 187–194 (2007)
47. Cheng, S.C., Gattrell, M., Guena, T., Macdougall, B.: J. Appl. Electrochem. **36**, 1317–1326 (2006)
48. Fletcher, S.: J. Electrochem. Soc. **141**, 18–23 (1994)
49. Darowicki, K.: Electrochim. Acta **40**, 439–445 (1995)
50. Kowalewska, M., Trzaska, M.: Influence of Si_3N_4 disperse ceramic phase on the corrosion resistance of micro- and nano-crystalline nickel layers. In: Proceedings of the VIII International Conference and Exhibition "Corrosion-2006," Lviv, Ukraine, 6–8 June 2006, pp. 615–619 (2006)
51. Trzaska, M., Lisowski, W.: Corros. Prot. **48**, 112–117 (2005)
52. Wyszynska, A., Trzaska, M.: Influence of the degree of dispersion of the ceramic phase on the corrosion resistance of Ni-P/Si_3N_4 composite coatings. In: Proceedings of the VIII International Conference and Exhibition "Corrosion-2006," Lviv, Ukraine, 6–8 June 2006, pp. 609–614 (2006)
53. Birbilis, N., Padgett, B.N., Buchheit, R.G.: Electrochim. Acta **50**, 3536–3540 (2005)
54. Macdoald, J.R.: Electrochim. Acta **35**, 1483–1492 (1990)
55. Lapicque, F., Storck, A., Wragg, A.A.: Electrochemical engineering and energy. The Language of Science. Springer, New York (1995)
56. Dobbelaar, J.A.L.: The Use of Impedance Measurements in Corrosion Research; The Corrosion Behavior of Chromium and Iron Chromium Alloys, Dissertation, Technical University, Delft (1990)
57. Gabrielle, C.: Identification of Electrochemical Processes by Frequency Response Analysis. Solartron Instrumentation Group (1980)
58. Miecinskas, P.: Russ. J. Electrochem. **141**, 731–735 (2005)

59. Walker, A.T., Wragg, A.: Electrochim Acta **22**, 1129–1134 (1977)
60. Mansfeld, F.: Electrochim. Acta **35**, 1533–1537 (1990)
61. Boukamp, B.A.: Data Electrochem. Syst. Solid State Ionics **20**, 31–44 (1986)
62. Trzaska, M.: Electrochemical impedance spectroscopy in corrosion studies of copper surface layers. In: Proceedings of the VII International Workshop 'Computational Problems of Electrical Engineering', Odessa, Ukraine, 27–30 Aug 2006, pp. 178–180 (2006)
63. Shinohara, K., Aogaki, R.: Electrochemistry **67**, 126–131 (1999)
64. Trzaska, Z., Marszalek, W.: Periodic solutions of DAEs with applications to dissipative electric circuits. In: Proceedings of the IASTED Conference on Modelling, Identification and Control (MIC06), Lanzarote, Spain, 6–8 Feb 2006, pp. 309–314 (2006)
65. Trzaska, Z.: Arch. Electr. Eng. **53**, 191–215 (2004)
66. Trzaska, Z.: Arch. Electr. Eng. **54**, 265–287 (2005)
67. Trzaska, Z., Marszalek, W.: IEE Proc. Pt. D Control Theory Appl. **140**, 305–308 (1993)
68. Trzaska, Z.W., Marszalek, W.: Arch. Electr. Eng. **55**, 255–271 (2006)
69. Farkas, M.: Periodic Motion. Springer, New York (1994)
70. Chian, H.D., Chu, C.C., Cauley, G.: Proc. IEEE **831**, 497–1529 (1995)
71. Chun-Lei, T., Wu, X.-P.: Proc. Amer. Math. Soc. **132**, 1295–1303 (2004)
72. Mhaskar, H.N., Prestin, J.: Adv. Comput. Math. **12**, 95–131 (2000)
73. Trenn, S., Willems, J.C.: Switched behaviors with impulses—a unifying framework. In: Proceedings of the 51st IEEE Conference on Decision and Control, Maui, pp. 3203–3208 (2012)
74. Haykin, S., Van Veen, B.: Signals and Systems. Wiley, New York (2003)
75. Oppenheim, A.V., Willsky, A.S., Young, I.T.: Signals and Systems. Prentice-Hall, Upper Saddle River (1983)
76. Sommariva, A.M.: Power analysis of one-ports under periodic multi-sinusoidal operation. IEEE Trans. Circuits Syst. **53**(9) (2006)
77. Cottone, F.: Introduction to Vibration Energy Harvesting. NiPS Energy Harvesting Summer School (2011)
78. Engel, T.G., Kontras, E.A.: Analysis and design of homopolar motors and generators. In: 17th International Symposium on Electromagnetic Launch Technology (EML), IEEE, La Jolla, CA, USA, 7–11 July (2014)
79. High Current Compact Generator, magniX Technologies Pty Ltd., Arundel, QLD 4214, Australia (2016), www.magniflux.com
80. Kim, J., Cottone, K.F., et al.: Energy scavenging for energy efficiency in networks and applications. Bell Labs Tech. J. **15**(2), 7–29 (2010)
81. Trzaska, Z.: Dynamic processes in the electric drive system in the urban traffic. Transport miejski i regionalny **11**, 26–31 (2013)
82. Trzaska, Z.: Experiments with disk Faraday generator. In: Proceedings of the Conference Progress in Applied Electrotechnics, Kościelisko, Poland, 23–27 June, pp. 251–260 (2003)
83. Chan, M.S.W., Chau, K.T., Chan, C.C.: Effective charging method for ultracapacitors. J. Asian Electr. Veh. **3**(2), 771–776 (2005)
84. Nezhad, M.B.: Study of Homopolar DC Generator, Ph.D. Thesis, The University of Manchester, Manchester (2012), https://www.escholar.manchester.ac.uk/uk-ac-man-scw:19941
85. Park, G.: Overview of Energy Harvesting Systems. Los Alamos National Laboratory, www.sciepub.com/reference/27519
86. Briand, D., Yeatman, E., Roundy, S. (eds.): Micro Energy Harvesting. Wiley-VCH, Weinheim (2015)
87. Boldea, I.: Electric generators and motors: an overview. CES Trans. Electr. Mach. Syst. **1**(1), 3–14 (2017)
88. Chyba, C.F., Hand, K.P., Thomas, P.J.: Energy conservation and Poynting's theorem in the homopolar generator. Am. J. Phys. **83**(72) (2015). https://doi.org/10.1119/1.4895389
89. van Essen, H., Kampman, B.: Impacts of Electric Vehicles—Summary Report. Publication number: 11.4058.26. Delft, April 2011. www.cedelft.eu
90. Eberhard, M., Tarpenning, M.: The 21st Century Electric Car. Tesla Motors Inc. (2006)

91. Demetgul, M., Guney, I.: Design of the hybrid regenerative shock absorber and energy harvesting from linear movement. J. Clean Energy Technol. **5**(1), 81–84 (2017)
92. Faraday Generator, eskola.hfd.hr/icm/.../14.%20%20Faraday%20Generator.pdf
93. Mazurkiewicz, A. (ed.): Nanosciences and Nanotechnologies. Present State and Development Perspectives. Institute of Exploitation Technology—State Research Institute, Radom (2007)
94. Trzaska, M., Trzaska, Z.: Nanomaterials produced by electrocrystallization method. In: Aliofkhazraei, M., Makhlouf, A.S.H. (eds.) Handbook of Nanoelectrochemistry. Springer, Basel (2015)
95. Stevic, Z., Radovanovic, I.: Energy efficiency of electric vehicles. In: Stevic, Z. (ed.) New Generation of Electric Vehicles. InTech, Rijeka (2012)
96. Bartle, R.G., Sherbert, D.: Introduction to Real Analysis, 2nd edn. Wiley, New York (1997)
97. Matak, M., Šolek, P.: Harvesting the vibration energy. Am. J. Mech. Eng. **1**(7), 438–442 (2013). https://doi.org/10.12691/ajme-1-7-57
98. Priya, S., Inamn, D. (eds.): Energy Harvesting Technologies. Springer, Berlin (2008)
99. Trzaska, Z.: Impact and chaotic phenomena in nonlinear nonsmooth electrical dynamical systems. KONES **91**(3), 77–85 (2015)
100. Wheel to Well Analysis of EVs. MIT Electric Vehicle Team. MIT. http://web.mit.edu/evt/summary_wtw.pdf (2008)
101. Perek, A., Kurnik, W.: Kinematically excited vibration of an asymmetric rotor/bearing system with magnetic lubricant. Mach. Dyn. Res. **39**(4), 5–19 (2015)
102. Trzaska, Z.: Experiments with Faraday disk generator. In: Proceedings of the Conference "Advances in Applied Electrotechnics" PES-4, Kościelisko, pp. 251–260 (2003)
103. Bolonkin, A.A.: New Concepts, Ideas and Innovations in Aero-space, Technology and Human Sciences. Nova, Hauppauge, NY (2007)
104. High Current Compact Generator, magniX Technologies Pty Ltd., Arundel, QLD 4214, Australia (2016). www.magniflux.com
105. Feng, F., Lu, R., Wei, G., Zhu, Ch.: Online estimation of model parameters and state of charge of $LiFePO_4$ batteries using a novel open-circuit voltage at various ambient temperatures. Energies **8**, 2950–2976 (2015). https://doi.org/10.3390/en8042950
106. Nezhad, M.B.: Study of Homopolar DC Generator. Ph. D. Thesis, University of Manchester, Faculty of Engineering and Physical Sciences (2012)
107. FaradayGenerator,eskola.hfd.hr/icm/.../14.%20%20Faraday%20 Generator.pdf
108. Hide, R., Skeldon, A.C., Acheson, D.J.: A study of two novel self-exciting single-disk homopolar dynamos: theory. Proc. Roy. Soc. Lond. **A452**, 1369–1395 (1996)
109. Moroz, I.: Self-Exciting Faraday disk homopolar dynamos. Int. J. Bifurc. Chaos **11**(12), 2961–2975 (2001)
110. Mazurkiewicz, A. (ed.): Nanosciences and Nanotechnologies. Present State and Development Perspectives. Institute of Exploitation Technology—State Research Institute, Radom, (2007)
111. Ivana, P., Vlcek, I., Ivanova, M.: The construction of the Pure homopolar generator reveals physical problem of Maxwell's equations. arXiv:1601.07750v2 (2016)
112. Kirschner, J., Moon, S.: An investigation of the homopolar generator: determining the impact of system characteristics on efficiency. Electromagnetism: A Modelling and Simulation Approach. Project 2: Final Paper, pp. 1–7 (2010)
113. Gilbert, A.D.: Dynamo theory. In: Handbook of Mathematical Fluid Dynamics, vol. II, pp. 355–441. North-Holland, Amsterdam (2003)
114. Struchtrup, H.: Thermodynamics and Energy Conversion. Springer, New York (2014)
115. Schupp, G.: Computational Bifurcation Analysis: An Efficient Method for Studying Stationary and Periodic Motions. SIMPACK User Meeting, Eisenach, 9–11 Nov 2004, pp. 1–9
116. Steenbergen, M.J.: Modeling of wheels and rail discontinuities in dynamic wheel-rail contact analysis. Veh. Syst. Dyn. **44**(10), 763–787 (2006). https://doi.org/10.1080/00423110600648535
117. Santamaria, J., Vadillo, E.G., Gomez, J.: Influence of creep forces on the risk of derailment of railway vehicles. Veh. Syst. Dyn. **47**(6), 721–752 (2009). https://doi.org/10.1080/00423110802368817

118. Iwnicki, S.: Handbook of Railway Vehicle Dynamics. Taylor & Francis, Boca Raton (2006)
119. Nwokah, O.D.I., Hurmuzu, Y.: The Mechanical Systems Design Handbook: Modeling, Measurements and Control. CRC Press, London (2002)
120. de Silva, C.W.: Vibrations. Fundamentals and Practice. CRC Press, London (2000)
121. Bachman, H., et al.: Vibration Problems in Structures. Birkhäuser-Verlag, Basel (1995). https://doi.org/10.1007/978-3-0348-9231-5
122. Benaroua, H.: Mechanical Vibrations. Analysis, Uncertain Control. Marcel Dekker, New York (2004)
123. Schupp, G., Weidemann, C., Mauer, L.: Modeling the contact between wheel and rail within multibody system simulation. Veh. Syst. Dyn. **41**(5), 349–364 (2004). https://doi.org/10.1080/00423110412331300326
124. Piotrowski, J., Kik, W.: A simplified model of wheel/rail contact mechanics for non-Hertzian problems and its application in rail vehicle dynamic simulations. Veh. Syst. Dyn. **46**(1–2), 27–48 (2008). https://doi.org/10.1080/00423110701586444
125. Bezin, Y., Iwnicki, S.D., Cavallett, M.: The effect of dynamic rail roll on the wheel-rail contact conditions. Veh. Syst. Dyn. **46**(1), 107–117 (2008). https://doi.org/10.1080/00423110701882348
126. von Wagner, U.: Nonlinear dynamic behavior of a railway wheelset. Veh. Syst. Dyn. **47**(5), 627–640 (2009). https://doi.org/10.1080/00423110802331575
127. Alonso, A., Giménez, J.G.: Non-steady state modeling of wheel-rail contact problem for the dynamic simulation of railway vehicles. Veh. Syst. Dyn. **46**(3), 179–196 (2008). https://doi.org/10.1080/00423110701248011
128. Zboinski, K., Dusza, M.: Bifurcation approach to the influence of rolling radius modeling and rail inclination on the stability of railway vehicles in a curved track. Veh. Syst. Dyn. **46**(1), 1023–1037 (2008). https://doi.org/10.1080/00423110802037255
129. Knothe, K., Stichel, S.: Schienenfahrzeugdynamik. Springer, Berlin (2007)
130. Feynman, R.P.: There's plenty of room at the bottom: an invitation to enter a new world of physics. Caltech Eng. Sci. **23**(5), 22–36 (1960). http://www.zyvex.com/nanotech/feynman.html
131. de Silva, C.W.: Vibration and Shock Handbook. Taylor & Francis, Boca Raton (2005)
132. Drexler, K.E.: Nanosystems: Molecular Machinery, Manufacturing, and Computation. Wiley, New York (1992)
133. Popov, V.L., et al.: Friction coefficient in 'rail-wheel'—contacts as a function of material and loading parameters. Phys. Mesomech. **5**(3), 113–120 (2002)
134. Shabanaa, A.A., Zaazaaa, K.E., Escalonab, J.L., Sany, J.R.: Development of elastic force model for wheel/rail contact problems J. . Sound Vib. **269**(1–2), 295–325 (2004). https://doi.org/10.1016/S0022-460X(03)00074-9
135. von Schwerin, R.: Multibody System Simulation: Numerical Methods, Algorithms, and Software. Springer, Berlin (1999)
136. Greenwood, D.T.: Advanced Dynamics. CUP, Cambridge (2003)
137. Misiak, J., Stachura, S.: Selected Problems of Static Stability and Dynamics of Rod and Shell Structures. Publishing Office of the Warsaw University of Ecology and Management, Warsaw (2010)
138. Blajer, W.: A geometrical interpretation and uniform matrix formulation of multibody system dynamics. Z. Angew. Math. Mech. **81**(4), 247–259 (2001)
139. Orlova, A., Boronenko, Y., Scheffel, H., Fröhling, R., Kik, W.: Tuning von Üterwagendrehgestellen Durch Radsatzkopplungen. ZEV-Glasers Annalen **126**(4), 270–282 (2002)
140. Evans, J., Iwnicki, S.: Vehicle Dynamics and the Wheel Rail Interface (2002). http://www.railtechnologyunit.com
141. Karnopp, D.: Vehicle Stability. Marcel Dekker Inc., New York (2004). https://doi.org/10.1201/9780203913567
142. Trzaska, Z., Marszalek, W.: Periodic solutions of DAEs with applications to dissipative electric circuits. In: Proceedings of the IASTED Conference Modeling, Identification and Control, Lanzarote, 6–8 Feb 2006, pp. 309–314

143. Trzaska, Z.: Straightforward method for studies of periodic non-harmonic states of linear systems. Arch. Electr. Eng. **53**(2), 191–215 (2004)
144. Trzaska, Z.: A new approach to shaping one-period energies of dynamical systems operating in non-sinusoidal states. Arch. Electr. Eng. **54**(3), 265–287 (2005)
145. Trzaska, Z., Marszalek, W.: Computing periodic solutions of linear differential-algebraic systems with nonsinusoidal excitations. Arch. Electr. Eng. **55**(3–4), 255–271 (2006)
146. Trzaska, Z.: One-period energy in dynamical systems with periodic discontinuous excitations. Elektronika **48**(3), 26–32 (2007)
147. Kisilowski, J., Knothe, K.: Advanced Railway Vehicle Dynamics. Technico-Scientific Publisher, Warsaw (1991)
148. Filipski, P.S., Baghzoul, Y., Cox, M.D.: Discussion of power definitions contained in the IEEE dictionary. IEEE Trans. Power Delivery **9**(3) (1994)
149. F. Z. Wang, "A Triangular Periodic Table of Elementary Circuit Elements," *IEEE Trans. Circuits and Systems—I: Regular Papers,* vol. 60, no. 3, pp. 616–623, March 2013.
150. Czamecki, L.S.: Powers in nonsinusoidal networks: their interpretation, analysis, and measurement. IEEE Trans. Instrum. Meas. **39**(2), 340–345 (1990)
151. Emanuel, A.: Powers in nonsinusoidal situations: a review of definitions and physical meaning. IEEE Trans. Power Delivery **5**(3), 1377–1389 (1990)
152. Marszalek, W., Trzaska, Z.: Properties of memristive circuits with mixed-mode oscillations. Electron. Lett. **51**(2), 140–141 (2015)
153. Corinto, F., Ascoli, A., Gilli, M.: Nonlinear dynamics of memristor oscillators. IEEE Trans. Circuits Syst. I: Reg. Pap. **58**(6), 1323–1336 (2011)
154. Trzaska, Z.: Mixed mode and chaotic oscillations in a system with a self-excited disk generator. In: Proceedings of the Conference "Modeling and Simulation" MiS-6, Kościelisko, pp. 154–161 (2010)
155. Zhang, L., Peng, H., Ning, Z., Mu, Z., Sun, Ch.: Comparative research on RC equivalent circuit models for lithium-ion batteries of electric vehicles. Appl. Sci. **7**, 1002 (2017). https://doi.org/10.3390/app7101002
156. Targosz, J.: Vibroisolation Systems in Railway and Automotive Transportation Applications. AGH University Science & Technology Press, Cracow (2007)
157. Zerzm, E.: Topics in Multidimensional Linear Systems Theory. Springer, London (2000)
158. Borrelli, R.R., Colleman, C.S.: Differential Equations: A Modeling Perspective. Wiley, New York (2004)

Chapter 7
Artificial Intelligence in the Service of Dynamical Systems Studies

7.1 Background of AI Modeling

Artificial intelligence (AI) appears with increasing intensity in our common day practice and increases human powers to solve many of the complex problems that labored people so far for a really long time. Artificial intelligence aspires at emulating human intelligence on machines to make them think and behave like human beings. Presently we are at the onset of a revolution of processing large amounts of data where human intelligence and machine intelligence coexist. Very often, AI is considered to simulate human intelligence, as opposed to natural intelligence, by machines, especially computer systems. The efficiencies of AI opened up new avenues for the non-conventional models which decisively have rise to a new research branch called computational intelligence (CI). The development of CI techniques follows a different path than that of the AI (Fig. 7.1).

In light of this relationship, AI is the exposure of intelligent behavior shown by machines with respect to the natural intelligence of human beings. It is a branch of computer science that is concerned with the progress of a technology that enables a machine or computer to think, behave, or act in a human-like manner. On the other hand, Computational Intelligence (CI) is more like a sub-branch of AI that emphasizes on the design, application and development of linguistically justified computational models. This involves the study of adaptive mechanisms to allow or facilitate the use of intelligent procedures in complex and evolving environments. The fundamental purpose of CI is to understand the computer paradigms that make intelligent behavior possible in natural or man-made systems in composite and variant environments. The practical applications of CI encompass intelligent household appliances, medical diagnosis, banking and consumer electronics, optimization applications, industrial applications, etc. The progress of CI techniques proceeds on a different path than that of the AI. Although, both AI and CI seek almost similar goals, they are quite different from each other by means of implementations. However the goal of both groups is the same: artificial systems powered by intelligence appear as strong

Z. Trzaska, *Mathematical Modelling and Computing in Physics, Chemistry and Biology*, Studies in Systems, Decision and Control 495,
https://doi.org/10.1007/978-3-031-39985-5_7

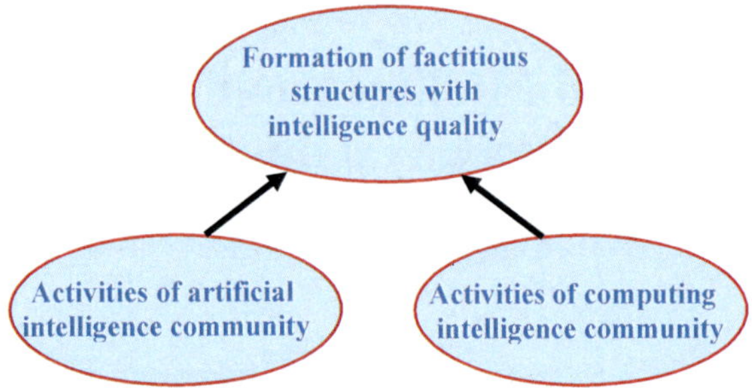

Fig. 7.1 Two distinctive communities of researches active in artificial intelligence and computational intelligence

tools for solutions of well difficult problems in real-life space. Moreover, we always want to achieve an overall objective, which is a better functioning of the intelligent system than one, that can be achieved without intelligent components. Note that the differences between AI and CI are important for scholars and should be complied in scientific papers for their proper classification. However, from the point of view of applications of intelligent systems this can be disregarded. Therefore, in what follows this chapter, it will simply be used a unique name, i.e. AI consisting of both AI and CI. It should be emphasized that in this chapter the readers are invited to enrich their knowledge or review and arrange it regarding AI taking into account the results of the original works published so far in relation to mathematical modeling and simulation of significant problems in physics, biophysics, medical and chemical domains.

7.2 Common Types of AI Algorithms

Artificial intelligence (AI) is a leading computer technology of the current age of the Fourth Industrial Revolution (Industry 4IR). It possesses the ability to imitate human behavior and intelligence by machines or special arrangements. Therefore, AI-based modeling is the key to build automated, intelligent, and smart systems according to current and forthcoming needs. Currently, there are available many AI procedures and related mathematical modeling and simulations being developed for scientific and medical applications such as nanotechnology, genomics, chatbot for inter-professional education, biomedical imaging, chemical reactions and processes, radiation treatment planning, molecular medicine, healthcare and prosecutorial investigations. These investigations are usually based on AI or mathematical model linked to a knowledge foundation for problem classification and solution. The main achievements of AI are usually rule based systems in which computers follow known methods of human thinking and try it achieve similar results as human being. Thus AI modeling takes into account the creation, teaching, training, and deployment of computer learning algorithms that emulate logical decision-making based on available data. AI models form a framework to encouragement advanced reference methodologies such as real-time and predictive analytics, as well as augmented analytics. Developing an effective AI model is a challenging task due to the dynamic nature and variation in real-world problems and data.

It should be noted that in this approach, validation of the model is difficult. Simulations for most of the systems studied require highly developed software components which pose challenges in proving the validity of the code. Furthermore, there are no substantive tools available to analyze the properties of the model and its dynamics. In general, there is also a lack of appropriate formalism for the comparison of models. For instance, if a simulation based on elements seems to be a simplification of another, then one would like to be able to connect their dynamics in a rigorous way. However, currently, it lacks a mathematically rich formal framework that simulates element-based models. This configuration should be based on a class of mathematical subjects that can be mapped to element-based simulation. These subjects should be characterized by a sufficiently general mathematical framework being able to gain key features of element-based simulations and, at the same time, should be sumptuous enough to get the substantial mathematical results.

In general the aim of AI models is to apply one or more algorithms to predict results or build decisions by trying to understand the relationship between multiple inputs of varying nature. AI models differ in how they approach this task, and AI developers working together can develop multiple algorithms to achieve a corresponding objective or function. The most commonly applied procedures utilize one of the following algorithms [1–10].

- Linear regression constitutes the linear relationship between one or more X input(s) and Y output(s), often mapped by a simple line graph.
- Logistic regression constitutes the relationship between a binary X variable (such as present or absent, true or false) and an output variable Y.
- Linear discriminant analysis operates like logistic regression except starting data being characterized by separate categories or classifications.
- Decision trees use branching patterns of logic to a set of input data till the decision tree attains a conclusion.
- Naive Bayes is a classification procedure taking into account that there are no relationships between starting inputs.
- K-nearest neighbor is a classification procedure that assumes that inputs with similar characteristics will be close together when their correlation is plotted (in terms of Euclidean distance).
- The learning vectorial quantization is similar to the neighbor closest to K, but instead of measuring the distance between individual data points, the model will converge as data points in prototypes.
- Support vector machine algorithms set up a divisor, called a hyperplane, which distinctly separates the data points for a more accurate classification.
- Bagging combines several algorithms to create a more accurate model, while random forest combines several decision trees to achieve a more accurate prediction.
- Deep neural networks refer to a structure of multiple layers of algorithms that inputs must cross, leading to a point of prediction or final decision.

In fact, AI models are becoming so large that quite data is required to effectively learn them, and the faster these data are loaded, the model is faster trained and deployed. In practice, learning ability is one of the most important factors for every intelligent system and the most intelligent activities should be based on the collection and use of recognized knowledge. In the realm of machine learning systems, we can find neural networks—the most well-known and valuable learning systems. It is worth to mention that the learning systems area does not cover all neural network set. This is because of some neural networks, which work without learning process. In the machine learning group there are also model recognition systems (which can sometimes be achieved as neural networks) and learning decision trees. Special components of artificial intelligence methods are Fuzzy Set and Fuzzy Logic system. They can be combined with Neural Networks and then are called neuro-fuzzy solutions which frequently also may be learned. Distinctive subareas of Artificial Intelligence area are Expert Systems, Intelligent agents and Genetic Algorithms. However, most strange part of Artificial Intelligence are Ant Colony methods and in general so called Swarm Algorithms inspired by the collective behavior of natural systems such as flocks of birds and swarms of bees.

7.3 Artificial Neural Network

7.3.1 Preliminary

It is useful to know that the oldest systems solving many problems through AI methods are neural networks. This highly intelligent and user-friendly technology is based on modeling small parts of a true neuronal system (for example, small pieces of the cerebral cortex), shown schematically on Fig. 7.2a. Moreover, artificial neural networks are estimated as the best computational tool for solving complex problems with total scarcity of information about rules governing the problem. They are able to solve practical problems through learning and are always state-of-the-art techniques of many practical applications [11–25].

The central nervous system (CNS) is made up of the brain and spinal cord. It is one of 2 parts of the nervous system. The other part is the peripheral nervous system, which consists of nerves that connect the brain and spinal cord to the rest of the body. The central nervous system is the body's processing centre. The brain controls most of the functions of the body, including awareness, movement, thinking, speech, and the 5 senses of seeing, hearing, feeling, tasting and smelling. The spinal cord is an extension of the brain. It carries messages to and from the brain via the network of peripheral nerves connected to it. Nerves also connect the spinal cord to a part of the brain called the brainstem.

If information from a biological neuron enters the neuron via dendrite, soma processes the information and transmits it via axon (Fig. 7.2b, c). Currently, as the best computational tool for solving complex problems with a complete lack of

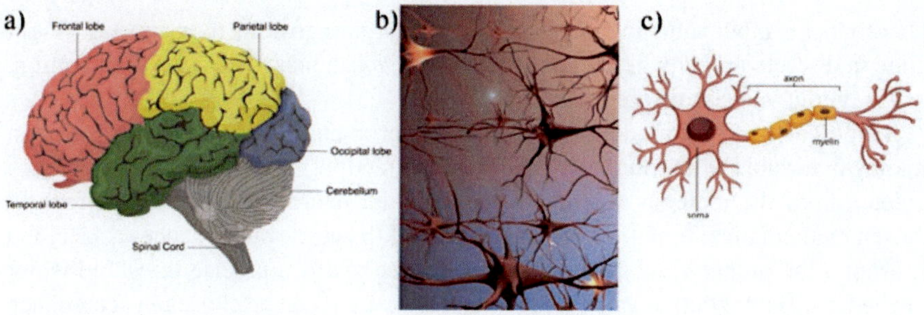

Fig. 7.2 Components of the human central nervous system: **a** brain, **b** bit of neural network, **c** single neuron

information about the rules governing the problem are evaluated artificial neural networks. This is very handy and user-friendly technology based on the modeling of small parts of real neural system (e.g., small pieces of the brain cortex) that are able to solve practical problems by means of learning. A scheme of an artificial neuron with multi inputs and single output is presented in Fig. 7.3. In such an artificial neuron information enters its body through x_i entries followed by weight $w_i = 1$, 2, …, n (each input can be individually multiplied by a weight). Then the body of an artificial neuron sums the weighted inputs and bias b by summator Σ and transforms the sum with a transfer function f. At the end an artificial neuron passes the processed information via output y. Thus, artificial neuron consists of four basic components—inputs with weights, summator, transfer function with threshold and output. The advantages of the simplicity of the artificial neuron model can be seen in his mathematical description, i.e.

$$y(k) = f\left(\sum_{i=0}^{n} w_i(k) \cdot x_i(k) + b\right) \tag{7.1}$$

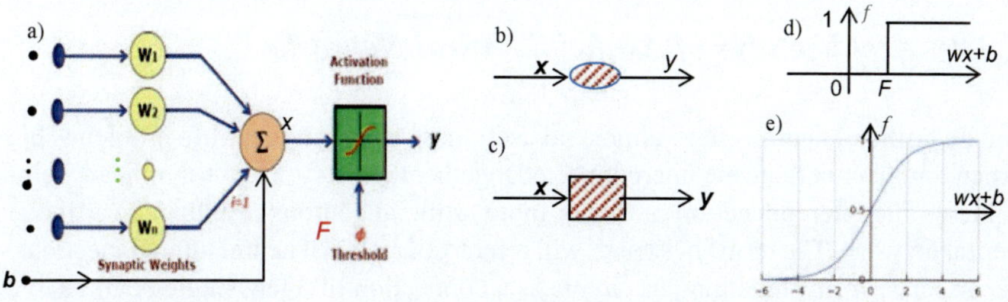

Fig. 7.3 An artificial neuron: **a** basic scheme, **b** integrated scheme with respect output–input, **c** modular block, **d** step function as a transfer function, **e** sigmoid transfer function

where $x_i(k)$ is input value in discrete time k with i going from 1 to n, $w_i(k)$ is weight value in discrete time k where i goes from 1 to n, b is a bias, f is a transfer function, $y(k)$ is output value in discrete time k.

It emerges from the model of an artificial neuron and its Eq. (7.1) that the main unknown variable in the model is its transfer function f exhibiting a threshold F. It determines the major properties of the artificial neuron and it can be properly chosen mathematical function. This means that if the entry value reaches a specified threshold, the output value translates into a value and if a precise threshold is not reached, a different output value will result. However, it can be chosen in accordance with the problem that the artificial neural network must solve but in most cases, it is selected from the following set of functions: Step function, linear function and Nonlinear function (often Sigmoid). In general the knowledge generated in neural networks is hidden and has no readable form, but can be collected automatically on the basis of examples forming the learning data set.

If the step function acts as a transition function, then, because it is a binary function, the output takes only two possible values (e.g. zero and one). Equation (7.2) provides an explanation of the condition

$$y = \begin{cases} 1 \; if \; x \geq threshold \\ 0 \; if \; x < threshold \end{cases} \tag{7.2}$$

When this type of transfer function is employed in the artificial neuron then it is known as ***perceptron***. In the case of the linear transfer function, the artificial neuron performs a simple linear transformation on the sum of the weighted entries and the bias. When using the non-linear transfer function, the ***sigmoid function*** is the most common.

However, the reader should be aware that a thorough description of artificial neural networks is far beyond the scope of this work. An in-depth approach to advanced issues is presented in numerous monographs in the field of artificial intelligence. Hereby, the set of concepts necessary to understand the work is reported.

7.3.2 Architectures of Artificial Neural Networks

Single artificial neurons have almost no usefulness in solving real-life problems, but when a number of them are interconnected together then they present significant helpfulness. The interconnection of two or more artificial neurons results in an artificial neural network. The term "network" will refer to all artificial neural interconnections. It can range from something as simple as a connection of a few single neurons to a vast collection of neurons in which everyone is connected to all the other neurons in the network. Neural networks consist of different layers placed on top of each other and interconnected as nodes to optimize and refine categorization or prediction. It should be pointed out that artificial neural networks (ANN) are a tool, or a model, rather than a method for implementing artificial intelligence in autonomous systems.

The learning process can be stated as the algorithm by which the free parameters of a neural network are adapted through a process of stimulation from the environment in which it works.

The most meaningful categorization of deep neural networks is within feedforward and recurring networks. Deep feedforward networks, often referred to as multi-layer perceptrons (LPMs), are the most common models of deep learning. The feedforward network is intended to be closer to a given function f. Depending on the task to be performed, the input is mapped to an output value. For instance, for a classifier, the N network matches an x-entry with a y-category. A feedforward network sets a mapping $y = N\ (x,\ w)$ and learns the values of the parameters w (weight) which give the best function approximation. These models are called feedforward because the information streams from the input layer, through intermediaries, to the output y. There are no feedback connections in which outputs of the model are sent back as input to the network itself. When neural feedforward networks are expanded to include feedback connections, they are referred to as recurrent neural networks.

The network is capable of solving complex real-life problems by processing information in their basic building blocks (artificial neurons) in a non-linear, distributed, parallel and local way. The idea leads to the combination of a data set for records, a cost function, an optimization process and a model. It is worth mentioning that a variety of artificial neural networks are developed and discovered so far by masses of researchers. Notice that neural networks are technologies definitively dedicated to digital computing (quantitative). The signals on the input, output and above all—each element in the neural network—are in the form of numbers even if their interpretation is of a qualitative type. This means that the neural network converts qualitative information into quantitative one. A popular type of such a conversion is named “one of N”. In general, artificial neural networks are assessed as the best computational tool for solving complex problems with total lack of information about rules governing the problem. If we have a problem and need to resolve it quickly and effectively, we can pick artificial neural networks as an easy-to-use tool, with many good programs available.

Of the many types of artificial neural networks discovered to date, the simplest yet effective tool for solving the most of practical problems is the Multi-Layer Perceptron (MLP) network [26–29]. With a thorough knowledge of both categories and examples of systems, this network can be used with a conjugate gradient learning rule. However, in the event that due to lack of data, the user of the neural network does not know the result to be expected, he can reach for another popular type of neural network namely, the Self-Organizing Map (SOM), also known as Kohonen network, which can learn without the teacher [30]. If an optimization problem is under investigations and the best solution in a complex situation is needed, then the recursive neural network, known as the Hopfield network, can be used. Experienced users and persons actively engaged in the concerned field can of course use also other types of neural networks, which are described in many publications, e.g. in [31–38].

How individual artificial neurons are interconnected is called architecture of an artificial neural network. The fact that interconnection can occur in many ways leads to many possible topologies that are split into two basic classes shown in schemes presented in Fig. 7.4a, b. The first scheme shows simple feed-forward architecture where information flows from inputs to outputs in a single direction only. To such an extent that some users recognize the expression "neural networks" as meaning only feed-forward networks. The main idea boils down to this, the nodes in the successively top layers condense immediately one after another the upper-level characteristics of the preceding layers. In the neural network literature, feed-forward has sometimes been used to refer directly to multilayer networks. The feed-forward network is intended to be closer to a given function f. Depending on the task to be performed, the input is mapped to an output value. For instance, for a classifier, the N network matches an x-entry with a y-category. Recall that a feed-forward network sets a mapping $y = N\ (x,\ w)$ and learns the values of the parameters w (weight) which give the best function approximation. Whereas the second diagram shown on the upper right of Fig. 7.4 represents a simple recurrent topology where some of the information flows from input to output, but also in the opposite direction. At the bottom of Fig. 7.4 are shown modular representation of the respective network. Note that these networks, which usually have no more than four layers, are some of the most common neural networks in use. They are referred to as recurrent neural networks.

For the purposes of analytical description, let us look at the simple three-layer artificial neural network. It is shown in Fig. 7.5. Following the signal flow direction (from input to output) we get the following description of the network

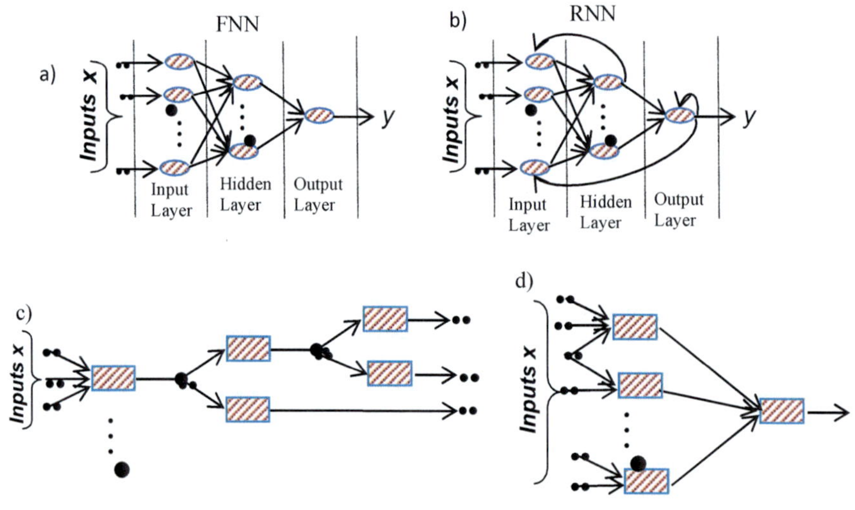

Fig. 7.4 Architectures of artificial neural networks: **a** feed-forward (FNN), **b** recurrent (RNN), **c** modular representation of a feed-forward network, **d** modular representation of a recurrent network

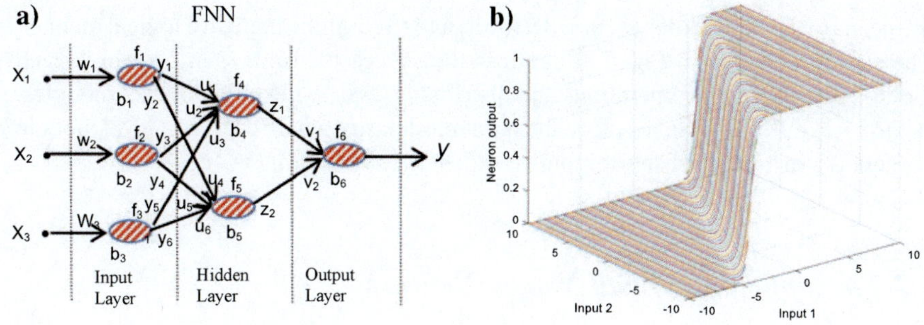

Fig. 7.5 Fed-forward artificial neural network: **a** three layer network: *x*, y, z—signals, b—biases, *u*, *v*, *w*—weights, *f*—transfer functions, **b** output-input relations in two-neuron layer network with sigmoid transfer function

$$\begin{aligned}
y_1 &= f_1(w_1x_1 + b_1) \\
y_2 &= f_2(w_2x_2 + b_2) \\
y_3 &= f_3(w_3x_3 + b_3) \\
y_4 &= f_4(w_4x_4 + b_4) \\
y_5 &= f_5(w_5x_5 + b_5) \\
y_6 &= f_6(w_6x_6 + b_6) \\
z_1 &= f_4(u_1y_1 + u_2y_3 + u_6y_5 + b_4) \\
z_2 &= f_5(u_1y_1 + u_5y_4 + u_6y_6 + b_5) \\
y &= f_6(v_1z_1 + v_2z_2 + b_6)
\end{aligned} \tag{7.3}$$

Because the fixed artificial neural network topology is known, it can therefore be used to resolve a given problem. Just as biological neural networks need to learn their proper responses to the given inputs from the environment, so the artificial neural networks need to do the same. The learning process can be defined as the algorithm by which the free parameters of a neural network are adapted through a process of stimulation by the environment in which it works. The type of learning is the set of instructions for how the parameters are changed. The next step is to learn the right answer from an artificial neural network and this can be achieved by learning (supervised, unsupervised or reinforcement learning). Regardless of the method used, the task of learning is to define the values of weight and bias on the basis of learning data to minimize the chosen cost function. Though the analytical description can be utilized on any complex artificial neural network, but in practice specialized computers and software are in use, which can help build, mathematically describe and optimize any type of artificial neuronal network.

Many problems are excellent solved using neural networks whose architecture consists of several modules, with sparse interconnections between modules. The modules can be connected in a variety of ways, some of which are shown in Fig. 7.4c, d. Modularity allows the neural network developer to solve smaller tasks separately by means of small modules (neural network) and to combine them in a logical manner. The structure shown in Fig. 7.4c realizes the successive refinement in which each module performs some operations and distributes tasks to next higher level modules. In the case of Fig. 7.4d takes place the input modularity where each first level module processes a different subset of inputs (not necessarily disjoint).

7.3.3 Hopfield Artificial Neural Network

If a problem of optimization and the best solution in a complex situation must be found, the recursive artificial neural network, known as the Hopfield network, may be used. Units in Hopfield networks are binary threshold units. They take only two different values for their states, and the value is determined by whether or not the unit's entry exceeds the F_i threshold. These networks, based on the idea of symmetric recursive connections, have made possible to apply the concept of the energy function. Thanks to this, it has become possible to translate the richness of the achievements of statistical physics into the language of neural networks. It is a type of recurrent artificial neural network that can be used to store one or more stable target vectors. These stable vectors represent memories that the network remembers when it comes with similar vectors that act as an indicator to the network's memory. Hopfield networks act as content-associative storage systems with binary threshold nodes, or with continuous variables. Hopfield's discrete networks describe the relations between binary neurons 1, 2, ..., i, j, ..., N. At a certain time, the state of the neural network is described by a V vector, which records neurons that shoot in a binary N-bit word. They also offer a template to understand human memory. Thus formed units with binary forms take only two different values for their states, which are determined by whether the units' entry exceeds their threshold or not. Binary units can be either 1 or 0, or 1 or -1, i.e.

$$y_i = \begin{cases} 0 & if \sum_j w_{ij} x_j > F_i \\ -1 & otherwise \end{cases}$$

$$y_i = \begin{cases} 1 & if \sum_j w_{ij} x_j > F_i \\ -1 & otherwise \end{cases} \tag{7.4}$$

Figure 7.6a shows the basic model of a single neuron, which is based on non-linear units denoted f_i with values from the set $\{-1, +1\}$ and the step activation function

$$s_i = \text{sign}\left(\sum_j w_{ij} s_j\right) \tag{7.5}$$

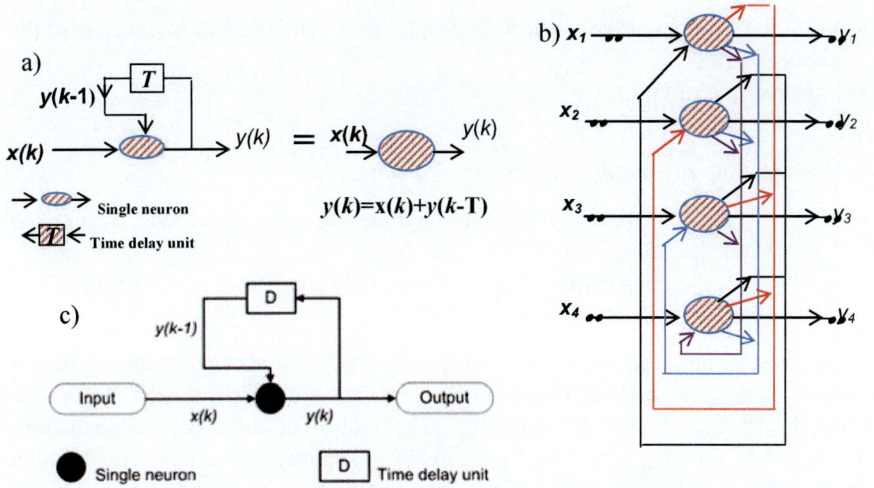

Fig. 7.6 Nonlinear networks: **a** a single neuron, **b** a four neuron Hopfield network, **c** neuron with delay feedback

with sign(0) = +1.

Interactions w_{ij} between neurons have units which usually take a value of + 1 or − 1, and this convention will be used throughout this chapter. These interactions are "learned" via Hebb's law of association, such that, for a certain state V^s

$$w_{ij} = V_i^s V_j^s; \quad w_{ii} = 0 \tag{7.6}$$

That is

- $w_{ii} = 0 \; \forall i$ (no unit has a connection with itself),
- $w_{ij} = w_{ji} \; \forall i, j$ (connections are symmetric).

The constraint that the weights are symmetrical ensures that the energy function decreases monotonically as well as following the activation rules. A Hopfield network with asymmetric weights may exhibit some periodic or chaotic behavior; but this behavior is confined to relatively small parts of the phase space and does not impair the network's ability to act as a content-addressable associative memory system.

Updating one unit (node in the graph simulating the artificial neuron) in the Hopfield network is performed using the following rule:

$$s_i \rightarrow \begin{cases} 1 & if \sum_j w_{ij} s_j \geq U_i \\ -1 & otherwise \end{cases} \tag{7.7}$$

where

- w_{ij} is the strength of the connection weight from unit j to unit i (the weight of the connection),
- s_i is the state of unit i,
- U_i is the threshold of unit i.

The updating rule implies that if $w_{ij} > 0$ then:

- at $s_j = 1$, the contribution of j in the weighted sum is positive. Thus, s_i is pulled by j towards its value $s_j = 1$,
- at $s_j = -1$, the contribution of j in the weighted sum is negative. In effect, s_i is pushed by j towards its value $s_j = -1$.

As a result, the values of neurons i and j will converge if the weight among them is positive. In the same manner, they will diverge if the weight is negative. The Hopfield network with the structure shown in Fig. 7.6 but with 12 neurons was used to identify the digit 4 as shown in Fig. 7.7a by means of the appropriate pixels. Based on the patterns shown in the figure, a training string is created (white boxes mean − 1, and black boxes mean 1). In this way, a 12-element test vector x(1) is obtained. In the next two trials, the shape of the digit changes to noisy forms, which are shown in Fig. 7.7b, c. They are represented by the vectors x(2) and x(3), respectively.

$$\begin{aligned} x(1) &= [1\ \ -1\ \ 1\ \ 1\ \ 1\ \ 1\ \ -1\ \ -1\ \ 1\ \ -1\ \ -1\ \ 1] \\ x(2) &= [1\ \ -1\ \ 1\ \ 1\ \ -1\ \ 1\ \ -1\ \ -1\ \ 1\ \ -1\ \ -1\ \ 1] \\ x(3) &= [1\ \ -1\ \ -1\ \ 1\ \ 1\ \ 1\ \ -1\ \ -1\ \ 1\ \ -1\ \ -1\ \ 1] \end{aligned} \tag{7.8}$$

Applying (7.6) gives the matrix W of neuronal weights as follows

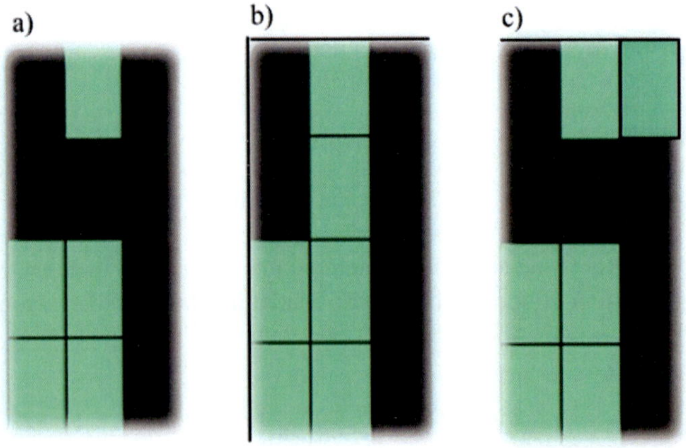

Fig. 7.7 Images of digit 4: **a** test, **b** and **c** noisy

$$W = \begin{bmatrix} 0 & -0.25 & 0.08 & 0.25 & 0.08 & -0.25 & -0.25 & -0.25 & 0.25 & -0.25 & -0.25 & 0.25 \\ -0.25 & 0 & -0.25 & -0.25 & -0.08 & -0.25 & 0.25 & 0.25 & -0.25 & 0.25 & 0.25 & -0.25 \\ 0.08 & -0.25 & 0 & 0.08 & -0.08 & 0.08 & -0.08 & -0.08 & 0.08 & -0.08 & -0.08 & 0.08 \\ 0.25 & -0.25 & 0.08 & 0 & 0.08 & 0.25 & -0.25 & -0.25 & 0.25 & -0.25 & -0.25 & 0.25 \\ 0.08 & -0.08 & -0.08 & 0.08 & 0 & 0.08 & -0.08 & -0.08 & 0.08 & -0.08 & -0.08 & 0.08 \\ -0.25 & -0.25 & 0.08 & 0.25 & 0.08 & 0 & -0.25 & -0.25 & 0.25 & -0.25 & -0.25 & 0.25 \\ -0.25 & 0.25 & -0.08 & -0.25 & -0.08 & -0.25 & 0 & 0.25 & -0.25 & 0.25 & 0.25 & -0.25 \\ -0.25 & 0.25 & -0.08 & -0.25 & -0.08 & -0.25 & 0.25 & 0 & -0.25 & 0.25 & 0.25 & -0.25 \\ 0.25 & -0.25 & 0.08 & 0.25 & 0.08 & 0.25 & -0.25 & -0.25 & 0 & -0.25 & 0.25 & 0.25 \\ -0.25 & 0.25 & -0.08 & -0.25 & -0.08 & -0.25 & 0.25 & 0.25 & -0.25 & 0 & 0.25 & -0.25 \\ -0.25 & 0.25 & -0.08 & -0.25 & -0.08 & -0.25 & 0.25 & 0.25 & 0.25 & 0.25 & 0 & -0.25 \\ 0.25 & -0.25 & 0.08 & 0.25 & 0.08 & 0.25 & -0.25 & -0.25 & 0.25 & -0.25 & -0.25 & 0 \end{bmatrix} \quad (7.9)$$

It turns out that when a vector identical to the pattern is given to the input of the network, then the network will not change its state, it also recognizes images that differ "slightly" from the patterns.

7.3.4 Equilibrium States in the Hopfield Network

Hopfield networks are characterized by a scalar value associated with each state of the network, called "energy" of the network, $E = E(\mathrm{t})$, where:

$$E = -\frac{1}{2}\sum_{i,j}^{n} w_{i,j} y_i y_j - \sum_i x_i y_i + \sum_i \theta_i y_i \quad (7.10)$$

This amount is referred to as "energy" because it decreases or remains the same when network units are updated. It either diminishes or remains unchanged on update (feedback) after every of iterations. In addition, in case of repeated updating, the network will eventually converge to a state that is a local minimum of the energy function which is regarded as a function of Lyapunov [39]. Therefore, if a state is a local minimum of energy function, this is a steady state for the network. Note that this energy feature belongs to a general class of models in physics called Ising models [40].

To determine if the network will converge to a stable configuration, it should suffice to verify that the energy function reaches its minimum, by

$$\frac{dE}{dt} \leq 0 \quad (7.11)$$

Therefore, the problem of selecting a specific "target" network state can be treated as a problem of selecting a state with minimal "energy" of the network (Fig. 7.8). The "energy" function is of course introduced here in a purely conventional way. In fact we should rather talk about the Lyapunov function, but most authors follow Hopfield in this energetic metaphor. It is worth emphasizing that this property makes possible to prove that the system of dynamical equations describing temporal evolution of neurons' activities will eventually reach a fixed point attractor state. It can be shown that all processes in a Hopfield network can change only in such a direction that the "energy" $E(t)$ can only decrease or stay the same—it can never increase. This means that knowing the energy function, we can predict the trajectories of signal changes in the entire network. The network is related to convergence if the total activity s_i of each neuron versus time is given by the following differential equation

$$\frac{ds_i}{dt} = -\frac{s_i}{\tau} + \sum_{j=1}^{n} w_{i,j} s_j + \theta_i \tag{7.12}$$

The w_{ij} coefficients, called synaptic weights, can be modified during the learning process, which is one of the essential distinguishing features of neural networks as adaptive information processing systems. The vector s of the sum of all neurons excitations in the network can then be related to the vectors of output signals Y from network elements and input (external) signals X by means of a matrix differential equation

$$\frac{d\boldsymbol{s}}{dt} = -\frac{\boldsymbol{s}}{\tau} + \boldsymbol{WY} + \boldsymbol{X} \tag{7.13}$$

supplemented with non-linear equations of the static characteristics of individual neurons

$$y_i = \varphi(s_i) \tag{7.14}$$

which is often represented in the form of a classic sigmoid

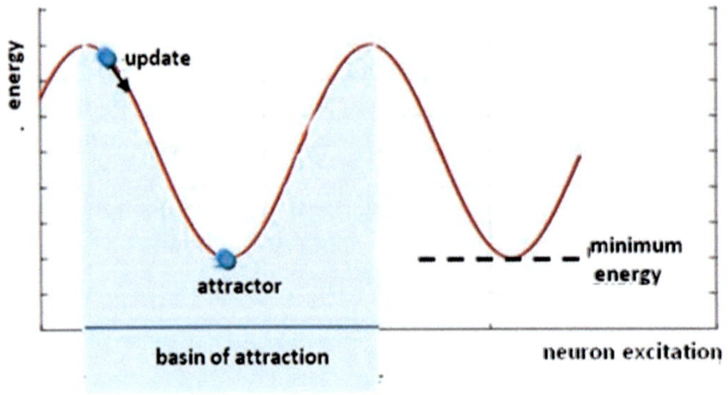

Fig. 7.8 Energy landscape of a Hopfield network, highlighting the current state of the network (up the hill), a state of attraction to which it will end up converging, a minimum energy level and a basin of attraction shaded in light blue. Note how the update of the Hopfield network is always going down in energy

$$y_i = \frac{1}{1 + \exp(-bs_i)} \tag{7.15}$$

where $i = 1, 2, \ldots, n$ and constant b must be appropriately adjusted.

The output signal y of the neuron will be the greater, the more the location of the input vector X in the space X resembles the location of the weights vector W in the space W. In this way, it can be said that the neuron recognizes the input signals, distinguishing those that are similar to its weight vector.

For the above non-linear dynamical system it is possible to define the Lyapunov function in the form

$$L = -\frac{1}{2}\boldsymbol{Y}^T\boldsymbol{W}\boldsymbol{Y} - \boldsymbol{X}^T\boldsymbol{Y} + \frac{1}{\tau}\sum_{i=1}^{k}\int_0^{y_i}\varphi^{-1}(\vartheta)d\vartheta \tag{7.16}$$

where upper index T denotes transposition.

In the case of the function φ with the shape of a unit step, the Lyapunov function takes the form

$$L = -\frac{1}{2}\boldsymbol{Y}^T\boldsymbol{W}\boldsymbol{Y} - \boldsymbol{X}^T\boldsymbol{Y} \tag{7.17}$$

It is easy to realize a nearest and direct relationship between the given function and the given above function of the "energy" minimized by the network. This is further argument for the fact that dynamic processes in the network will always course on the way to "energy" minimization, although this course may not be monotonic, and there may even be difficulties convergence guaranteed.

The Lyapunov function forms foundations for the universal approximation theorem which establishes the following. Let φ: $\boldsymbol{R} \rightarrow \boldsymbol{R}$ be a non-constant, bounded and continuous function representing the activation function. Let I_m denote the m-dimensional unit hypercube $[0, 1]^{\mathrm{m}}$. The space of real-valued continuous functions on I_m is denoted by $C(I_m)$. Then, given any $\varepsilon > 0$ and any function $f \in C(I_m)$, there exist an integer N, real constants v_i, $b_i \in R$ and real vectors $w_i \in R^m$ for $i = 1, \ldots, N$, such that

$$F(x) = \sum_{i=1}^{N} v_i\varphi\left(w_i^T x + b_i\right) \tag{7.18}$$

is an approximate realization of the function f with

$$|F(x) - f(x)| < \varepsilon \tag{7.19}$$

for all $x \in I_m$.

7.4 Learning Neural Networks

The learning process can be defined as the algorithm for adjusting the free parameters of a neural network by a stimulating process from the environment in which it operates [41]. The procedure consists in estimating the parameters of neurons (usually weights) so that the whole network can perform a particular task. In other words, the foundations of network learning are algorithms which have the ability to learn according to the data provided. A human provides the computer with many examples (input data) and solutions to a given problem (output data—results), and the computer learns and adapts to a given problem through analysis, self-education and inference. The following sequence is typically used:

- the environment stimulates the neural network,
- the neural network modifies any parameter,
- the neural network responds in a new way in accordance with the new structure.

The most prevalent artificial neural network architectures and the associated training methods illustrate the advantages and disadvantages of their employment for solutions of specific problems. Note that the network has to solve the task with known answer and we correct parameters of neurons in such a way—the system answer to be consistent with this answer. Most frequently learning neural network means: change of weights. The ANN is trained in the process of iteratively testing large amounts of data and verifying the results to ensure the accuracy and correct operation of the model. During this process respective specialists should be easy available to modify and improve the AI model as it learns.

A subset of artificial intelligence (AI) and machine learning (ML) is what is known as deep learning. It involves neural networks with multiple layers that attempt to simulate the behavior of the human brain, although it is still far from reaching its capabilities. In general, there would be a given experiment $\mathcal{E}$, which can easily be considered as a dataset $\mathcal{D} = (x_1, x_2, \ldots, x_n)$ .The information required to complete the task is collected and drawn from both structured and unstructured data. The learning algorithms can be split into three distinct approaches:

- Supervised learning in which, taking into account the known outputs $Y = (y_1, y_2, \ldots, y_N)$, we learn to generate the correct output seen as targets when new data sets are supplied. In most cases, targets are delivered by a human supervisor. There are also approaches in which target states are automatically retrieved by the machine learning model. The supervised learning find applications in solutions of numerous problems, both offline [42, 43] and online [34, 44].
- Unsupervised learning when algorithms use the regularity of data to generate an alternative representation used for reasoning, forecasting, or clustering.
- Reinforcing learning which produces actions $A = (a_1, a_2, \ldots, a_n)$ that affect the environment and receives awards $R = (r_1, r_2, \ldots, r_n)$. It makes possible to determine, how to map observations to actions, in order to maximize a superior digital signal. The trainer is not informed of the steps to be taken, but rather needs to find out which steps are most effective by trying them [45]. This means that the

reinforcement learning consists of learning what to do (i.e. mapping situations versus actions) in order to maximize a digital premium.

Although deep learning fascinates many engineers and practitioners and is used in many industries, it is still seen as an anxious technology. In the case of neural networks, it is unclear what the model is learning. We can only judge whether it does its job well. Therefore, it can seem like magic in how deep learning algorithms work.

7.5 Fuzzy Logic Models

7.5.1 Concise Overview

Recently, the so-called "intelligent" modeling and control methodologies, which employ techniques motivated by biological systems and human intelligence to develop models and controllers for dynamic systems, are investigated. Typical examples of techniques that make use of human knowledge and deductive processes are fuzzy modeling and control. Fuzzy set is a natural way to deal with the imprecision. Many real world representations rely on significance rather than precision. General observation is that as complexity of reality rises, its precise descriptions lose meaning and meaningful descriptions lose precision. It is, however, often possible to describe the functioning of systems by means of natural language, in the form of if–then rules. Fuzzy rule-based systems can be used as knowledge-based models constructed by using knowledge of experts in the given field of interest. Such approaches explore alternative representation schemes, using, for instance, natural language, rules, semantic networks or qualitative models, and possess formal methods to incorporate extra relevant information. On the other hand artificial neural networks realize learning and adaptation capabilities by imitating the functioning of biological neural systems on a simplified level. By contrast, in Boolean logic, the truth values of variables may only be the integer values 0 or 1 and nothing in between. Fuzzy logic had, however, been studied since the 1920s, as infinite-valued logic—notably by Jan Łukasiewicz, Professor of Mathematics at the University of Warsaw, Poland, in the years 1915–1945, and Professor of Mathematical Logic at the Royal Irish Academy in the years 1946–1956. Thanks to investigations of Łukasiewicz a number of axiomatization of classical propositional logic was appeared. An especially elegant axiomatization comprises only three axioms (True, False, Possible), and is still invoked to this day. He was a pioneer investigator of multi-valued logics; his three-valued propositional calculus, introduced in 1917, was the first explicitly axiomatized non-classical logical calculus [46].

Now, investigations on the fuzzy set theory and its applications are progressing at a great rate. This is caused by the increasing universal consciousness that the theory of fuzzy sets is a tool for applications in both theoretical fields as well as a technique for enrichment their foundations.

Nowadays and upcoming days, it will be stimulated a breadth of the disciplines of Information Sciences with intelligent systems, expert and decision support systems, approximate reasoning, self-adaptation and self-organizational systems, information and knowledge, modeling and computing as seen in the following studies like decision making.

For systems involving non-linearity and the lack of an accessible analytical model, fuzzy logical control emerged as one of the most promising approaches. There can be no doubt that fuzzy inference is a step toward simulation of human thought.

7.5.2 *Fuzzy Systems*

The term *fuzzy logic* was introduced with the 1965 proposal of fuzzy set theory by Iranian Azerbaijani mathematician Lotfi Zadeh [26]. According to Zadeh the fundamental goal of fuzzy logic is to form the theoretical foundation for reasoning about fuzzy propositions; such reasoning has been called to as approximate reasoning. He has proposed systematic route of modeling imprecision of the data using fuzzy sets. He also presented strong mathematical setting to support his statement. Fuzzy logic is based on the observation that people make decisions based on imprecise and non-numerical information. On the basis of fuzzy logic, fuzzy set theory arose, which permits the progressive appreciation of the membership of elements in a set what is described with the aid of a membership function valued in the real unit interval [0, 1]. In fuzzy set theory, classical bivalent sets are usually called *crisp sets*. For instance, fuzzy controls may be unreliable sensor readings ("noisy" data), or amounts associated with human perception, such as comfort, beauty, etc. For a large number of practical problems, the collection of an acceptable level of knowledge necessary for physical modeling is difficult, time-consuming and costly, if not impossible. In most systems, the underlying phenomena are only partially understood and accurate mathematical models cannot be derived or are too complex to be helpful. However, fuzzy systems can process such information, whereas conventional (crisp) systems do not. It is a computational approach based on the "degree of truth" rather than the usual "right or wrong value" (0 or 1). Fuzzy models or sets are mathematical means of representing vagueness and imprecise information (hence the term fuzzy). The origin of applying the description of systems using fuzzy models was that in the majority of systems, the underlying phenomena are understood only partially and crisp mathematical models cannot be derived or are too complex to be useful. Hence, fuzzy logic itself is not fuzzy, rather it deals with the fuzziness in the data. And this fuzziness in the data is best described by the fuzzy membership function. This value quantifies the grade of membership of the element in X to the fuzzy set A. It is defined by the membership function $\mu \in [0, 1]$. Figure 7.9 shows a particular case of the so-called trapezoidal membership function. It also shows all key parameters that characterize this function. Support for a fuzzy set A is all $x \in X$ points such as $\mu_A(x) > 0$. It is easy to see that the fuzzy membership function presented in the graphic way visualizes convincingly the degree of affiliation of any value in a given fuzzy set. In the chart, the X axis represents the universe of discourse and the Y axis represents the degree of belonging in the interval [0, 1].

Often, in the real world, a given element can belong to several sets having different values of memberships, specifically, when concepts using linguistic variables are under investigations. For instance, consider three linguistic representations for the temperature—called *cool*, *pleasant* and *warm*, then depending upon value, temperature 'τ' has different membership in all three sets (Fig. 7.10). The temperature value $\tau = 5$ °C might have membership value 0.7 in cool set, 0.3 in pleasant set and 0 in warm set. Because, 5° is considered quite cool but it is not warm at all, so its membership value in cool set would be high, where is it will be very low in warm set.

Various fuzzy membership functions may be applied in studies of fuzzy systems. The most frequently used are the following membership functions: triangular, trapezoidal, sigmoid, Gaussian, Chauchy also called bell function, etc. Let's note that these functions are mathematically relatively very simple. However, it is worth emphasizing that fuzzy logic is meant to deal with the fuzziness, so use of complex membership functions would not add much more precision in the final output. It should also be mentioned that the core of any fuzzy logic system is a fuzzy inference system in which fuzzification is the first stage of its surveys. A meaningful dose of information

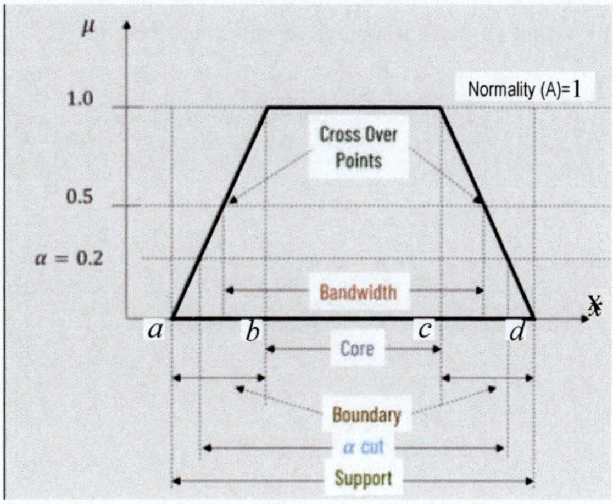

Fig. 7.9 Parameters characterizing a given fuzzy set with trapezoidal membership function

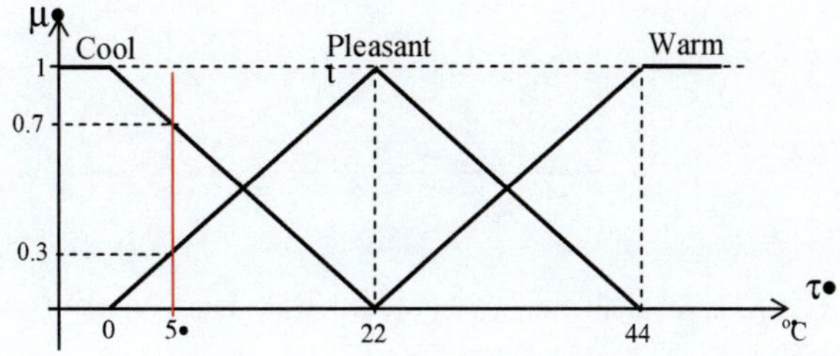

Fig. 7.10 Temperature range representation using fuzzy function

about these systems is available as the knowledge of proper experts, process operators and designers. The properties of fuzzy sets are in line with fuzzy terminologies defined in terms of parameters shown in Fig. 7.10 in the case of trapezoidal membership function. Similar characterizations belong to the other types of membership functions. Note that fuzzy systems are suited to uncertain or approximate reasoning, particularly for the system with a mathematical model that is hard to derive. These models have the capability of recognizing, representing, manipulating, interpreting, and using data and information that are vague and lack certainty. Presently, fuzzy logic penetrates many fields, from control theory to artificial intelligence. Applications extend to consumer goods such as cameras, camcorders, washing machines and microwave ovens through industrial process control, medical instrumentation, decision support systems and portfolio selection.

In neural network learning widely used is sigmoid function. It is used specifically in solutions of problems, where it suppresses the input and maps it between 0 and 1. It is controlled by parameters a and b, where a controls the slope of the membership function plot at the crossover point $x = b$. Mathematically it is defined as

$$\mu_{\text{sigm}}(x, a, b) = \frac{1}{1 + e^{-a(x-b)}} \tag{7.20}$$

where $x \in X$ represents the input from the process. Graphically, this membership function can be represented as indicated in Fig. 7.11.

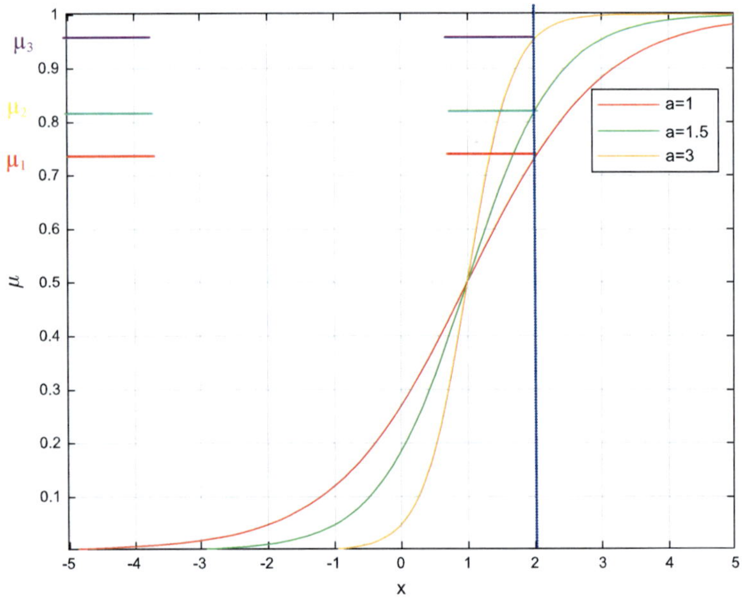

Fig. 7.11 Sigmoid membership function for three values of a and $b = 1$ with particular values corresponding to the input $x_0 = 2$

For instance, if it is necessary to determine μ_{sigm} corresponding to $x = 2$ and three successive values of parameter $a = [1; 1.5; 3\}$ with $b = 1$, then using the equation of a sigmoid membership function (7.10) gives

$$\mu_{\text{sigm}}(x, a, 1) = \frac{1}{1 + e^{-a(x-1)}}$$

$$\mu_{\text{sigm1}}(x, 1, 1) = \frac{1}{1 + e^{-1(2-1)}} = 0.7311,$$

$$\mu_{\text{sigm2}}(x, 1.5, 1) = \frac{1}{1 + e^{-1.5(2-1)}} = 0.8176,$$

$$\mu_{\text{sigm3}}(x, 3, 1) = \frac{1}{1 + e^{-3(2-1)}} = 0.9526$$

One of the important properties of fuzzy sets is convergence. It can be defined as follows. Fuzzy set A is convex if $\mu_{\underline{A}}(\lambda x_1 + (1 - \lambda)\, x_2)) \geq \min\,(\mu_{\underline{A}}(x_1), \mu_{\underline{A}}(x_2))$, where $x_1, x_2 \in X$. In more detail we get: for any elements x, y and z in a fuzzy set A, the relation $x < y < z$ implies that: $\mu_{\underline{A}}(y) \geq \min\,(\mu_{\underline{A}}(x), \mu_{\underline{A}}(z))$. If this condition holds for all points, the fuzzy set is called convex fuzzy set. Figure 7.12 gives the graphical illustration. Convex fuzzy sets are strictly increasing and then strictly decreasing. A set that is not convex is called a *non-convex set*. Concluding, we get that A is convex if all its α-level sets are convex.

Various theories of convex fuzzy mapping defined by the order "fuzz-max" have been extensively and profoundly discussed in the literature by many researchers [46–51]. Note that semi-strict convex fuzzy mappings are a significant class of fuzzy mappings that are different from convex fuzzy mappings.

The criteria for a superior or inferior semi-continuous fuzzy mapping defined on a non rainfall convex subset R_n are fuzzy mapping (semi-convex, strictly convex,

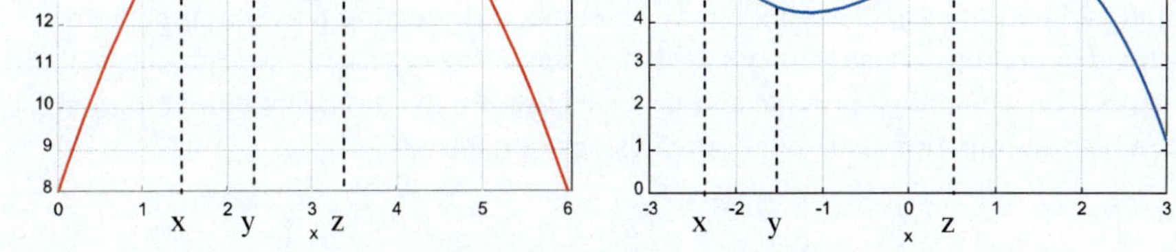

Fig. 7.12 Illustration of fuzzy set properties: **a** convex fuzzy set, **b** nonconvex fuzzy set

respectively). They can be carried out by a check of intermediate convexity (semi-strict convexity, strict convexity, respectively). It is useful to look for that kind of practical criterion for all kinds of convexity in fuzzy optimization.

7.5.3 Building Fuzzy Models

Recently, a great deal of research activity has focused on the development of transparent rule based fuzzy models which can accurately predict the quantities of interest, and at the same time provide insight into the system that generated the data. The attention is concentrated on the selection of appropriate model structures in terms of the dynamic properties, as well as the internal structure of such fuzzy rules as linguistic, relational, or Takagi–Sugeno type. Consequently, modeling that is traditionally associated with better understanding of the underlying mechanisms of action of the real and designed systems is seen as a combination of a thorough understanding of the nature and behavior of the system, and an appropriate mathematical treatment that leads to a useable model. The prior knowledge and data (process measurements) are two common sources of information serving for building fuzzy models. Usually "experts" i.e. process designers, operators, etc. originate the prior knowledge which can be of a rather approximate nature (qualitative knowledge, heuristics). For these reasons fuzzy models can be regarded as simple fuzzy expert systems. From the point of view of system identification procedures, a fuzzy model is regarded as a set of local submodels. The fuzzy sets naturally provide smooth transitions between the submodels, and enable the integration of various types of knowledge within a common framework. Accordingly, the modeling framework is based on rule-based fuzzy models, which describe relationships between variables by means of *if–then* rules, such as:

If on the Beach the Sunshine is Greater, then the Air Temperature Rises Faster.

These rules establish logical relationships between system variables by connecting the qualitative values of one variable (sunshine is greater) to the qualitative values of another variable (temperature rises faster). Observe that qualitative values are generally clearly interpreted from a linguistic point of view and are referred to as linguistic terms (labels, values). The meaning of the linguistic terms with regard to the input and output variables which may be numerical (heating power, temperature) is defined by suitably chosen fuzzy sets. In this sense, fuzzy sets, or more precisely, their membership functions, provide an interface between the input and output numerical variables and the linguistic qualitative values in the rules.

a)

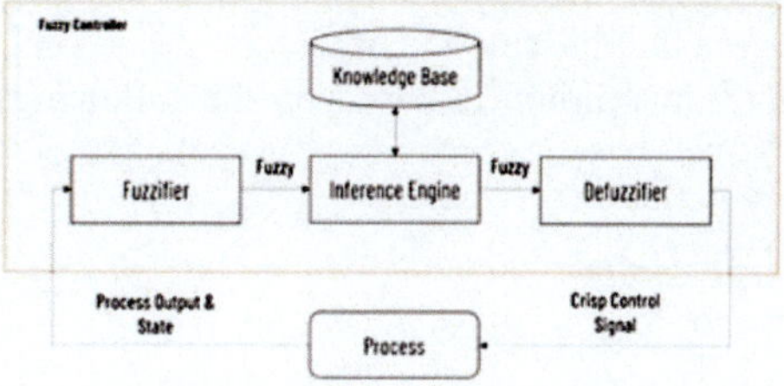

b)

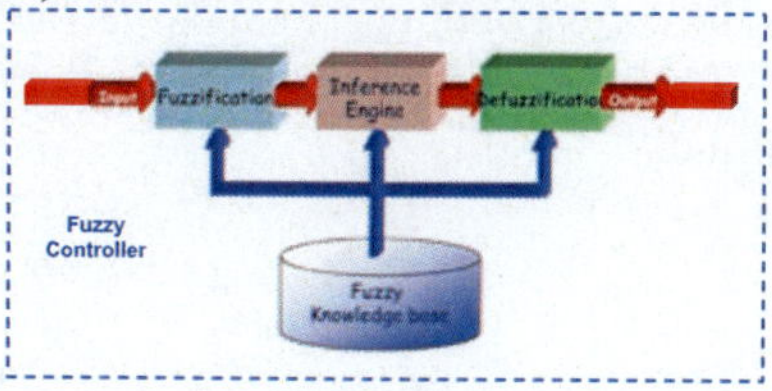

Fig. 7.13 System with fuzzy controller: **a** block structure, **b** fuzzy controller structure

7.5.4 Defuzzification

In the final stage of calculations of the studied system with the use of fuzzy sets is a defuzzification process where the fuzzy output is translated into a single crisp value. In other words, defuzzification is performed by a decision-making algorithm that selects the best net worth from a fuzzy package. Therefore, defuzzification can be considered as an inverse transformation compared with the fuzzification process, because in such process, the fuzzy output is converted into crisp values to be applied to the investigated system model. It is obvious that final decision regarding the correct operation of the system has to be taken only on crisp values.

An illustration of the use of defuzzification in effective control of a given system is presented in Fig. 7.13.

The fuzzy results generated in computing cannot be used in an application. Thus defuzzification must convert the fuzzy output of fuzzy inference engine into crisp value, so that it can be fed, for instance, to the controller input. Note that the controller can only understand the crisp input. So it is necessary to convert the fuzzy result into crisp value.

Several forms of defuzzification exist, including the centre of gravity (COG), the mean value of the maximum (MOM), and the centre of the largest area (COA) methods. The COG method returns the value of the middle of the region under the curve and the MOM approach may be considered as the point at which the balance is achieved on a curve. The centre of the largest area method applies to the fuzzy set with multiple sub-regions. Then, the center of gravity of the sub-region with the largest area can be used to calculate the defuzzified value.

The centroid defuzzification procedure is the most prevalent and intuitively appealing among the defuzzification methods. According to it in order to produce the value u of the control variable sensitive to all rules it should be assumed the center of gravity of the final fuzzy space. It is expressed by the following formula

$$u = \frac{\sum_{i=1}^{n} \alpha_i \mu_i}{\sum_{i=1}^{n} \alpha_i A_i} \tag{7.21}$$

where μ_i is the value of the membership function of the output fuzzy set of rule i, A_i is the corresponding area and α_i is the degree that the rule i is fired.

The procedure of the mean value of the maximum is based on the following formula

$$x = \frac{\sum_{i=1}^{n} \alpha_i H_i x_i}{\alpha_i H_i} \tag{7.22}$$

where x is the control (output) value to be applied, n is the number of rules in a multi input single output system, H_i is the maximum value of the membership function of the output fuzzy set, which corresponds to rule I_i, x_i is the corresponding control (output) value, and α_i is the degree that the rule i is fired.

The centre of the largest area method is simple, computationally effective and widely used *defuzzification*. In the first step a sub-region of the fuzzy set with the largest area A_m is fixed. Next x', the center of gravity of A_m, is determined. In result, the defuzzified value of the control signal is calculate as follows

$$x^* = \left(\int \mu_{Am}(x)x'dx\right)/\left(\int \mu_{Cm}(x)dx\right) \tag{7.23}$$

It should be mentioned here such defuzzification methods as: lambda cut method, maxima methods, weighted average method have also been developed, and are applicable in specific cases of fuzzy sets. However, there is no systematic procedure for choosing a good defuzzification strategy. The selection of defuzzification procedure depends on the properties of the application.

Defuzzification is the process of obtaining a single number from the output of the aggregated fuzzy set. It is used to transfer fuzzy inference results into a crisp output. In other words, defuzzification is realized by a decision-making algorithm that selects the best crisp value based on a fuzzy set.

7.6 Real-Life Artificial Intelligence Applications

The popularity, social significance and function of Artificial Intelligence are soaring by the day. AI applications have significantly expanded over the past few years and have found their applications in almost every sector of human activity. Below is the list of the top 17 applications of AI, namely:

1. E-Commerce—create referral engines that enable to better co-operate with clients.
2. Education—helps increase productivity among faculties and helps them concentrate more on students than office or administration work.
3. Lifestyle—influences on the lifestyle of consumers by shaping tastes for autonomous vehicles, spam filters, facial recognition, and recommendation system.

4. Navigation—based on GPS technology can provide users with accurate, timely, and detailed information to improve safety.
5. Robotics—robots powered by AI use real-time updates to sense obstacles in its path and pre-plan its journey instantly.
6. Human resource—applications based on specific parameters can scan job candidates' profiles.
7. Healthcare—finds diverse applications to build sophisticated machines that can detect diseases and identify cancer cells.
8. Agriculture—identifying defects and nutrient deficiencies in the soil had done using computer vision, robotics, and machine learning applications.
9. Gaming—creating smart, human-like NPCs to interact with the players.
10. Automobiles—building self-driving vehicles.
11. Social Media—considers likes and the followed accounts to determine what posts are shown on explore tab.
12. Marketing—marketers, using AI, can deliver highly targeted and personalized ads with the help of behavioral analysis, and pattern recognition in ML, etc.
13. Chatbots—can comprehend natural language and respond to people online who use the "live chat" feature that many organizations provide for customer service.
14. Finance—can help to significantly improve a wide range of financial services.
15. Astronomy—warns everyone by weather storm in this beautiful world of technology.
16. Data Security—industry leaders employ AI for a variety of purposes, including providing valued services and preparing their companies for the future.
17. Travel and Transport—creating a new market for firms and entrepreneurs to develop innovative solutions for making public transportation more comfortable, accessible, and safe.

In general, Artificial Intelligence appears as revolutionizing industries with its applications and helping solve complex problems. The focus of successful achievements is on the algorithms' progress and development. The new technologies being developed to automate machine learning pipelines and greatly speed up the development process are also fascinating—and deserving of corporate focuses.

Figure 7.14 presents an iterative process in which we build on previous training results to figure out how to approach the successful training problem. We always want to achieve an overall objective, which is a better functioning of the intelligent system than one, that can be achieved without intelligent components.

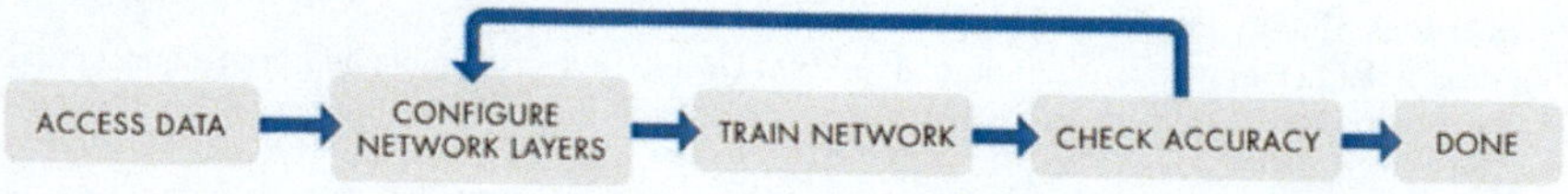

Fig. 7.14 Scheme of activities leading to successful applications of AI

References

1. Rosenblum, M., Pikovsky, A., Kuhn, A.A., Busch, J.L.: Real-time estimation of phase and amplitude with application to neural data. Sci. Rep. **11**(1), 1–11 (2021)
2. Belega, D., Petri, D., Dallet, D.: Amplitude and phase estimation of real-valued sine wave via frequency-domain linear least-squares algorithms. IEEE Trans. Instrum. Meas. **67**(5), 1065–1077 (2018)
3. Brownlee, J.: Machine Learning Mastery with Python: Understand Your Data, Create Accurate Models, and Work Projects End-to-End. Machine Learning Mastery (2016)
4. Buitinck, L., Louppe, G., Blondel, M., Pedregosa, F., Mueller, A., Grisel, O., Niculae, V., Prettenhofer, P., Gramfort, A., Grobler, J., Layton, R., VanderPlas, J., Joly, A., Holt, B., Varoquaux, G.: API design for machine learning software: experiences from the scikit-learn project. In: ECML PKDD Workshop: Languages for Data Mining and Machine Learning, pp. 108–122 (2013)
5. Akponeware, A.O., Adamu, Z.A.: Clash detection or clash avoidance? An investigation into coordination problems in 3D BIM. Buildings **7**(3), 75 (2017)
6. Baldwin, E.: AI creates generative floor plans and styles with machine learning at Harvard (2019). https://www.archdaily.com/918471/ai-creates-generative-floor-plans-and-styles-with-machine-learning-at-harvard/
7. Bejnordi, B.E., Veta, M., Van Diest, P.J., Van Ginneken, B., Karssemeijer, N., Litjens, G., Van Der Laak, J.A., Hermsen, M., Manson, Q.F., Balkenhol, M., et al.: Diagnostic assessment of deep learning algorithms for detection of lymph node metastases in women with breast cancer. JAMA **318**(22), 2199–2210 (2017)
8. Brady, T.F., Yellig, E.: Simulation data mining: a new form of computer simulation output. In: Proceedings of the Winter Simulation Conference, p. 5 ff. IEEE (2005)
9. De la Rosa, E., Yu, W.: Randomized algorithms for nonlinear system identification with deep learning modification. Inf. Sci. **364**, 197–212 (2016)
10. dos Santos, F.A., Rodrigues, A., Micheletti, A.: Design and experimental testing of an adaptive shape-morphing tensegrity structure, with frequency self-tuning capabilities, using shape-memory alloys. Smart Mater. Struct. **24**(10), 105,008 (2015)
11. Baymani, M., Effat, S., Niazmand, H., Kerayechian, A.: Artificial neural network method for solving the Navier-Stokes equations. Neural Comput. Appl. **26**(4), 765–773 (2015)
12. Raja, M.A.Z., Samar, R., Haroon, T., Shah, S.M.: Unsupervised neural network model optimized with evolutionary computations for solving variants of nonlinear MHD Jeffery-Hamel problem. Appl. Math. Mech. **36**(12), 1611–1638 (2015)
13. Raja, M.A.Z., Samar, R.: Numerical treatment for nonlinear MHD Jeffery-Hamel problem using neural networks optimized with interior point algorithm. Neurocomputing **124**, 178–193 (2014)
14. Brunton, S.L., Kutz, J.N.: Data-Driven Science and Engineering: Machine Learning, Dynamical Systems, and Control. Cambridge University Press, Cambridge (2019)
15. Burrows, S., Stein, B., Frochte, J., Wiesner, D., Müller, K.: Simulation data mining for supporting bridge design. In: Proceedings of the Ninth Australasian Data Mining Conference, vol. 121, pp. 163–170 (2011)
16. Aelenei, D., Aelenei, L., Vieira, C.P.: Adaptive façade: concept, applications, research questions. Energy Procedia **91**(Supplement C), 269–275 (2016)
17. Akadiri, P.O., Chinyio, E.A., Olomolaiye, P.O.: Design of a sustainable building: a conceptual framework for implementing sustainability in the building sector. Buildings **2**(2), 126–152 (2012)
18. Akeyoshi, K., Kanai, E.: Mechanical Properties of Tempered Glass. VII International Glass Congress (paper 80) (1965)
19. Amarante dos Santos, F., Bedon, C., Micheletti, A.: Explorative study on adaptive facades with superelastic antagonistic actuation. Struct. Control Health Monit. **27**(4), e2463 (2020)
20. Aznar, F., Echarri, V., Rizo, C., Rizo, R.: Modelling the thermal behaviour of a building facade using deep learning. PLoS ONE **13**(12), e0207616 (2018)

21. Badue, C., Guidolini, R., Carneiro, R.V., Azevedo, P., Cardoso, V.B., Forechi, A., Jesus, L., Berriel, R., Paixão, T., Mutz, F., et al.: Self-driving cars: a survey (2019). arXiv:1901.04407
22. Baumeister, T., Brunton, S.L., Kutz, J.N.: Deep learning and model predictive control for self-tuning mode-locked lasers. JOSA B **35**(3), 617–626 (2018)
23. Beckmann, R., Knackstedt, W.: Process for the production of modified, partially acetalized polyvinyl alcohol films. U.S. Patent No. 4,144,376, 13 Mar (1979)
24. Benoit, M., Ispas, S., Jund, P., Jullien, R.: Model of silica glass from combined classical and ab initio molecular-dynamics simulations. Eur. Phys. J. B-Condens. Matter Complex Syst. **13**(4), 631–636 (2000)
25. Bhowmik, A., Castelli, I.E., Garcia-Lastra, J.M., Jørgensen, P.B., Winther, O., Vegge, T.: A perspective on inverse design of battery interphases using multi-scale modelling, experiments and generative deep learning. Energy Storage Mater. (2019)
26. Bishop, C.M.: Pattern Recognition and Machine Learning, Information Science and Statistics, 1st edn. Springer, Berlin (2006)
27. Bostrom, N.: Superintelligence. Dunod, Paris (2017)
28. Brownlee, J.: Clever Algorithms: Nature-Inspired Programming Recipes. Machine Learning Mastery (2011)
29. Adeli, H.: Neural networks in civil engineering: 1989–2000. Comput. Aided Civ. Infrastruct. Eng. **16**(2), 126–142 (2001)
30. Binkhonain, M., Zhao, L.: A review of machine learning algorithms for identification and classification of non-functional requirements. Expert Syst. Appl. X **1**, 100001 (2019). https://doi.org/10.1016/j.eswax.2019.100001
31. Effati, S., Buzhabadi, R.: A neural network approach for solving Fredholm integral equations of the second kind. Neural Comput. Appl. **21**(5), 843–852 (2012)
32. Raja, M.A.Z., Khan, J.A., Shah, S.M., Samar, R., Behloul, D.: Comparison of three unsupervised neural network models for first Painlevé transcendent. Neural Comput. Appl. **26**(5), 1055–1071 (2011)
33. Sabouri, J., Effati, S., Pakdaman, M.: A neural network approach for solving a class of fractional optimal control problems. Neural Process Lett. **45**(1), 59–74 (2017)
34. Boschert, S., Rosen, R.: Digital twin—the simulation aspect. In: Mechatronic Futures, pp. 59–74. Springer, Berlin (2016)
35. Bukieda, P., Lohr, K., Meiberg, J., Weller, B.: Study on the optical quality and strength of glass edges after the grinding and polishing process. Glass Struct. Eng. (2020). https://doi.org/10.1007/s40940-020-00121-x
36. Chang, N.B., Bai, K.: Multisensor Data Fusion and Machine Learning for Environmental Remote Sensing. CRC Press, USA (2018)
37. Barbosa, F., Woetzel, J., Mischke, J., Ribeirinho, M.J., Sridhar, M., Parsons, M., Bertram, N., Brown, S.: Reinventing Construction: A Route to Higher Productivity. McKinsey Global Institute (2017)
38. Chui, M., Manyika, J., Miremadi, M.: What AI can and can't do (yet) for your business. McKinsey Quarterly (2018)
39. Bunker, S.: Artificial Intelligence: Building Smarter Machines (2018)
40. Chollet, F., et al.: Keras (2015). https://github.com/fchollet/keras; Chowdhary, K.R.: Introducing artificial intelligence. In: Linnhoff-Popien, C., Zaddach, M., Grahl, A. (eds.) Fundamentals of Artificial Intelligence, pp. 1–23. Springer, Berlin (2020)
41. Bzdok, D., Altman, N., Krzywinski, M.: Points of significance: statistics versus machine learning (2018). https://doi.org/10.1038/nmeth.4642
42. Bluyssen, P.M., Oostra, M.A., Meertins, D.: Understanding the indoor environment: how to assess and improve indoor environmental quality of people? In: Proceedings of CLIMA 2013: 11th REHVA World Congress & 8th International Conference on IAQVEC "Energy Efficient, Smart and Healthy Buildings", Prague, Czech Republic, 16–19 June 2013. CiteSeer (2013)
43. Borgstein, E., Lamberts, R., Hensen, J.: Mapping failures in energy and environmental performance of buildings. Energy Build. **158**, 476–485 (2018)

44. Borrmann, A., König, M., Koch, C., Beetz, J.: Building Information Modeling: Technologische Grundlage und Industrielle Anwendungen. Springer, Berlin (2015)
45. Effati, S., Pakdaman, M.: Artificial neural network approach for solving fuzzy differential equations. Inf. Sci. **180**(8), 1434–1457 (2010)
46. Łukasiewicz, J.: Aristotle's Syllogistic from the Standpoint of Modern Formal Logic, 2nd edn enlarged. Oxford University Press, Oxford (1963)
47. Bostrom, N.: Superintelligence. Dunod, Paris (2017)
48. Chen, S., Wang, H.: SAR target recognition based on deep learning. In: 2014 International Conference on Data Science and Advanced Analytics (DSAA), pp. 541–547. IEEE (2014)
49. Crevier, D.: AI: The Tumultuous History of the Search for Artificial Intelligence. Basic Books Inc., USA (1993)
50. Daily, M., Medasani, S., Behringer, R., Trivedi, M.: Self-driving cars. Computer **50**(12), 18–23 (2017)
51. Day, J.J.K., O'Brien, W.: Oh behave! Survey stories and lessons learned from building occupants in high-performance buildings. Energy Res. Soc. Sci. **31**, 11–20 (2017). https://doi.org/10.1016/j.erss.2017.05.037